Lecture Notes in Networks and Systems

Volume 240

The series "Lecture Notes in Networks and Systems" publishes the latest developments in Networks and Systems—quickly, informally and with high quality. Original research reported in proceedings and post-proceedings represents the core of LNNS.

Volumes published in LNNS embrace all aspects and subfields of, as well as new challenges in, Networks and Systems.

The series contains proceedings and edited volumes in systems and networks, spanning the areas of Cyber-Physical Systems, Autonomous Systems, Sensor Networks, Control Systems, Energy Systems, Automotive Systems, Biological Systems, Vehicular Networking and Connected Vehicles, Aerospace Systems, Automation, Manufacturing, Smart Grids, Nonlinear Systems, Power Systems, Robotics, Social Systems, Economic Systems and other. Of particular value to both the contributors and the readership are the short publication timeframe and the world-wide distribution and exposure which enable both a wide and rapid dissemination of research output.

The series covers the theory, applications, and perspectives on the state of the art and future developments relevant to systems and networks, decision making, control, complex processes and related areas, as embedded in the fields of interdisciplinary and applied sciences, engineering, computer science, physics, economics, social, and life sciences, as well as the paradigms and methodologies behind them.

Indexed by SCOPUS, INSPEC, WTI Frankfurt eG, zbMATH, SCImago.

All books published in the series are submitted for consideration in Web of Science.

More information about this series at https://link.springer.com/bookseries/15179

David Scaradozzi · Lorenzo Guasti ·
Margherita Di Stasio · Beatrice Miotti ·
Andrea Monteriù · Paulo Blikstein
Editors

Makers at School, Educational Robotics and Innovative Learning Environments

Research and Experiences from FabLearn Italy 2019, in the Italian Schools and Beyond

Springer

Editors
David Scaradozzi
Dipartimento di Ingegneria
dell'Informazione (DII)
Università Politecnica delle Marche
Ancona, Italy

Margherita Di Stasio
Istituto Nazionale di Documentazione,
Innovazione e Ricerca Educativa (Indire)
Florence, Italy

Andrea Monteriù
Dipartimento di Ingegneria
dell'Informazione (DII)
Università Politecnica delle Marche
Ancona, Italy

Lorenzo Guasti
Istituto Nazionale di Documentazione,
Innovazione e Ricerca Educativa (Indire)
Florence, Italy

Beatrice Miotti
Istituto Nazionale di Documentazione,
Innovazione e Ricerca Educativa (Indire)
Florence, Italy

Paulo Blikstein
Teachers College
Columbia University
New York, NY, USA

ISSN 2367-3370 ISSN 2367-3389 (electronic)
Lecture Notes in Networks and Systems
ISBN 978-3-030-77042-6 ISBN 978-3-030-77040-2 (eBook)
https://doi.org/10.1007/978-3-030-77040-2

This Springer imprint is published by the registered company Springer Nature Switzerland AG
The registered company address is: Gewerbestrasse 11, 6330 Cham, Switzerland

Preface

This open access book contains observations, outlines, and analyses of educational robotics methodologies and activities, and developments in the field of educational robotics emerging from the findings presented at FabLearn Italy 2019, the international conference that brought together researchers, teachers, educators, and practitioners to discuss the principles of making and educational robotics in formal, non-formal, and informal education.

Our analysis of these extended versions of papers presented at FabLearn Italy 2019 highlights the latest findings on learning models based on making and educational robotics. We investigate how innovative educational tools and methodologies can support a novel, more effective, and more inclusive learner-centered approach to education.

The following key topics are the focus of discussion: makerspaces and fab labs in schools, a maker approach to teaching and learning; laboratory teaching and the maker approach, models, methods, and instruments; curricular and non-curricular robotics in formal, non-formal, and informal education; social and assistive robotics in education; and the effect of innovative spaces and learning environments on the innovation of teaching, good practices, and pilot projects.

We would like to extend a special massive thanks to Arianna Pugliese, for her immeasurable work in coordinating the book project and in organizing the FabLearn Italy 2019 event, and to Lorenzo Calistri for his technical support during the event. We also thank Silvia Modena and Althea Muirhead for their painstaking and qualified support in the linguistic revision of every page of this book.

Florence, Italy — Lorenzo Guasti
Ancona, Italy — David Scaradozzi
Florence, Italy — Beatrice Miotti
Florence, Italy — Margherita Di Stasio
Ancona, Italy — Andrea Monteriù

Introduction

Education is crucial for equipping men and women with the critical competencies that will enable them to understand the world, find employment, and participate in future society. Understanding how a student learns (and what she needs to learn) is essential for enhancing the ability of teachers to guide her toward the desired outcome. New technologies can support this process by gathering information seamlessly and providing hints automatically. In addition, new technologies are an important part of the school renewal process when associated with innovative methodologies.

Maker pedagogy [1, 2] and Robotics in Education (RiE) are a broad area of technology applications in education [3, 4]. Within RiE, educational robotics (ER) helps students explore powerful ideas and authentic learning in open-ended STEM environments. Although it is fundamental to assess, evaluate, and identify such learning in pedagogical terms, it is still unclear how students' learning achievements should be measured and certified within a normal curriculum. The qualitative method seems to be the principal evaluation methodology used in pedagogical practice to provide a rich and in-depth analysis of students' learning. Although the qualitative methods are well suited to representing the open-ended nature of the learning environment, they may be heavily subject to observer bias and external factors.

Moreover, they are less effective in synthesizing information, which is a required ability when evaluating and supporting a whole class of students at the same time. In this respect, educational data mining (EDM) and learning analytics (LA) can help by identifying and modeling students' learning and by assessing maker pedagogy and robotics in education activities in a natural classroom environment. To successfully apply EDM and LA in a real classroom scenario, a modeled system (environment) that automatically and systematically collects data is needed. When reliable data are collected, they can be analyzed employing EDM and LA techniques to provide teachers with clear and understandable results [5].

The studies concerning maker pedagogy are in depth and specifically take a fresh look at student-centered experiential education. Papert [6, 7] expanded the scientific viewpoint and included a theory that different types of tools, media, and materials can shape our cognition, viewing learning as a process of building and correcting theories in close connection with constructions in the world. The theory behind the maker movement, called "Constructionism," is geared toward understanding what students learn when they are immersed in learning environments with materials and tools in which they can build objects, share them with peers and teachers, and refine their understanding of the world based on multiple cycles of constructing and debugging. Maker education is a natural continuation of the Constructionism theory.

Maker Dimension

The craft dimension of "making" gives meaning and perspective for understanding a digital world and society.

In this publication, which collects most of the contributions from FabLearn Italy 2019, we have been able to study exciting projects involving many fields in pedagogy and engineering. Specifically, it is very interesting to see how the maker movement has involved ecology, environmental sustainability, and gender equality. Many of the projects presented concern solutions to real-life problems, especially regarding the socially weak or everyday situations in which there is a distinct need for a "solution" to a problem. This allows teachers and students to immerse themselves in real-life problems and study by confronting the world around them.

From the contributions, it is clear that the "maker philosophy" has made a forceful entry into didactics, especially as regards laboratory work. Today, it is no longer simply a matter of buying a 3D printer for the classroom. It is about overhauling the science curriculum to include microelectronic platforms like Arduino and Raspberry, and the combined use of coding, robotics, electronics, and printing.

No less important is the reflection on learning spaces. The papers presented reveal how important it is that schools and lesson plans evolve from this perspective (see also the FabLearn Italy program in Fig. 1). For example, the presence of a modern "makerspace" laboratory within the school building is a potential lever for innovating all laboratory teaching, giving teachers a place (and the tools) to design new modern and practical teaching units.

Trends and Perspectives: Modeling the Learning Environment as a Holistic Automation System

"School has to change" is the common thread that runs through all the contributions in this book. School and all elements that make it up—the education system, learning approaches, building design, organization of time, assessment methodologies—are centuries old.

It can be deduced from the contributions and from the researchers who took part in FabLearn Italy 2019 that school will only change once we change the way we research school innovation. It will only happen if the entire educational environment

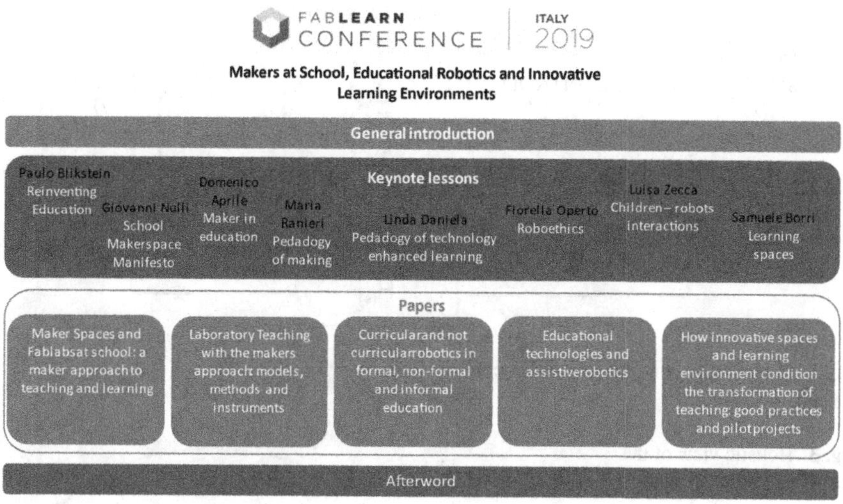

Fig. 1 Content in this book from the FabLearn Italy 2019 program

is studied and developed holistically (Fig. 1), and this should include all the subjects and skills that have been considered crucial until now: pedagogy and engineering, architecture, robotics and automation, making artifacts, coding, virtual and digital 3D design, etc. The ultimate aim is to create a rich augmented environment in which to acquire and consolidate problem-solving skills and the ability to observe a scientific phenomenon in depth, including through modeling, as often occurs in science [8–10].

Following the automation system theory, such an environment can be viewed as a process where the information flows from teachers to learners. A supervisor has to fine-tune the learning tools (curricula, devices, and spaces) to optimize the learning outcomes. In this scenario, learners can be modeled as a system, with its dynamics and its state–space representation. Teachers are the controllers that need feedback from classical assessment methods (class assignments) or modern sensors (EDM and LA) (see Fig. 2, the feedback control system).

To design the learning tools correctly and fine-tune them, researchers and authorities need to look at the learning environment holistically. They have to identify the think–make–improve core strategy and consider the background noise (influence of the environment). To obtain a parametric model as a reference for modern sensors (EDM and LA), a digital twin of the process must be identified as a dynamic complex system. This design method's primary outcome will be new instruments for identifying errors and obtaining feedback, key performance indicators, and ethics involvement indicators (see Fig. 3).

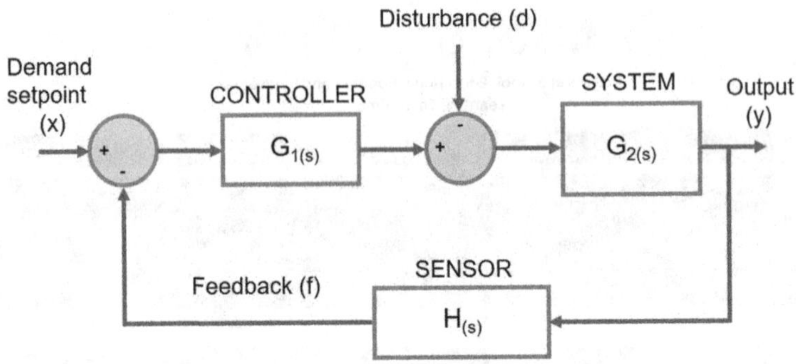

Fig. 2 General diagram of the feedback control system

Fig. 3 Holistic view of the learning complex system

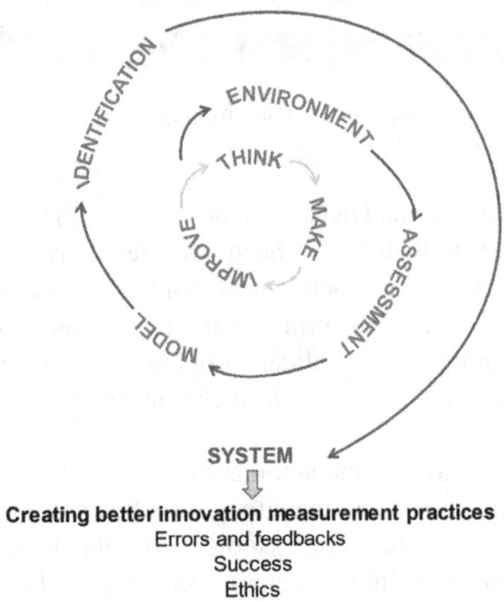

Conclusions and Book Index

We can conclude from the topics covered at FabLearn Italy 2019 that research in the field of school innovation is thriving and active. Each section of the book presents a different area of research and corresponds to a different cultural area featured at the event (see Fig. 1).

Lorenzo Guasti
l.guasti@indire.it

David Scaradozzi
d.scaradozzi@univpm.it

References

1. Martinez, S.L., Stager, G.: Invent to learn. Making, Tinkering, and Engineering in the Classroom. Construting Modern Knowledge, Torrance, Canada (2013)
2. Guasti, L., Rosa, A.: Maker@ Scuola. Stampanti 3D nella scuola dell'infanzia. Firenze, Assopiù Editore, 91–106 (2017)
3. Rossi, P.: La scienza e la filosofia dei moderni: aspetti della rivoluzione scientifica. Bollati Boringhieri (1989)
4. Scaradozzi, D., Screpanti, L., Cesaretti, L.: Towards a definition of educational robotics: a classification of tools, experiences and assessments. In Smart Learning with Educational Robotics (pp. 63–92). Springer, Cham (2019)
5. Benitti, F.B.V.: Exploring the educational potential of robotics in schools: a systematic review. Comput. Educ. **58**(3), 978–988 (2012)
6. Papert, S.: Mindstorms: Children, Computers, and Powerful Ideas. Basic Books, New York (1980)
7. Floridi, L.: Philosophy and Computing: An Introduction. Psychology Press (1999)
8. Blikstein, P., Wilensky, U.: Bifocal modeling: a framework for combining computer modeling, robotics and real-world sensing. In annual meeting of the American Educational Research Association (AERA 2007), Chicago, USA (2007)
9. Fuhrmann, T., Schneider, B., Blikstein, P.: Should students design or interact with models? using the bifocal modelling framework to investigate model construction in high school science. Int. J. Sci. Educ. **40**(8), 867–893 (2018)
10. Fuhrmann, T., Greene, D., Salehi, S., Blikstein, P.: Bifocal biology: the link between virtual and real experiments. In Proceedings of the Constructionism 2012 Conference, Athens, Greece (2012). Retrieved from http://etl.ppp.uoa.gr/constructionism2012dailly/constructionism2012_proceedings.zip.

Contents

How Innovative Spaces and Learning Environment Condition the Transformation of Teaching: Good Practices and Pilot Projects

Introduction to the Main Topics

Perspectives for School: Maker Approach, Educational Technologies and Laboratory Approach, New Learning Spaces

Margherita Di Stasio and Beatrice Miotti

Abstract Technology has become an ordinary, constituent element of today's world. Therefore, digital skills are not only important, but ultimately necessary for all. This makes it mandatory for school and educational research to construct learning paths with an educational rather than a training approach to technology. The purpose of school is not to prepare students to be users of an instrument, but, above all, to kindle in them an awareness of the process that can stem from creation. This chapter aims to highlight some of the topics emerging from international discussion on school and learning methodologies. First, one of the interesting topics is the Maker approach and productive partnerships with schools, including in relation to the original concept of craftsmanship, forerunner of the term "maker". Second, educational robotics and the laboratory as tools for improving deep learning, abstract thinking and problem-solving skills, without forgetting to look at the ethical and social issues of introducing robots into our lives. Third, learning spaces and new school buildings have to meet the changing paradigms of learning approaches. There are countless opportunities for improving digital skills and methodological approaches. Teachers should be aware of all these possibilities and be ready to get training and apply them in their everyday lessons.

Keywords Makers · Educational robotics · Learning spaces · Constructivism · Teacher training

Although this chapter has been jointly conceived, Margherita Di Stasio wrote Section 1, 2 and 4; Betrice Miotti developed Section 3.

M. Di Stasio (✉) · B. Miotti
Istituto Nazionale Documentazione Innovazione Ricerca Educativa (Indire), via M. Buonarroti 10, 50122 Florence, Italy
e-mail: m.distasio@indire.it

B. Miotti
e-mail: b.miotti@indire.it

D. Scaradozzi et al. (eds.), *Makers at School, Educational Robotics and Innovative Learning Environments*, Lecture Notes in Networks and Systems 240,
https://doi.org/10.1007/978-3-030-77040-2_1

1 Introduction

The evolution of technology is not limited to itself: as the history of science and technology—*sub specie philosophica*—have taught us, periods of true revolution involve technological, cultural and social factors [1].

It is clear that, in recent years, our world has undergone a never-before-seen technological acceleration [2]. Our reality is marked by highly pervasive technologies that have us facing dimensions like the *infosphere* [3] and the *internet of things* [4] as parts of a world in which our life is characterized as *onlife* [5]. One emerging and characteristic element of this world is so-called "digital craftsmanship", Makers, that is finding its own dimension in society and in the economy.

An educational and non-training approach to technology aims not only to produce users of instruments, but, above all, to kindle in them an awareness of the process that can stem from creation.

Education and pedagogy should work actively to make students, children and young people aware of the world in which they live and be able to be active in it.

Digital craftsmanship is also finding a distinctive role in education [6], in a perspective in which fab labs are "democratizing the maker movement for students" [7].

2 Maker Dimension

The craft dimension of "doing" is what gives meaning and perspective for understanding a digital world and society. That is not a contradiction in terms. This is why.

The spread of personal computers and networks has led to the dissemination of technology in everyday personal and work life. Web 2.0 and the emergence of social networking have led to the creation of participatory cultures [8] and the development of media education.

Even if the instrument created by Bernard Lee[1] for a community of select scientists seemed to be available to everyone, the relationship between hardware and software and the Logic behind technology has remained the prerogative of a narrow group of specialists.

The desire to share what is produced by man with technology, and to have technology serving the needs of humans distinguishes the open source and free software[2] community. H*acker ethic* comes from, develops and is spread, both in the literature and indeed, by people like Himanen [9] and Torvald [10], who gave his name to the Linux operating system.

The hacker movement and the maker movement share this desire to make technology accessible and understandable for more people.

---

[1] The first website can be browsed at http://info.cern.ch/.

[2] On this topic see https://www.fsf.org/ and https://www.fsf.org/.

As stated [7], the maker movement draws inspiration from the hackerspace for improvement; it also presents its activities as being accessible to everybody, not only to people who are prepared and specialized.

The very first statement of the *Maker Movement Manifesto* claims that "making is fundamental to what it means to be human. We must make, create, and express ourselves to feel whole. There is something unique about making physical things. Things we make are like little pieces of us and seem to embody portions of our soul" [11]. The value of experience emerges in affective terms. On one hand, it restores the human element to the handling of technology. On the other, it is related to the learning by doing tradition of Freinet and Papert.

Some making dimensions are meaningful in the school context. The addition of makerspaces can be of great significance when it comes to redesigning teaching and learning spaces. Making is a structured and complex process and may require active and real collaboration in which the co-construction of knowledge is encouraged. It leads to a deeper understanding of the mechanisms that underlie interaction with technology, opening up opportunities for active media education. It also helps the development of design and reflection processes that can be modelized and applied in different social and educational contexts.

3 Trends and Perspectives

"School has to change" is the common thread that runs through all the contributions in this track. School and all of the elements that make it up—educational system, learning approaches, building design, organization of time—are centuries old.

It is rooted in the lecture-based teaching model, which is no longer suitable [12, 12] because the youth of today "are connected to everything, and they have a sense of urgency because of climate change at our doorstep, the dying oceans, and social issues such as systemic racism coming to a head. All of this is happening in front of their eyes. This is a generation that feels the urgency of changing the world." [14].

And so, the most important question research tries to answer is: "How and what do we want those children and teens to learn, so that in the future they can become more like the inspiring young people mentioned above? [Cf. Boyan Slat, Greta Thunberg and the Parkland survivors]" [14].

It is clear how school should change. Instead of the lecture-based school model a "[…] well-thought-out laboratory approach helps children to learn, plan, communicate and collaborate, act independently and responsibly, solve problems, identify connections and interpret information." [15] Moreover, "The education system should not be designed for a standard (and its related standardized tests), but should bring out the divergent thinking and creativity of each student." [16]. Technology and educational robotics can be used to improve the learning process: "pedagogical factors that make technology-enhanced learning successful: learning motivation, cognitive development, cognitive load, and knowledge development in the discourse of technology-enhanced learning" [17] and "educational use of robots stimulates productive, creative and divergent thinking and is useful for understanding how students learn." [15].

Thus the maker approach can be seen as a virtuous example and a model to imitate, because "contemporary craftsmen are those who know how to use digital technologies with mastery, and consider quality, innovation and social cooperation to be fundamental values in their work" [18], "if you redesigned schools from scratch, they would probably look like a well-functioning makerspace" [14]. However, in order for this collaboration between schools and the maker approach to be worthwhile, a School Makerspace manifesto is needed "Schools and makers are two very broad, diverse and elaborate entities. Yet, as long as they can find common ground, they can create a lab where they can work together but also separately, according to their individual needs." [13].

But school cannot change if teachers are not ready to reinvent the way they teach, by moving from a face-to-face lesson model to a variety of teaching methods and learner-centered strategies, including with technological support. So, in the new school model, it is extremely important to "strengthen the professionalism of teachers, through research-training and participatory action research," [15] and to ensure "teachers have the appropriate digital competence to use technology, which in turn affects the extent to which the opportunities created by technology will be used in the learning process" [17].

Looking at the use of robotics or artificial intelligence at school but also in everyday life, we cannot help but consider "the ethical, legal and societal issues (ELS) inherent to that field" that we are going face. Roboethics is a new science that studies "the protection of privacy, the defense of human dignity, distributive justice and the dignity of work" [19] in relation to ELS issues.

3.1 Experiences and Points of View

The urgent need for change in the school system clearly emerges in [14], which puts forward a number of interesting points of view for building an education system from scratch. The school model has been the same for centuries and only recently have new methodologies been put forward; however the renewal process is too slow for the needs of young people today. Blikstein states that schools should do more than teach arithmetic and other subjects; it should also equip young people to deal with the big issues of society and the world. Blikstein also argues that the best approach to learning can be found in the makerspace model, where project-based learning, learning that is meaningful to kids, student agency, inclusive environments, new types of skills, using technology as an expressive material are already happening.

According to [18] the maker approach stems from the concept of craftsmanship, although in a broader sense that is not only related to manual labor, but also to intellectual activities, such as art, mastery, and computer programming. Ranieri states that contemporary craftspeople are experts at using technology; they openly share information, promote the circulation of ideas, and are candid about problems and how to solve them. This maker approach should also be applied in educational settings, by bending technology to students' needs and promoting new learning paths.

The same ideas are presented by Aprile in [16]. As a teacher, Aprile agrees with Blikstein [14] that the current vision of school is outdated, especially given the challenges of a globalized, hyperdynamic world. Aprile compares schools to a factory production line where the pedagogical models, the buildings and the way time is organized limit the ability to take account of every student's specific skills and awareness. The author proposes the term "edumaker" to represent a teacher–maker in education with a proactive attitude and a desire to get involved personally in the learning process (analytical thinking, creativity, problem-solving, collaborative working).

Setting up a makerspace in a school is a very challenging process, but it demands a vision that is shared by both the school and the maker to evaluate opportunities and establish terms that are beneficial to both. In [13], the author proposes a "Makerspace Manifesto" which schools and makers can use immediately to organize a collaboration. The arrangement ensures benefits for both partners: schools provide the furniture, and makers can use it and the makerspace outside of school hours. Makers can train teachers in laboratory practices, to empower them to offer maker activities during curricular lessons.

In order to get to a place where it can distance itself from the face-to-face lesson model and embrace a variety of teaching methods and learner-centered strategies, including with technological support, school needs to re-think its very foundations. In [12], Borri refers to the "1 + 4 Learning Spaces for a New Generation of Schools" framework, whereby learning spaces should be designed for a range of different activities, featuring task-oriented zones fitted with flexible furniture and a variety of educational materials and tools. Each 1 + 4 space is suitable for a different type of student learning activity, whether it is individual work or teacher-supervised: Group Space, Exploration Lab, Agora, Individual Area, Informal Area.

The laboratory is also central to the research reported in [15]. Here it is an educational space and a privileged setting, where research and training can take place and new knowledge can be developed. Zecca describes the results of research in primary school students where robots are used to develop research skills, such as observation, explanatory hypothesis formulation, hypothesis testing, and review of hypotheses in light of the results observed, using a dialogical approach. In this research, children did not have to program robots but rather to explain the robot's behavior, in order to develop abstract thinking and problem-solving skills.

In [17], Daniela suggests we should reflect on technology-enhanced learning and the place and role of technology in education. The author identifies five different objectives for learning technology in educational models: learning technology that supports the knowledge acquisition process; learning technology that provides access to knowledge; learning the principles of developing new technologies and finding new creative solutions; the interrelationships of learning motivation and learning achievements; evaluating learning achievements.

The use of technology, especially robotics, in educational settings and for social purposes has led the research community to question its impact on ethical, legal and social aspects. In [19] Operto describes robotics as a very powerful tool for studying and increasing our knowledge, not only of the universe around us—space, oceans, our bodies—but also our brains/minds. On the other hand, robotics is going to be pervasive in the world, and ethical issues need to be addressed so we can define (the extent and limits of) human responsibility and machine autonomy, in cases of damage caused by a learning robot. Roboethics, the field that studies ELS issues, is going to have an impact on the design, programming, shape and use of robots.

4 Conclusions

> The information society is a neo-manufacturing society, in which data, information and digital processes are raw materials. We need to teach the languages of digital design; in other words, not only Italian and English, but also mathematics, statistics, industrial design, computer science, law, the language of science, architecture, art, music, and dead languages. We need to be able to read and write digitally so we can be critical and informed creators and caretakers of the world around us.[3]

We can refer to these statements made by Floridi a few years ago in an Italian newspaper, and use them as an expression of our expectations and hopes for a school that creates citizens of the world and not consumers.

The practice of a maker education can contribute to this aim in two ways: first, by providing a different technological awareness; second, but no less important, by setting out a methodological approach that allows the adoption of new and different attitudes to fundamental issues, such as equity, gender issues sustainability and many more.

References

1. Rossi, P.: La scienza e la filosofia dei moderni: aspetti della rivoluzione scientifica. Bollati Boringhieri (1989)
2. Floridi, L.: Philosophy and Computing: An Introduction. Routledge, London and New York (1999)
3. Floridi, L.: The Fourth Revolution: How the Infosphere is Reshaping Human Reality. OUP, Oxford (2014)
4. New Media Consortium, Horizon Report: 2017 Higher Education Edition. https://www.unmc.edu/elearning/_documents/NMC_HorizonReport_2017.pdf (2017)
5. Floridi, L.: The Onlife Manifesto: Being Human in a Hyperconnected Era, p. 264. Springer Nature (2015)
6. Schön, S., Ebner, M., Kumar, S.: «The Maker Movement Implications from Modern Fabrication, New Digital Gadgets, and Hacking for Creative Learning and Teaching». eLearning Papers Special Edition, 86–100 (2014)
7. Blikstein, P.: Maker movement in education: History and prospects. Handbook Technol. Educ. pp. 419–437 (2018)

[3] https://www.ilsole24ore.com/art/il-design-futuro-sara-gruppo--riscoperta-sapere-umanistico-fa-evolvere-l-uomo-tecnologico-AEw2R0GF.

8. Jenkins, H.: Confronting the Challenges of Participatory Culture: Media Education for the 21st Century, p. 145. The MIT Press (2009)
9. Himanen, P.: The Hacker Ethic. Random House (2001)
10. Torvalds, L., Diamond, D.: Just for Fun: The Story of an Accidental Revolutionary. Harper Audio (2001)
11. Hatch, M.: The Maker Movement Manifesto, p. 11. McGraw-Hill Education (2013)
12. Borri, S.: From the classroom to the learning environment. In: Scaradozzi, D., Guasti, L., Di Stasio, Miotti, B., Monteriù, A., Blikstein, P. (eds.) Makers at School, Educational Robotics and Innovative Learning Environments—FabLearn Italy 2019. Springer (in press)
13. Nulli, G.: School makerspace manifesto. In: Scaradozzi, D., Guasti, L., Di Stasio, Miotti, B., Monteriù, A., Blikstein, P. (eds.) Makers at School, Educational Robotics and Innovative Learning Environments—FabLearn Italy 2019. Springer (in press)
14. Blikstein, P.: If we could start from scratch, what would schools look like in the 21st century? Rethinking Schools as a Locus for Social Change. In: Scaradozzi, D., Guasti, L., Di Stasio, Miotti, B., Monteriù, A., Blikstein, P. (eds.) Makers at School, Educational Robotics and Innovative Learning Environments—FabLearn Italy 2019. Springer (in press)
15. Zecca, L.: The game of thinking. Interactions between children and robots in educational environments. In: Scaradozzi, D., Guasti, L., Di Stasio, Miotti, B., Monteriù, A., Blikstein, P. (eds.) Makers at School, Educational Robotics and Innovative Learning Environments—FabLearn Italy 2019. Springer (in press)
16. Aprile, D.: Makers in education: teaching is a hacking stuff. In: Scaradozzi, D., Guasti, L., Di Stasio, Miotti, B., Monteriù, A., Blikstein, P. (eds.) Makers at School, Educational Robotics and Innovative Learning Environments-FabLearn Italy 2019. Springer (in press)
17. Daniela, L.: Pedagogical considerations for technology-enhanced learning. In: Scaradozzi, D., Guasti, L., Di Stasio, Miotti, B., Monteriù, A., Blikstein, P. (eds.) Makers at School, Educational Robotics and Innovative Learning Environments—FabLearn Italy 2019. Springer (in press)
18. Ranieri, M.: Making to learn. The pedagogical implications of making in a digital binary world. In: Scaradozzi, D., Guasti, L., Di Stasio, Miotti, B., Monteriù, A., Blikstein, P. (eds.) Makers at School, Educational Robotics and Innovative Learning Environments—FabLearn Italy 2019. Springer (in press)
19. Operto, F.: Elements of roboethics. In: Scaradozzi, D., Guasti, L., Di Stasio, Miotti, B., Monteriù, A., Blikstein, P. (eds.) Makers at School, Educational Robotics and Innovative Learning Environments—FabLearn Italy 2019. Springer (in press)

Making: Laboratory and Active Learning Perspectives

Margherita Di Stasio

Abstract In a world where technology is an ordinary part of everyday life, it is particularly important for making, coding, and educational robotics to feature in school programs. On one hand, the construction elements that are peculiar to digital fabrication make the pedagogical perspective active; on the other, they can support the development of active teaching practices. Although it appears to be a kind of revolution, something that is absolutely new, on deeper reflection, it is clear that making and coding can play a role in school as part of an important pedagogical tradition.

Keywords Active learning · Making · Coding · Educational robotics · Active pedagogical perspective · Laboratory

1 Introduction

A fundamental trend in public opinion and in 21st-century policy-making worldwide has been the call for the digital and technological development of school.

From the Recommendation of the European Parliament and of the Council on key competences for lifelong learning, published in 2006 and revised in 2018, to the Every Student Succeeds Act passed in 2015–2016, technology has an important role to play in lifelong education. Beginning as a separate competence for the purpose of achieving confident use of computers and the opportunities offered by the web, technological competence has gradually become a multidimensional competence, not only oriented to the use of tools. The technological evolution of the world has brought the need for all people, and particularly the young generations, to develop another kind of digital consciousness based on a creative and constructive commitment.

From an educational perspective, it is no longer sufficient to teach the use of computers or technological artifacts for everyday life and work proficiency; it is also

M. Di Stasio (✉)
Istituto Nazionale Documentazione Innovazione Ricerca Educativa (Indire), Florence, Italy
e-mail: m.distasio@indire.it

D. Scaradozzi et al. (eds.), *Makers at School, Educational Robotics and Innovative Learning Environments*, Lecture Notes in Networks and Systems 240,
https://doi.org/10.1007/978-3-030-77040-2_2

necessary to foster engagement with technology through creative and constructive paths, such as coding and making.

From this point of view, making can be seen as an active pedagogical approach.

2 Making as a Bridge Between Pedagogical Tradition and Technological Innovation

Since the beginning of the twentieth century, many key authors have stressed the importance of the learning process and the role that experience can play, with a particular focus on experience and learning by doing.

According to Dewey (1938), it is important for students to face a problem emerging from a real situation related to their experience, in order to stimulate thought. It is important for students to handle the elements necessary for solving the problem; it is also important for them to be able to develop solutions that they can test in practice.

In order to acquire persistent and significant knowledge, students need to be involved actively, in a learning process that is linked to their reality and their experiences.

Technology and technological instruments can be included in students' experience and are part of their surroundings. Technology, machines and techniques can also be a functional part of the learning environment. Célestin Freinet gives us an in-depth analysis of this point.

The classroom becomes a learning environment when it evolves through technologies. Every new technology or technique presupposes that it will be introduced into the work cycle of a new instrument that is better suited to meeting the needs of the technical and social environment. According to Freinet (1967), for school to be improved, school equipment needs to be modernized. This does not mean merely buying new materials. It demands a profound change in pedagogical terms in order to construct what Freinet describes as the school of the future, centered on children as community members (Freinet 1957).

In this brief historic excursus, we discover Seymour Papert's constructionism to be the closest precursor to maker pedagogy. Concerning the role of technologies in education, as noted, "Papert advocates technology in schools not as a way to optimize traditional education, but rather as an emancipatory tool that puts the most powerful construction materials in the hands of children—again, another idea that inspired the resurgence of the 'maker' sensibilities" (Blikstein 2013). The deep connection between constructionism and making is also evident in Papert's (1991) words when he suggests that constructionism should be thought of as "learning-by-making." He defines the constructivist view of learning as "building knowledge structures."

This leads us to think of *making* as a manipulative version of knowledge-building processes and to reflect on the tradition of the collaborative construction of knowledge (Scardamalia and Bereiter 2006; Brown and Campione 1994).

In the maker approach we can find the main features outlined in this excursus: experience and learning by doing; technologies as a basis for rethinking the learning environment; manipulation and construction of knowledge as a social and shared activity.

3 Technology, People, Society

As we can see in the contributions collected in this section, some features of making in the educational context can manifest in a specific way. The technical aspect can be read in terms of the relationship between people and technology, the importance of prototyping for building knowledge, the innovation of teaching practices through technology, method and environment; the confidence that the maker approach can build through technology can finally make way for the specific aspect of digital citizenship.

3.1 Experiences and Point of View

Considering Freinet, Malaguzzi and Munari's pedagogical models, in her contribution *The Maker Movement: From the Development of a Theoretical Reference Framework to the Experience of DENSA Coop. Soc*, Valentina Costa emphasizes the role of techniques as a basis for sharing knowledge construction as a distinctive aspect of the maker movement approach.

The relationship between the technical aspects and the educational impact of the use of the 3D printer is the focus of Di Tore et al. in *Learning by Making: 3D Printing Guidelines for Teachers*. Specifically, they analyze the impact of 3D printer features, such as resolution, type of materials and timings, in relation to their use in subject teaching.

In *Furniture Design Education with 3D Printing Technology*, Meltem Eti Proto and Ceren Koc Saglam consider the use of the 3D printer in the construction of a design method based on the experience of a university furniture design course.

In *Makerspaces for Innovation in Teaching Practices*, Giuseppe Alberghina approaches the theme of rapid prototyping as the basis for the integration between fab labs, makerspaces and school, and for innovating the education system through traditional teaching practices and making.

With a focus on the potential connection between making, coding and other methods, in *Service Learning: a Proposal for the Maker Approach*, Frazzarin and Leonori present an experience of service learning during which coding activities take place.

Also, organizational changes are needed for a real innovation in practices: this is the perspective with which Ricciardi et al. view a of teacher training experience for tinkering activities in their contribution *Officina degli Errori: An Extended Experiment to Bring Constructionist Approaches to Public Schools in Bologna.*

In *Fab the Knowledge*, Scataglini and Busciantella Ricci focus on the collaborative dimension that making can support through participatory design processes and underline the possibilities that making can open up in terms of project-based and inquiry-driven education.

In *Chesscards: Making a Paper Chess Game with Primary School Students, a Cooperative Approach*, Agnese Addone and Luigi De Bernardis underline motivation and the engagement of students involved in a learning process based on a collaborative and active approach.

Also, the construction, implementation and setup of a fab lab can be a process that strengthens a community by connecting different realities in the territory.

In *Montessori Creativity Space: Making a Space for Creativity*, Fattizzo and Vania describe and analyze the process of implementing a makerspace in a school based on Malaguzzi's concept of atelier, as an experience of collaboration by the entire school community.

In *Museum Education Between Digital Technologies and Unplugged Processes. Two Case Studies*, Carlini describes an experience of museum education in which the setup of a school fab lab is part of a process that involves several institutions in the territory: school, museum, and municipality.

The introduction of a maker approach in the educational context involves a new kind of relationship between people and technology.

On one hand, from a maker perspective it is possible to gain a deeper understanding of technology.

Going in this direction is Anatasia Pyrini's contribution, *Teaching Environmental Education Using an Augmented Reality World Map*, which discusses the theme of digital citizenship and competence, and looks at how this can be developed at school, with students and with teachers.

In *Roboticsness—Gymnasium mentis*, Lisimberti and Aprile present an experience in which making and educational robotics are the paths to bring school closer to everyday digital and technological reality, in which students acquire skills to become prosumers who can take part in and navigate a complex world.

On the other hand, technology tends to adapt to the needs of school, placing itself in a human-centered perspective. This kind of process is described in *A New Graphic User Interface Design for 3D Modeling Software for Children* where Giraldi et al. put forward a concept for a graphic user interface designed for preschool children.

4 Conclusions

The making approach can bring a new way of achieving several didactic aims in school. On one hand, maker pedagogy can update to the digital era what we can consider the *magna pars* of an active approach to contemporary education. On the other, the awareness of technology that is gained from making can build—in addition to technological competence—the technological literacy that is needed for full digital citizenship.

In this evolution of the relationship between school, technology and lifelong learning, making and coding take their place in education, not only as education technology and practical activities, but also as basic ideas for defining "clusters of innovative pedagogical approaches" (Paniagua and Istance 2018).

References

Blikstein, P.: Digital fabrication and 'making' in education: the democratization of invention. FabLabs: Mach. Makers Inventors **4**(1), 1–21 (2013)

Brown, A.L., Campione, J.C.: Guided discovery in a community of learners. In: McGilly, K. (ed.) Classroom Lessons: Integrating Cognitive Theory and Classroom Practice. MIT Press/Bradford Books, Cambridge, MA (1994)

Dewey, J.: Experience and Education. Macmillan, New York (1938)

Freinet, C.: L'école moderne française: guide pratique pour l'organisation matérielle, technique et pédagogique de l'Ecole Populaire. Editions Rossignol, Vienne (1957)

Freinet, C.: Les techniques Freinet de l'école moderne. Armand Colin, Paris (1967)

Paniagua, A., Istance, D.: Teachers as Designers of Learning Environments: The Importance of Innovative Pedagogies, Educational Research and Innovation, OECD Publishing. Paris (2018). https://doi.org/10.1787/9789264085374-en

Papert, S., Harel, I.: Situating constructionism. Constructionism **36**(2), 1–11 (1991)

Scardamalia, M. & Bereiter, C.: 'Knowledge Building: Theory, Pedagogy, and Technology', The Cambridge Handbook of the Learning Sciences, 97–115, (2006).

Robotics in Education: A Smart and Innovative Approach to the Challenges of the 21st Century

Laura Screpanti, Beatrice Miotti, and Andrea Monteriù

Abstract Robotics in Education (RiE) is a broad term that refers to a variety of applications. Robots can enhance learning and teaching, but they can also help overcome impairments, whether physical or social. Even though the advantages of bringing new technologies into schools are clear, the lack of a well-established set of good practices, assessment of experiences, and tools slows down their adoption. This chapter aims to highlight the key points that emerge from the recent enhancements in RiE. First, the market and research are continuously developing new tools for school to try to meet needs and tailor products. Second, there is a wealth of formal and non-formal experiences, both in the literature and in school activities. Third, research is still validating tools and methodologies that will assess the impact of introducing robotics into education. Despite the wide availability of tools and experiences, there is still a certain degree of uncertainty about how to cope with technology in education and how to evaluate the outcomes of such activities. The increasing cross-pollination between schools and researchers from different fields is producing valuable experiences that will soon close the gap.

Keywords Educational robotics · Social robotics · Assistive robotics · STEM · Digital skills

L. Screpanti (✉) · A. Monteriù
Dipartimento di Ingegneria dell'Informazione, Università Politecnica delle Marche, Ancona, Italy
e-mail: l.screpanti@univpm.it

A. Monteriù
e-mail: a.monteriu@univpm.it

B. Miotti
Istituto Nazionale di Documentazione, Innovazione e Ricerca Educativa (Indire), Florence, Italy
e-mail: b.miotti@indire.it

D. Scaradozzi et al. (eds.), *Makers at School, Educational Robotics and Innovative Learning Environments*, Lecture Notes in Networks and Systems 240,
https://doi.org/10.1007/978-3-030-77040-2_3

1 Introduction

Technology has always been the key to any human's achievement and progress. Modern technologies are becoming increasingly complex and connected. Dealing with such technologies requires a lot more than numeracy and literacy skills. Local and national governments are taking measures to empower citizens with the key competencies they need to live and work in the society of today and tomorrow. Thus, one challenge is to train people in enabling technologies to ensure a workforce that is skilled in them. Another challenge is to transform education so that everyone masters technology rather than simply using it. In order to solve this last issue, governments have set up more comprehensive strategies. For example, Italy's National Plan for Digital Education (PNSD) is a long-term plan aimed at introducing digital infrastructure, digital skills and robotics into schools.

Supported by governments and welcomed by teachers with a mixture of enthusiasm and mistrust, technologies and robotics have filtered out of research laboratories and industry and have finally entered schools. Some of these technologies are intended to be used in the classroom to overcome impairments and to assist the teaching–learning process. Others are intended to be "objects to think with" [1].

This chapter aims to highlight the key points that emerge from the recent enhancements in Robotics in Education (RiE).

2 Robotics in Education

Many fields have greatly benefitted from achievements in the field of robotics. Education has a variety of challenges that tools and methods from robotics can help solve [2]. Specifically, robotics provides a unique learning experience, with helpful solutions for every student's learning needs. As such, RiE encompasses a variety of well-established sub-areas of robotics. Considering the applications of robotic devices in the field of education, it is possible to identify four main areas: assistive robotics, social robotics, socially assistive robotics, and educational robotics (ER) [3]. Each of these areas is applied in the field of education for different purposes. Assistive robotics in education can help overcome physical impairments that might prevent learning and teaching, thereby contributing to well-being and inclusiveness [4, 5]. Social robots in education are either students' tutors or companions, engaging their interest and transforming lessons into interactive and connected learning environments. Socially assistive robots help students reduce a social impairment, assisting users through social rather than physical interaction [6]. Founded on constructionism and using re-programmable robotic kits—usually consisting of several unassembled components—ER helps students develop many technology- and subject-related competences [7]. Hard skills and soft skills (i.e., communication, teamwork) can both be developed through ER activities.

The wealth of robotic devices brings many opportunities but also new challenges. An effort should be made to evaluate their effectiveness in engaging learners and developing competences [2, 8] and to train teachers in pedagogical and technological aspects [9]. Technological transparency may be regarded as an advantage in social and assistive robots, as it enables end users to easily manage and seamlessly accept their device. On the contrary, it can be a disadvantage when it comes to ER [3, 10].

3 Trends and Perspectives

Parts V and VI of this book propose many interesting ideas and perspectives about RiE, emphasizing three key points: good practices, assessment of experiences, and development of tools.

3.1 Good Practices

Good practices for robotics as an educational methodology should be a reference for other experiences. They should consist of a set of common features that can be compared (i.e., age of participants, learning environment, etc.). Many of the activities described in parts V and VI of this book have been done at schools as an occasional or short-term project [11–13] or as a comprehensive, organic experience [14, 15].

The authors in [12] describe how primary school students built a game about escaping safely from an earthquake and saving their town's cultural heritage. Robotic kits, coding platforms, and QR codes can all help to boost school performance, motivation, interest in science, technology, engineering, and mathematics (STEM), and teamwork. Most importantly, students learned how to behave during an earthquake and got to know their local culture. Not only can robotics be used to gamify learning, it can also introduce students to other school subjects. The authors in [11] describe how to introduce primary school students to music, science, mathematics, geography, critical thinking and technology through Nintendo Labo technology and many other materials from the real world. This helped enhance the "spiral of creative learning," not only through immersive technology or prefabricated robotic kits, but also through specially designed authentic learning and tinkering activities that take place in innovative learning environments. Similarly, [13] describes an ER activity for developing logic, mathematics and physics skills. Starting from an analysis of a story by Dino Buzzati, students are asked to use the information retrieved from the text to model the behaviors of characters and to solve simple prediction problems. Interestingly, the author observes three approaches to the problem: empirical (simulating on paper the possible strategy to solve the problem), algebraic (looking for a relationship between two variables), and physical-algebraic (using notions from both mathematics and physics).

Many schools are enlisting robotics in their usual activities to bring a new approach to teaching STEM subjects, and to increase knowledge and competitiveness in scientific and technological fields. Notably, the authors in [14] promoted the culture of technology and science across Europe, improving achievement and motivation, right from primary school. To this end, secondary school students developed the mobile "Robot for Geometry" (R4G) for teaching mathematics, geometry, and fractions to younger students. Furthermore, the authors in [15] report a unique experience in Italy in which robotics is introduced as a primary school subject, and describe the learning objectives, lesson plans, and robotic devices involved. Notably, this curriculum has already been through its first phase of validation and has also been improved. Moreover, topics like the Internet of Things and distributed control were added to update the educational path.

Outreach activities complement formal learning in a variety of ways. Examples in parts V and VI of this book include a coding club [16], a global robotics contest [17], and an international project [18].

The author in [16] describes the experience of the Pomezia CoderDojo, where the MBot is used not only for educational purposes, but also for improving collaborative and social skills. The experience looks at the Seymour Papert learning-by-doing methodology as the best way to deploy the potential of coding and ER.

The authors in [17] aim to fill the gap between the world of scientific and educational competitions, focusing on participants, their knowledge of robotics and the hardware involved. The RoboCup@Home Education Project has created a new set of organization rules and standardized courseware and platforms that users have access to. Notably, organizers scaled up the project around the world and are now getting deeper into schools, providing self-paced training for teachers and setting up a community for the exchange of good practices.

Finally, the authors in [18] present the preliminary results of a comparative research on robotics education from a media education point of view. High school students from Italy and the Democratic Republic of the Congo performed a SWOT analysis of robotics after a lecture they received on the European recommendations to the Commission on civil law rules on robotics. This analysis is the starting point for a more comprehensive look into what robotics means for their lives and for society at large.

3.2 Assessment

Robotics was employed in several projects to develop knowledge of a wide range of subjects, hard and soft skills, and even complex competencies, or to raise awareness about social and environmental issues. Despite this wealth of robotic applications in education, many stakeholders at different levels advocate for measurable indexes of the effectiveness of RiE. The authors in [19] begin by acknowledging the results corroborating the enthusiasm with which new learning theories and methodologies have been met in schools. They then move on to highlight the lack or underuse of models and tools for assessing relatively new educational activities, like robotics, coding, making, tinkering. Hence, teachers and educators should be trained not only in pedagogical matters and educational paradigms but also in assessment models, to empower them with suitable tools for the digital context. Furthermore, the study carried out by the authors in [20], whose aim is to evaluate the effect of learning platforms, highlights that modern pedagogy should look for ways to bridge the gap between how students learn and how teachers teach. New and innovative teaching methods and learning environments are necessary for preparing students for life in a constantly changing society.

An automated evaluation system is the proposal of the authors in [21], to support teachers by seamlessly collecting data from an ER activity, and autonomously assessing how students cope with it. Exploring a machine learning approach and data mining techniques, they analyze real-world data and find three main behaviors: "Mathematical strategy", "Incremental Tinkering approach" and "Irregular Tinkering approach", which are closely linked to the "planner scientist" and "bricoleur scientist" [22] and echo the results observed by [13].

Teachers can also benefit from collaboration with universities. Cross-pollination between school education and academic research can lead to sound research design and a stronger educational experience. This is the case for the authors in [23], who report an example of a cross-pollination project in a lower secondary school. The Bricks for Kidz (B4K) methodology and ER are brought into a curricular setting and used to propose didactic and cross-curricular targets and define criteria for assessing student performance. The results of the research highlight the strong need for teacher training in pedagogical and didactic methodologies. The authors in [24] show how the number and the quality of social relationships can be analyzed in a formal learning environment during ER using quantitative techniques, like sociograms and questionnaires. The findings of this study report the effectiveness of short ER activities for stimulating social skills, but they also highlight that more studies are needed to explore the complex world of student relations.

Assessing students' relationships and expectations is very important when it comes to the gender issue. Based on data collected during ER and coding activities throughout Europe, the authors of [25] find that only a small percentage of female students pursue a career or studies in computer science and technology. According to their analysis, the barriers to such careers can be found in girls' self-limiting beliefs. These, in turn, stem from the low expectations of parents, peer groups, the school system, and society, and eventually lead many to choose secondary schools on the basis of perceived possibilities for working while taking care of their families. A job in a technical or scientific field is seen as being in conflict with the ability to care for a family. Coding and robotics activities might not be the solutions to the problem, but they can be used to build self-confidence and problem-solving abilities, which in turn can help to evaluate one's inner strengths.

3.3 Technological Development

Industry 4.0, Society 5.0, Education 2.0 are all broad terms that point to the deep impact of technology on every aspect of human life. Technological developments are at the core of the change in perspective that many fields are witnessing. Low costs and wide availability allow teachers to bring affordable instruments into schools for innovative activities at different levels.

Flying robots, for example, and quad-rotor micro aerial vehicles (MAVs), in particular, are the most appealing vehicles recently introduced as inexpensive, straightforward educational platforms for teaching the behavior and control of MAVs to students in primary schools (Tynker and Blocky), secondary schools and universities (JavaScript and Python, MATLAB & Simulink). As shown in [26], it can even support a robust nonlinear control algorithm (sliding mode control), thus demonstrating the flexibility and versatility of this novel educational ecosystem.

Mobile robots like the one described in [27] bring several applications into the classroom, enabling students to explore several aspects of technology, from the design of functional hardware to the development of sophisticated software. The fully open source and open hardware robot based on the Arduino platform and Robotic Operating System can explore the environment and can track and recognize objects via machine learning software.

The degree of complexity and transparency of the tools brought to students can vary according to age, grade, competence level, learning objective, and funding availability. The same open-source low-cost single-board microcontroller, Arduino, can improve laboratory practice in upper secondary schools and change students' attitudes towards STEM subjects. The authors in [28] report a series of experiments using this platform as an interface for data acquisition and for building a simple prototype of a rover.

Teachers need to keep track of all the different tools available, to choose the most suitable for their needs, and create an activity that can target the intended outcome, whether it is computational thinking or teamwork, communication, and mathematics. All of this work requires teachers to improve their digital skills and to enhance pedagogical and didactic innovation. The Weturtle.org platform presented in [29] draws on the concept of "Community of Practice" and the Technological Pedagogical Content Knowledge (TPCK) model to promote active use of the community, the training of teachers, and to enable more experienced teachers to become trainers themselves.

Both teachers and students increase their interest in robotics when they understand that it can help themselves or others to reduce a barrier. In fact, students can use robots not only as platforms on which to learn robotics, but also for assistance during the learning process. Assistive robotics provides particular benefits for education, as it can reduce the impact of impairments and increase autonomy. The authors in [30] present a proof of concept for a smart wheelchair with localization and navigation capabilities, which can be integrated with an academic management system to enable students who cannot walk on their own to reach any academic building or room on a university campus autonomously.

4 Conclusions

The picture that emerges of Robotics in Education is one of a wide variety of applications available to help students and teachers in their everyday lives. Plenty of technological tools and systems are available to help meet the needs of different tasks and age levels.

This means that teachers can create student-tailored and student-centered activities, often with a hands-on learning approach. In order to learn how to exploit technology in education, teachers need to be trained, not only in the technology itself, but also in integrating technology into their educational practice, for both instructional design and docimology. Some support comes from the technology itself, with platforms for online learning and communities of practice, where materials can be shared without time and geographical restrictions. Moreover, cutting-edge research that combines learning analytics and educational data mining is providing further insight into the learning process, and, eventually, it can provide teachers with a decision support system grounded in evidence-based education.

Despite the wide availability of tools, there is still a certain degree of uncertainty about how to deal with technology in classrooms and how to evaluate the outcomes of such activities. In fact, even if the technological developments are driving the digital revolution in education, the lack of guidelines on how to integrate robotics into education, and the difficulty assessing the complex competencies that such new practices bring forth are preventing teachers and schools from fully exploiting and exploring the benefits that technology could bring. On the other hand, the increasing cross-pollination between schools and researchers from different fields is producing valuable experiences in technology-enhanced teaching.

In conclusion, schools and stakeholders in education are still exploring how best to exploit the benefits of new technological developments. Tech developers, in turn, are working closely with pedagogical experts to meet the needs of teachers, students, and educators. In today's hyper-connected, digitized society, education cannot neglect to include new forms of literacy that will increase students' knowledge, skills, attitudes and values helping them to fulfil their potential and contribute to the well-being of their communities and environment.

References

1. Papert, S.: Mindstorms Brighton. Harvester Press (1980)
2. Benitti, F.V.B.: Exploring the educational potential of robotics in education: a systematic review. Comput. Educ. **58**(3), 978–988 (2012)
3. Scaradozzi, D., Screpanti, L., Cesaretti, L.: Towards a definition of educational robotics: a classification of tools, experiences and assessments. In: Daniela, L. (ed.) Smart Learning with

Educational Robotics—Using Robots to Scaffold Learning Outcomes, pp. 63–92. Springer, Berlin (2019)
4. Ciuccarelli, L., Freddi, A., Longhi, S., Monteriù, A., Ortenzi, D., Pagnotta, D.P.: Cooperative robots architecture for an assistive scenario. In: 2018 Zooming Innovation in Consumer Technologies Conference (ZINC), Novi Sad, 2018, pp. 128–129
5. Foresi, G., Freddi, A., Monteriù, A., Ortenzi, D., Pagnotta, D.P.: Improving mobility and autonomy of disabled users via cooperation of assistive robots. In: 2018 IEEE International Conference on Consumer Electronics (ICCE), Las Vegas, NV, USA (2018)
6. Pivetti, M., Di Battista, S., Agatolio, F., Simaku, B., Moro, M., Menegatti, E.: Educational robotics for children with neurodevelopmental disorders: a systematic review. Heliyon **6**(10) (2020)
7. Prist, M., Cavanini, L., Longhi, S., Monteriù, A., Ortenzi, D., Freddi, A.: A low cost mobile platform for educational robotic applications. In: 10th IEEE/ASME International Conference on Mechatronic and Embedded Systems and Applications, Conference Proceedings, Senigallia, Italy, September 10–12, 2014
8. Scaradozzi, D., Cesaretti, L., Screpanti, L., Mangina, E.: Identification of the students learning process during education robotics activities. Front. Robot. AI **7**, 21 (2020)
9. Scaradozzi, D., Screpanti, L., Cesaretti, L., Storti, M., Mazzieri, E.: Implementation and assessment methodologies of teachers' training courses for STEM activities. Technol. Knowl. Learn. **24**, 247–268 (2019)
10. Alimisis, D., Alimisi, R., Loukatos, D., Zoulias, E.: Introducing maker movement in educational robotics: beyond prefabricated robots and "black Boxes". In: Daniela, L. (eds.) Smart Learning with Educational Robotics, pp. 93–115. Springer, Cham (2019)
11. Gagliardi, M., Bartolucci, V., Scaradozzi, D.: Nintendo Labo for educational robotics at the primary school. In: Scaradozzi, D., Guasti, L., Di Stasio, Miotti, B., Monteriù, A., Blikstein, P. (eds.) Makers at School, Educational Robotics and Innovative Learning Environments—FabLearn Italy 2019. Springer (in press)
12. Pazzaglia, P., Scaradozzi, D.: Escape from Tolentino during an earthquake saving more lives and cultural heritage objects as you can. In: Scaradozzi, D., Guasti, L., Di Stasio, Miotti, B., Monteriù, A., Blikstein, P. (eds.) Makers at School, Educational Robotics and Innovative Learning Environments—FabLearn Italy 2019. Springer (in press)
13. Torre, M.: Buzzati robots. In: Scaradozzi, D., Guasti, L., Di Stasio, Miotti, B., Monteriù, A., Blikstein, P. (eds.) Makers at School, Educational Robotics and Innovative Learning Environments—FabLearn Italy 2019. Springer (in press)
14. Cantarini, M., Polenta, R.: Good educational robotics practices in upper secondary school in European projects. In: Scaradozzi, D., Guasti, L., Di Stasio, Miotti, B., Monteriù, A., Blikstein, P. (eds.) Makers at School, Educational Robotics and Innovative Learning Environments—FabLearn Italy 2019. Springer (in press)
15. Valzano, M., Vergine, C., Cesaretti, L., Screpanti, L., Scaradozzi, D.: Ten years of educational robotics in primary school. In: Scaradozzi, D., Guasti, L., Di Stasio, Miotti, B., Monteriù, A., Blikstein, P. (eds.) Makers at School, Educational Robotics and Innovative Learning Environments—FabLearn Italy 2019. Springer (in press)
16. Cannone, L.: Educational robotics in informal contexts: an experience at CoderDojo Pomezia. In: Scaradozzi, D., Guasti, L., Di Stasio, Miotti, B., Monteriù, A., Blikstein, P. (eds.) Makers at School, Educational Robotics and Innovative Learning Environments—FabLearn Italy 2019. Springer (in press)
17. Iocchi, L., Tan, J.C.C., Castro S.: RoboCup@Home Education: a new format for educational competitions. In: Scaradozzi, D., Guasti, L., Di Stasio, Miotti, B., Monteriù, A., Blikstein, P. (eds.) Makers at School, Educational Robotics and Innovative Learning Environments–FabLearn Italy 2019. Springer (in press)
18. Todino, M.D., De Simone, G., Kidiamboko, S., Di Tore, S.: European recommendations on robotics and their issues on education in different countries. In: Scaradozzi, D., Guasti, L., Di Stasio, Miotti, B., Monteriù, A., Blikstein, P. (eds.) Makers at School, Educational Robotics and Innovative Learning Environments—FabLearn Italy 2019. Springer (in press)

19. Tegon, R., Labbri, M.: Growing deeper learners. In: Scaradozzi, D., Guasti, L., Di Stasio, Miotti, B., Monteriù, A., Blikstein, P. (eds.) Makers at School, Educational Robotics and Innovative Learning Environments—FabLearn Italy 2019. Springer (in press)
20. Rudolfa, A., Daniela, L.: Learning platforms in the context of education digitization as strong innovative character with respect to education methodologies applied—Experience of Latvia. In: Scaradozzi, D., Guasti, L., Di Stasio, Miotti, B., Monteriù, A., Blikstein, P. (eds.) Makers at School, Educational Robotics and Innovative Learning Environments—FabLearn Italy 2019. Springer (in press)
21. Cesaretti, L., Screpanti, L., Scaradozzi, D., Mangina, E.: Analysis of educational robotics activities using a machine learning approach. In: Scaradozzi, D., Guasti, L., Di Stasio, Miotti, B., Monteriù, A., Blikstein, P. (eds.) Makers at School, Educational Robotics and Innovative Learning Environments—FabLearn Italy 2019. Springer (in press)
22. Turkle, S., Papert, S.: Epistemological pluralism and the revaluation of the concrete. J. Math. Behav. **11**(1), 3–33 (1992)
23. Vitti, E.V., Parola, A., Sacco, M.M., Trafeli, I.: Learning technologies for curricular STEAM skills. In: Scaradozzi, D., Guasti, L., Di Stasio, Miotti, B., Monteriù, A., Blikstein, P. (eds.) Makers at School, Educational Robotics and Innovative Learning Environments—FabLearn Italy 2019. Springer (in press)
24. Screpanti, L., Cesaretti, L., Storti, M., Scaradozzi, D.: Educational robotics and social relationship in the classroom. In: Scaradozzi, D., Guasti, L., Di Stasio, Miotti, B., Monteriù, A., Blikstein, P. (eds.) Makers at School, Educational Robotics and Innovative Learning Environments—FabLearn Italy 2019. Springer (in press)
25. Bagattini, D., Miotti, B., Operto, F.: Educational robotics and gender perspective. In: Scaradozzi, D., Guasti, L., Di Stasio, Miotti, B., Monteriù, A., Blikstein, P. (eds.) Makers at School, Educational Robotics and Innovative Learning Environments—FabLearn Italy 2019. Springer (in press)
26. Corradini, M.L., Ippoliti, G., Orlando, G., Terramani, S.: Study and development of robust control systems for educational drones. In: Scaradozzi, D., Guasti, L., Di Stasio, Miotti, B., Monteriù, A., Blikstein, P. (eds.) Makers at School, Educational Robotics and Innovative Learning Environments-FabLearn Italy 2019. Springer (in press)
27. Di Dio Bruno, G.: Erwhi Hedgehog: a new learning platform for mobile robotics. In: Scaradozzi, D., Guasti, L., Di Stasio, Miotti, B., Monteriù, A., Blikstein, P. (eds.) Makers at School, Educational Robotics and Innovative Learning Environments—FabLearn Italy 2019. Springer (in press)
28. Marzoli, I., Rizza, N., Saltarelli, A., Sampaolesi, E.: Arduino: from Physics to Robotics. In: Scaradozzi, D., Guasti, L., Di Stasio, Miotti, B., Monteriù, A., Blikstein, P. (eds.) Makers at School, Educational Robotics and Innovative Learning Environments—FabLearn Italy 2019. Springer (in press)
29. Storti, M., Mazzieri, E., Cesaretti, L.: Weturtle.org: a web-community for teachers' training and resource sharing on educational technologies. In: Scaradozzi, D., Guasti, L., Di Stasio, Miotti, B., Monteriù, A., Blikstein, P. (eds.) Makers at School, Educational Robotics and Innovative Learning Environments—FabLearn Italy 2019. Springer (in press)
30. Freddi, A., Giaconi, C., Iarlori, S., Longhi, S., Monteriù, A., Proietti Pagnotta, D.: Assistive robot for mobility enhancement of impaired students towards a barrier-free education: a proof of concept. In: Scaradozzi, D., Guasti, L., Di Stasio, Miotti, B., Monteriù, A., Blikstein, P. (eds.) Makers at School, Educational Robotics and Innovative Learning Environments–FabLearn Italy 2019. Springer (in press)

Innovative Spaces at School. How Innovative Spaces and the Learning Environment Condition the Transformation of Teaching

Giovanni Nulli, Gianluigi Mondaini, and Maddalena Ferretti

Abstract This paper introduces the contributions to Track C1 of the symposium, which explored the link between architectural space and learning processes, while trying to outline their connection and mutual influence. The paper also aims to outline major trends and innovative approaches in the field of school design. Specifically, it refers to the relationship between the architecture of school, the users' spatial perception, and the capacity to increase learning skills through the experience of comfort and quality spaces. Also, the relationship with the urban structure is investigated as a crucial aspect of school architecture.

Keywords School design · Innovative spaces · Learning environments · Spatial experience · Architecture and city · Architecture and pedagogy

1 Introduction

The assumption that space works as a "third teacher" addresses the importance of the learning environment in contemporary pedagogy, highlighting how space can contribute to the learning process.

Malaguzzi (in Edwards et al. 2012), the founder of Reggio Children, used the words "third teacher" because of the importance of furnishing a space for a particular purpose, and the importance of having someone with an artistic background to design it. This professional should work with teachers to create a pedagogically oriented space.

G. Nulli (✉)
Istituto Nazionale Documentazione Innovazione Ricerca Educativa (INDIRE), Florence, Italy
e-mail: g.nulli@indire.it

G. Mondaini · M. Ferretti
Department of Civil and Building Engineering, and Architecture (DICEA), Università Politecnica delle Marche, Ancona, Italy
e-mail: g.mondaini@univpm.it

M. Ferretti
e-mail: m.ferretti@univpm.it

D. Scaradozzi et al. (eds.), *Makers at School, Educational Robotics and Innovative Learning Environments*, Lecture Notes in Networks and Systems 240,
https://doi.org/10.1007/978-3-030-77040-2_4

In the National Curriculum Guidelines (Italian Ministry of Education, University and Research, 2012), the Italian curriculum for primary and lower secondary schools has different sections focused on the importance of the learning environment and how it can promote awareness in self-learning, learning by doing, collaboration and the acquisition of a scientific method.

Active pedagogy is based on the previous statement, as both Malaguzzi and the National Curriculum Guidelines have in mind a student who is free to move, experiment, and able to try out his ideas with the teacher's support.

What kind of impact does this pedagogy have on the design of a space?

Building new schools is a long and expensive process, in which different aspects converge: laws, lack of models, decision-makers who know nothing about either pedagogy or architecture. Finally, not many new schools are built every year. It can be hard to design a new school from scratch. The process should be a collaboration between motivated decision-makers, a well-informed school head, and a well-intentioned local community working together to find an architect who is willing to design alongside a pedagogy expert.

There are also increasing cases of schools reimagining old classrooms and existing buildings, in which the first step is producing a pedagogic idea that will be the starting point for designing the space. Classrooms can be clustered together and feature specialist corners, hallways (usually very large), and old labs, closet or storage rooms can be opened up to become larger spaces where students can take part in teamwork, experiential learning and be freer to move.

2 The Topic: A Dialogue Between Architecture and Pedagogy

The main theme that Track C1 tackles is the relationship between architecture and education, especially the positive effect the quality of designed spaces has on the learning experience.

Designing a school building today means experimenting with new models of space that rethink the type of building we became accustomed to, at least in Italy, between the Second World War and the end of the twentieth century. The space will no longer be created by walls alone, that is, a connecting system and repeatable modules in the "classroom cell", but also by the people who animate, interpret and modify it with their presence and creativity in the didactic process. The beauty and empathic attractiveness of the school environment must return to being the center of focus, as an integral and inseparable part of the educational process; it must be a priority for those who design schools, a vision where the inhabitants of the building come first, and the perspective of the dwelling takes precedence over that of the building (Ingold 2000). A new idea of school is pursued that interprets the current needs of teachers and students and, with the quality of its spaces, is able to adhere to contemporary teaching and learning models.

The principal aspect that is foundational to the architecture of the educational space of tomorrow is its open dimension. School can no longer be rigid, featuring the traditional classroom–corridor model; instead it must consist of dynamic spaces that are both useful and stimulating, and can adapt and engage physically and empathically with those who live and work there.

The definition of new spatial models is certainly conditional on a dialog between architecture and pedagogy and the possibilities offered by new technologies. Today's pedagogy increasingly emphasizes action as the foundation of learning. "Learning by doing" must become the slogan that informs every new project, imagining active settings in the relationship between space and the user. There is a growing move away from the abstract approach to knowledge in favor of personalized, multi-perspective learning (Attia and Weyland 2016).

Architecture should therefore offer a radical shift in the typological flattening of the past, in favor of multi-functional, dynamic and fluid open spaces, experimenting with new conditions of transparency and multi-sensory space for education. The design of physical spaces acts as a strategic tool for updating learning environments and for mobilizing a significant capacity for growing motivated learners and creative problem solvers (Bosch 2018).

3 Trends and Perspectives

In Northern Europe, in particular, we find avant-garde schools, where the traditional classroom concept has been superseded by open-space learning environments, with integrated ad hoc furniture solutions that bring students and teachers together, in a way that supports the contemporary pedagogical needs we have outlined. We also find creative interior architectures, real "learning landscapes" formed by open, multi-functional spaces. In addition to offering physical and psychological well-being, welcoming open spaces have a positive effect on learning processes, and contribute to establishing a spirit of community between children and teachers and developing social awareness.

In the Italian context, in recent years, we have also seen a change in the design of training spaces, which, finally, is evidence of a cultural shift. Architecture is proposing new ideas, with buildings designed with a greater awareness of the relationship between itself and learning.

3.1 Experiences and Points of View

In line with these premises, Track C1 explores the relationship between architecture and education through good practices and pilot projects. The session dealt with the importance of the quality of space in the education experience and it investigated the interactions between a school's architecture and its urban context, with a view to finding new opportunities for mutual exchange and enhancement. Different perspectives were collected during the symposium, as outlined in the following paragraphs.

Rubino considers architecture's role in influencing the learning process, or whether architecture needs to adapt to new teaching methods, building on the rich literature and a series of good practices that demonstrate possible interactions between pedagogy and architecture. Relevant to this point is the recent development in European schools that makes use of hybrid spaces (e.g., staircases, open air gardens) as new learning areas. Similarly, Ferrari et al. ask whether school architecture should reflect different teaching models, by providing multifaceted and adaptable spaces that are modifiable for future needs. The focus is the classroom of the future, for which a group of ten architects was asked to make a contribution. Yet, some pioneering examples of school architecture from the past already displayed similar design attempts. They focused on children as the protagonists of the space, using architecture to enhance their first community experience and their creative skills. Cabras et al. explore the link between space and dynamic perception and connect it to the idea of "affordance" where, by experiencing innovative spaces, children can learn new concepts and ideas. The reverse approach is also true and the school space designed by the authors and proposed in the paper is shaped around the psychosomatic perception of the child, thus features a continuous architectural space where the experience of learning takes place through action and movement. For example, the school's "water room", which recycles the historic building's old water tank, has an important pedagogic function, as it draws children's attention to sustainability and environmental issues. In Faiferri et al. a stronger and more reciprocal relationship between the school building and the city is aimed for, as an essential part of the future of education. Projects developed within the framework of the International Scientific School's "Innovative Learning Spaces", held yearly in Alghero, explore the possibilities of an increased connection between school and urban structure. In Mondaini et al., school design is directly associated with the need to respond to the current technological evolution of communication processes and educational practices. The classroom is no longer suited to teaching activities and its boundaries must open to a more adaptable and flexible concept of educational environment. Some northern European references and some of the authors' design experiences show innovative design approaches trying to break the constrictions of school architecture of the past and propose transparent, fluid learning environments that establish a continuity with the outdoor space. Finally, D'Annuntiis et al. report the experience of a university course held by the authors in collaboration with UNICEF, one of the first of its kind in Italy. Particular care was taken to understand children's rights to education and their need to acquire the skills for new tools and technologies. Also, architecture students were asked to elaborate on good practices in school architecture and to test a participatory planning process involving children in Grottammare and aimed at co-creating more suitable learning and living spaces in the town.

4 Conclusions

The contributions in this track present different solutions with different origins. Some of them focus on the necessary link between the architectural design process, the teaching methods, and the pedagogic needs: some methods work effectively when space is designed to support the learning experience. For example, flexible areas with specific furniture can boost new visions and creations. By adhering to the concept of affordance and respecting the rights of young people, different spaces can stimulate new meanings in students and suggest new learning paths to teachers. Also, as social stakeholders, schools are part of the city and should be connected to important amenities, making them more accessible to the neighborhood.

The track highlighted that different approaches to space and pedagogy, and new design processes already exist in academic research and have been experimented in several schools. So what could be next?

On the academic side, we noted an important connection between the two sciences involved in this process: architecture and pedagogy should work more closely together to provide frameworks and guidelines to governing institutions. Secondly, it is important to communicate the pedagogic role of space to the public stakeholders that are responsible for school buildings. At this time of pandemic, it is especially important for all stakeholders to collaborate on creating solutions that combine safety, wellbeing and new methods. Furthermore, it is important to provide school models that include new pedagogic ideas and innovative school environments. That is how a new design approach can be implemented. One possible model is the one created by Indire (Tosi 2019), based on five different kinds of spaces plus one (the classroom), which offers both a flexible framework that can be studied and adapted to different schools, and a tool for teachers for designing new activities for a new space.

In conclusion, a closer collaboration between architecture and pedagogy is needed to help school and politics reimagine school buildings and their internal space. It is also important to keep an open mind to experimental and innovative school initiatives, and define models arising in different contexts as opportunities for upgrading the learning environment and the overall educational experience.

References

Attia, S., Weyland, B.: Body and Mind: How the Corporeal Can Serve as a Contact Point Between Architecture and Pedagogy. FAMagazine, Parma (2016)

Bosch, R.: Designing for a Better World Starts at School. Saxo Publish, Copenhagen (2018)

Cabras, L., Pusceddu, F.: UP school: motion, perception, learning. In: Scaradozzi, D., Guasti, L., Di Stasio M., Miotti, B., Monteriù, A., Blikstein, P. (eds.) Makers at School, Educational Robotics and Innovative Learning Environments—FabLearn Italy 2019, vol. 1. Springer, Heidelberg (2020)

D'Annuntiis, M., Cipolletti, S.: Child friendly architectures. Design spaces for children and adolescents. In: Scaradozzi, D. et al. (op.cit.)

Edwards, C., Gandini, L., Forman, G.: The hundred languages of children: The Reggio Emilia experience in transformation, 3rd edn. Prager Press, Santa Barbara, CA (2012)

Faiferri, M., Bartocci, S.: Landscapes of knowledge and innovative learning experiences. In: Scaradozzi, D. et al. (op.cit.)
Ferrari, M., Tinazzi, C.: School space ideas for the future. In: Scaradozzi, D. et al. (op.cit.)
Ingold, T.: The Perception of the Environment. Routledge, London (2000)
MIUR. (2012). Indicazioni Nazionali per il Curricolo, Roma, Le Monnier
Mondaini, G., Rosciani, M.: Adaptive environments. New spaces for learning. In: Scaradozzi, D. et al. (op.cit.)
Rubino, A.C.: Multipurpose learning environments for a flexible didactic. In: Scaradozzi, D. et al. (op.cit.)
Tosi, L. (Ed.).: Fare didattica in spazi flessibili: progettare, organizzare e utilizzare gli ambienti di apprendimento a scuola. Giunti Scuola (2019)

Keynotes

Makers in Education: Teaching is a Hacking Stuff

Domenico Aprile

Abstract Several changes and challenges affect contemporary society: public administration, business, society at large. The lowest common denominator is digitization through IT tools. And what about school systems? Do they play a decisive role in enabling the skills needed to these challenges? We send our children to school to prepare them for a fast-changing world. But have our schools (and teachers) changed in the last 100 years? Modern teachers are being asked to change their teaching and learning practices, and to implement methodologies (including the use of IT tools) and make way for new teaching practices.

Keywords Teaching · Making · IT · Methodology · Skills

1 Problems and Goals

1.1 Troubleshooting

Currently, school systems have more than one problem with which to contend. A video [1] by Next School [2] (almost 9 million views on YouTube) identifies six problems with the world's education systems:

- Industrial age values
- Lack of autonomy
- Inauthentic learning
- No room for passion
- No differences between children's learning styles
- Too much lecturing.

Seth Godin [3] and Ken Robinson [4] agree that the current education system was designed in the Industrial Age to churn out factory workers. Yet, while modern

D. Aprile (✉)
Liceo Scientifico "Fermi-Monticelli" - European High School, Brindisi, Italy

D. Scaradozzi et al. (eds.), *Makers at School, Educational Robotics and Innovative Learning Environments*, Lecture Notes in Networks and Systems 240,
https://doi.org/10.1007/978-3-030-77040-2_5

society is no longer based on the values of mass production, Taylor's principles of scientific management still run deep in schools.

A few years ago, I asked my students to do some research and represent the values of Industry 4.0. I asked them to work in teams. One of the teams produced an excellent video, in which they highlight the differences between Fordism and modern organizations: working time, organization, tools, skills (operational and educational), process ownership). Everything has changed.

Our education system, however, sends several dangerous messages to our children: they should follow the established order, rather than taking responsibility and making the most of their lives; they should only memorize concepts and a standard set of information against which what they know will be measured (not evaluated); everyone has to learn the same things, at the same time in the same way; if you take a bit longer to learn something, you are a failure; do not use digital resources.

As a result, children are bored and not engaged, and most of what they learn has already faded the day after their exams. All this is proof that our education systems are outdated and ineffective.

1.2 Changing the Paradigm

Education systems seek to meet the challenges of a globalized and extremely dynamic world by offering static teaching/learning models.

The educational model that forms the basis for modern education systems was designed for a society that no longer exists. It was strongly influenced by the deductive reasoning that developed during the Enlightenment and is based on classical thought, rooted predominantly in academic learning. Schooling is organized according to the Aristotelian units of time, space and action. Curiously, its organization resembles a factory production line: bell rings (time); separate facilities (space); children grouped by age (batch processing); one entity intervenes with another entity, so that the first modifies the second (action).

First, there needs to be a shift away from batch schooling to custom education. The education system should not be designed for a standard (and its related standardized tests), but should bring out the divergent thinking and creativity of each student. As a computer science teacher, I tell my students that solutions often exist in multiples and are the result of heuristic, action-research processes. This means taking account of every student's specific skills and awareness, rather than evaluating them against a standard.

A radical change is needed; one that takes us from a vertical approach to teaching/learning to one that is horizontal. In this paradigm the aesthetic—as opposed to the anesthetic—should be taught, stimulating children to share, collaborate and learn together.

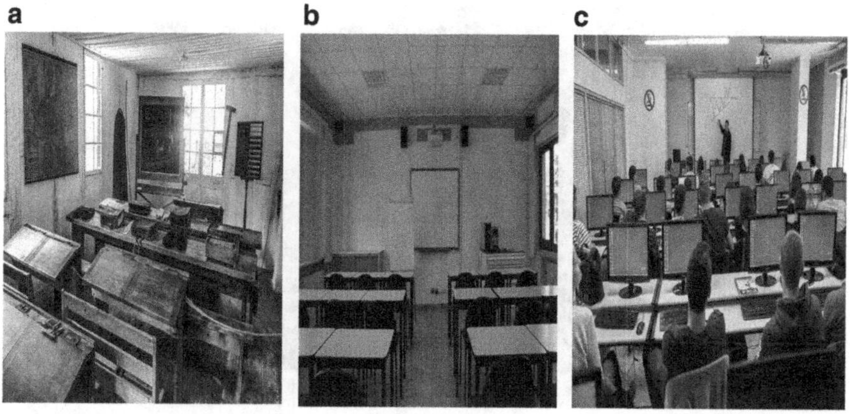

Fig. 1 A classroom in 1920 (**a**), a classroom in 2020 (**b**), a computer lab (**c**)

2 A Maker in Education

2.1 A Quantum Leap

A visitor from 100 years ago would struggle to make sense of today's world.

But if he went into a school, he would see something familiar: the layout of classrooms is almost identical, with a teacher's desk, students' desks arranged in a row, a board (at that time it would have been a blackboard, now it is an IWB). Even the computer labs retain this layout, which was typical of the Fordist-Taylorist factory (Fig. 1a–c).

Classroom activities are mainly based on the top-down transmission of content by a teacher sitting in a chair.

What we need, instead, is a physically active teacher engaged in a horizontal relationship with her students (Fig. 2).

We need to switch from a premise of "*Let me explain how to do it*" to one of "*Let's do it together!*"

2.2 What is an Edumaker *(*Maker in Education*)?*

A maker in education is a hybrid between a hacker and a maker.

The term *hacking* dates back to the early 1950s, when MIT students in Boston, driven by a thirst for knowledge, coined the term "*unnel hacking*" to define their raids in underground tunnels.

Therefore, hacking arises from a desire for knowledge.

What is a *maker*? Borrowing the definition from Dale Dougherty, editor of MAKE magazine: "*Makers want to hack this world the same way we used to hack computers.*"

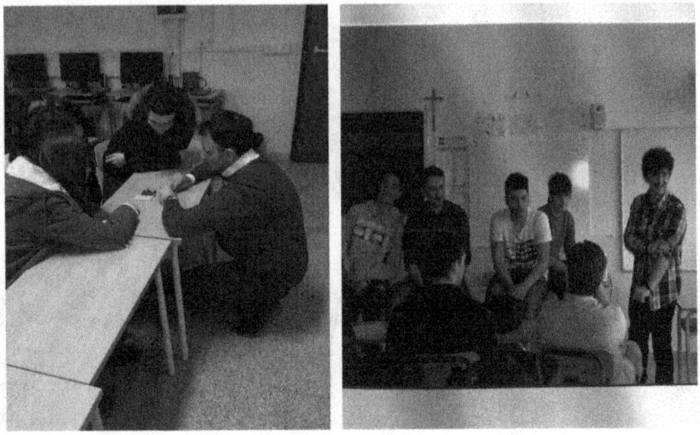

Fig. 2 Education activities

This is why I believe makers and hackers are very similar.

A maker in education has a proactive attitude, he does not say "*no! You don't care about this stuff!*", but rather, "*let's try to do it together*" and asks himself "*what can the kids learn*?"

A maker in education upturns the system from learning by transmission to one of working in collaboration; from the *global* (all the same things) to the *glocal* (all the same opportunities).

A maker in education builds a knowledge network that:

- develops a business idea with students and promotes entrepreneurship;
- builds relationships between schools and local authorities;
- exports his teaching model to other schools in the area;
- experiments with an open didactic model and favors "other" methods over traditional methods;
- shares his own teaching experiences which are replicable.

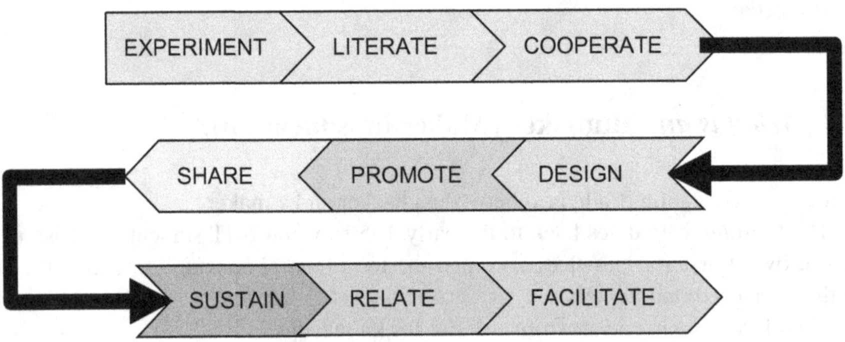

Table 1 draws parallels between a maker and a maker in education.

Table 1 Maker versus maker in education

Maker	Maker in Education
Rethinking production and business models	Rethinking the learning model
Manual skills	Robotics, 3D printing
Cooperation	Cooperative learning
Design	Redesigns learning content (less standardized "production" and more customization), while moving beyond the idea that owning content leads to competence. "Designs and redesigns the curriculum"
Sustainability	Physics: Dematerialization; Virtual: focus on skills
Virtual and physical communities	Virtual and physical learning community using several kinds of channels (I.M., learning platforms…)
Sharing knowledge and technology	Sharing knowledge and technology

A maker in education overturns a paradigm. He is an *edumaker* [5].

3 Experience of a Maker in Education

3.1 Co-m@king_LAB

Co-m@king_LAB (*Coding and making laboratory*) [6] is a knowledge hub for creative design and prototype realization, which uses innovative practices and technologies related to *making* and *IoT*.

Activities combine a "*hands-on minds-on*" approach with research/action-edge content: Third Industrial Revolution; *Europe 2020; Digital Agenda;* Open Data (and, more generally, Open Innovation); Internet of Things; Makers, Industry 4.0.

This is pursued using open-source hardware and software platforms (e.g., Arduino, Raspberry). The aim is to achieve the operating competences of the maker's new digital frontier techniques, as well as educational and cognitive competences (*analytical thinking, creativity, problem-solving, collaborative working*).

Students are involved in designing, creating robot prototypes (including 3D printing components) and developing hard skills (programming, designing, using electronic boards) and soft skills (cooperation, team building and leading, creativity, problem-solving).

Co-m@king_LAB aims to update computer science teaching in Italian *Licei Scientifici* (science-oriented high schools), through the use of unconventional methods and tools (for example, the mobile laboratory). It seeks to go beyond the physical space of the classroom to foster a school based on relationships, skills and Papertian constructionism.

4 Conclusions

In the paradigm still in force in today's school systems, only students learn. Yet this model no longer works: teachers also learn and learning methods are synchronous (in terms of space and time).

We should remember that technology is not neutral: we need to "*think with machines*" [7] and teaching is not exempt from this requirement.

Through technology, children get used to taking part and being proactive. In the same way that the Web has evolved from 1.0 (internet of information) to 2.0 (internet of people) and 3.0 (internet of things), teaching cannot stay the same. In a world where 20 years is a temporal abyss: Twenty years ago, computer operating systems were installed on 12 floppy disks; today we can download them from the internet.

The school system is stuck in a 1.0 model and struggles even to understand that change is necessary.

References

1. Problems with our School System. https://youtu.be/okpg-lVWLbE. Published on 15 Dec 2016
2. Next School Homepage. https://www.nextschool.org/. Last access 18 Sept 2020
3. Seth Godin Homepage. https://www.sethgodin.com/. Last accessed 14 Sept 2016
4. In memory of Ken Robinson, Homepage. http://sirkenrobinson.com/. Last access 16 Sept 2020
5. Edumakers Homepage. http://www.edumakers.eu/index.php/en/. Last access 15 Sept 2020
6. Co-m@king_LAB Homepage. http://www.comakinglab.education/. Last access 15 Sept 2020
7. Moriggi, S.: Connessi. Beati quelli che sapranno pensare con le macchine. Ed. San Paolo Cinisello Balsamo (MI) (2013)

If We Could Start from Scratch, What Would Schools Look like in the Twenty-First Century? Rethinking Schools as a Locus for Social Change

Paulo Blikstein

Abstract Today's debate about education is prone to focusing on system optimization, test score improvement, and budgetary concerns. However, education is much more: it is primarily about a vision for our societies. As we think about a new vision, it has to speak to the ethos of our time. Today's youth are heavily focused on social change, addressing global problems such as climate change, systemic racism, and economic inequality. This requires new content and pedagogies. Thus, schools should be rebuilt to support such endeavors, emphasizing ways of learning in which students have more agency and learning is more relevant. Currently, the schools where this work is possible are most typically located in affluent countries and regions. We should work to democratize the possibility of "learning how to change the world," making public schools a viable locus for fostering social change.

Keywords Educational change · Constructionism · Maker education · Critical pedagogy · Social justice

1 Introduction: How Do Educational Systems Get Built?

Today's debate about education is prone to focusing on system optimization, test score improvement, and budgetary concerns. However, education is much more than that: it is primarily about a vision for our societies. It is humanity's way of encoding culture and knowledge into a civilization's DNA and transferring it to future generations. As such, it is a society's way of shaping itself and paving its own future. It is high time we treat it as such.

When talking about education, we should thus start from a vision of what a school, city, or nation wants from it, and then work towards engineering and implementing that vision. Unfortunately, today's educational debate is dominated by a

---

P. Blikstein (✉)
Teachers College, Columbia University, New York City, NY, USA
e-mail: paulob@tc.columbia.edu

D. Scaradozzi et al. (eds.), *Makers at School, Educational Robotics and Innovative Learning Environments*, Lecture Notes in Networks and Systems 240,
https://doi.org/10.1007/978-3-030-77040-2_6

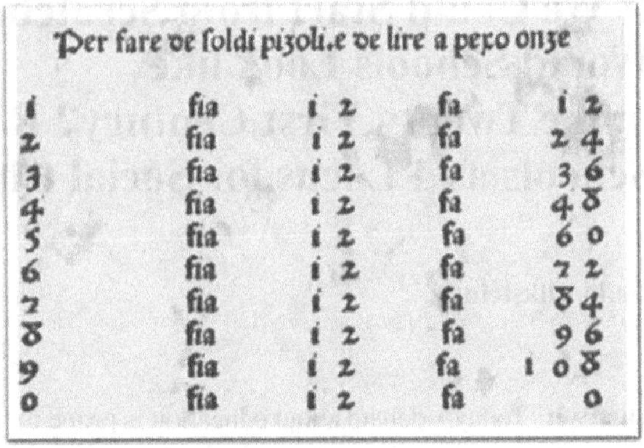
Per fare de soldi pizoli.e de lire a pezo onze

1	fia	12	fa	12
2	fia	12	fa	24
3	fia	12	fa	36
4	fia	12	fa	48
5	fia	12	fa	60
6	fia	12	fa	72
7	fia	12	fa	84
8	fia	12	fa	96
9	fia	12	fa	108
0	fia	12	fa	0

Fig. 1 Page from Treviso Arithmetic textbook from 1478 (Swetz 1987)

bean-counting discourse that, instead of considering what young people are interested in learning or who they want to become, looks at schools to ensure countries' competitiveness on the international market.

Education systems are built by starting with societal needs, and then working backwards from there: building national standards, then creating textbooks, educational materials, and professional development for teachers—all before learning makes it into the classroom.

An example of this process in practice is depicted in Fig. 1, which shows an excerpt from the Treviso Arithmetic textbook from 1478 (Swetz 1987)—specifically, a section on multiplication table algorithms. The book also covers addition, subtraction, and long division. The author explicitly starts with the rule of three and then moves on to problems concerning discounts, partnerships, bartering, and coin alloys—all very relevant for 15th-century Venetians, who needed the skills to do quick math operations for business purposes.

Incredibly, the same content is still being taught in schools today, well over 500 years later! This is in spite of the world having changed immensely in the centuries since. For several years now, machines have replaced humans working on routine and/or manual tasks (see Fig. 2). On the other hand, the need for professionals capable of dealing with tasks such as unscripted communication or unstructured problem-solving is ever growing. Simple arithmetic skills are no longer in high demand, as we can easily delegate such tasks to our ubiquitous smartphones.

Therefore, one piece of our educational puzzle is to figure out the societal needs of the twenty-first century. They are undoubtedly different from those of 15th-century Venice, even though we are still teaching the same content. One way to determine our needs is the type of quantitative research that Autor et al. (2003) did.

However, the other puzzle element cannot be measured, as it is not something economists can study by running a regression and observing the impact of a given variable. That second aspect is about a vision for the future and the kind of society we want to work towards becoming. One example is how we arrived at universal suffrage. One hundred years ago, when women still did not have the right to vote, many movements worldwide began demanding change. The required legislative modifications

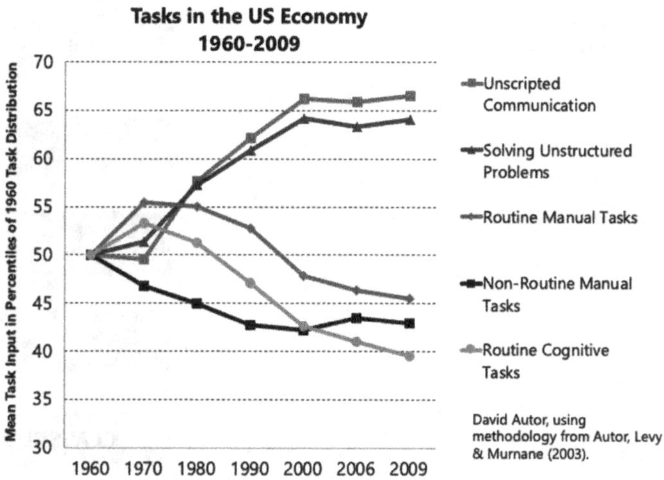

Fig. 2 Source: Autor, based on methodology from Autor et al. (2003)

were not preceded by studies proving this would be economically advantageous—it was simply a matter of principle, a vision of society that people believed in, where everybody had equal voting rights.

2 What is Our Vision for the Future?

When designing an education system, we should thus heed not only the economics, but also the vision: what do we want kids to be like now and in the future? When we look at how we build education systems, we should consider both the societal needs and the societal *ethos*. This we see reflected in Fig. 3, where societal needs and ethos inform the curriculum.

Today's society is changing at a swift pace. Ten years ago, data science and artificial intelligence were still in their infancy, many professions did not even exist, and many issues were not yet a big concern for our societies. When things keep evolving so rapidly, national standards cannot be written in stone and expected to last 30 years or more.

We should advocate for systems that will allow national standards to be reviewed and adjusted every five years or so, removing outdated/obsolete content (e.g., spending three years learning arithmetic) and replacing it with more relevant topics. Given the limited time frame available in a school day, it is all about making informed choices: after all, we cannot cram all the content from both the fifteenth and twenty-first centuries into the same 24-hour day.

Once we accept that national standards should evolve regularly, it becomes apparent that we should stop focusing on designing lesson plans to be followed

Fig. 3 How our current educational systems got built

precisely in the classroom. Instead, we should design learning spaces, toolkits, activities, and guidelines that are flexible and enable different kinds of learning, particularly project-based learning.

Teacher professional development should also be reviewed. Instead of training teachers as technicians that will teach a curriculum and deliver a lesson plan exactly as prescribed in the textbook, we should encourage teachers to be intellectuals and designers in the classroom.

By now, some countries have realized that fundamental changes are needed. In Canada's British Columbia, a new course, "Applied Design, Skills and Technology," is mandatory for all students in Kindergarten through Grade 12 (British Columbia Ministry of Education 2016). This means that, alongside the usual classes that teach math, science, languages, arts, and so on, there is a class where students learn design, computer science, and maker content. Similarly, US public policy—in the form of the Next Generation Science Standards (NGSS)—deems it mandatory for all students to build prototypes and design and optimize solutions (NGSS Lead States 2013). In doing so, it encourages students to do engineering and project-based learning in science classrooms.

3 Sobral, Brazil: Examples of Possible Change

Along these same lines, in the northeastern Brazilian city of Sobral, the science program is being completely revamped, from curriculum redesign to makerspace implementation and teacher training. The municipality is implementing an approach to learning sciences where students learn how to build hypotheses, use models to explain scientific phenomena and apply their knowledge to solve issues in their community.

Figure 4 shows a lesson plan on vertebrates redesigned by a local teacher in Sobral. Traditionally, this would involve drawing maps, with the teacher listing the different species' characteristics on the whiteboard, and the students then being expected to memorize them. Instead, the teacher created a laser-cut toolkit in the makerspace that allowed them to create their own made-up vertebrates. Students thus created ocean giraffes that could swim and imaginary dinosaurs. Not only was this activity significantly more engaging than the traditional approach, but it also had more to it than creating made-up animals. Part of the unit also required the students to come up with their animal's various features: what kinds of food do they eat, how do they breathe, how do they reproduce, how do they move? All these characteristics also had to be justified. For instance, one student who presented a made-up marine dog-like animal explained that, while it would still swim with its tail above water, it would need thorns all over to help it escape from underwater predators. In this case, being creative is closely related to the subject matter—the animals were made up, but students had to use science to justify their choices.

One of the unit's main learning goals was to match form and function—a fundamental idea in biology. But instead of simply memorizing the course content, the students actively designed an animal, matching form and function in a very authentic way. This was followed by a large school fair where the students shared their imaginary animals along with their reasoning for the choices they had made. The unit was redesigned by a teacher we had previously trained at a normal public school in Sobral under no particularly special conditions.

Another public school teacher in Sobral redesigned a unit introducing the nervous system and how our body's internal sensors send signals to the brain, thus allowing us to control our limbs. Here the teacher had students "build neurons" using robotics kits that would mimic the human nervous system, instead of just memorizing facts about the human body.

In a school in California, a teacher reimagined the in-class microscope experience, finding issue with how students only got three minutes each with one of the

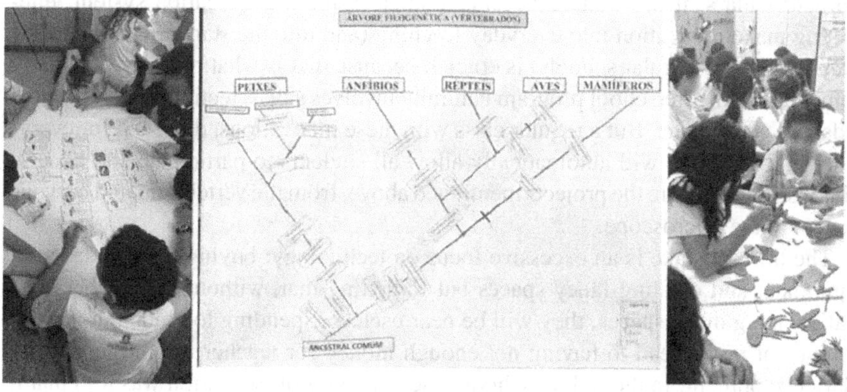

Fig. 4 Lesson plan redesign: Vertebrate animals

microscopes available in the science lab. Instead, each student built their own five-dollar laser-cut wooden microscope, complete with plastic lens. Not only did each student now have their own microscope in class, but they could also all take their microscopes home and conduct all kinds of investigations.

A further real-life project was found in Thailand, where students living in a rural community were faced with an irrigation problem: they never knew what the ideal time was for irrigation. They decided to build a robotic system that would detect air temperature and soil humidity and use this data to water the plants at the optimal time, thereby minimizing evaporation. Such projects are being created worldwide, even in places with limited resources.

There are therefore many examples of how it is possible to build this new kind of structure, where instead of fixed lesson plans followed precisely by teachers, we are building toolkits and environments that allow rich learning to happen.

However, while it is true that we have all these new possibilities in classrooms, no matter where we go, we find that three fundamental mistakes are almost always being made.

4 Three Mistakes in Progressive Education

The first mistake is the lack of a national plan. It is tempting to think that all you need is a few great teachers and plenty of positive energy. However, without a national plan, and national standards, one cannot build a sustainable system that includes project-based learning and maker education, nor can one create the incentives and policies to make such an approach universal. This is why we ought to look towards the countries that have already started to develop such standards—like Canada and the United States, as mentioned previously.

The second mistake is that many attempts fail to truly integrate making into the curriculum. It is not enough to have afterschool programs, maker fairs, and other extracurriculars, if most students remain stuck in the old education system. Integrating maker education into everyday teaching (and into the standard 45-min time frame typical of regular schools) is crucial, because that is what makes it genuinely democratic. An afterschool program naturally involves self-selection, and only some kids will participate. But a regular class with these innovations, designed for everybody in the school, will automatically allow all students to participate and benefit. This also happened in the projects mentioned above, from the vertebrates and nervous system to the microscope.

The third mistake is an excessive focus on technology: buying an abundance of equipment and creating fancy spaces but forgetting that, without the teachers and staff running those spaces, they will be near useless. Spending too much money on equipment might lead to having not enough money for teacher training or curricular development. In the Sobral schools, the Secretary of Education was told that it was necessary to hire one additional lab teacher per school for an effective maker education program, because you could not thrust yet another responsibility upon the existing teachers. Thanks to a robust training program and a support system in place for both new and established teachers, incredible projects are now coming out of Sobral—but the key to success was having well-prepared people running those programs, not just the equipment.

5 The Future of Education Looks like the Present of Makerspaces

We often talk about maker education as something new and separate from regular schooling. However, if we could start building an education system from scratch, how would we go about it? If we think about what current research says about how humans learn and how to encourage that in a school environment, we realize that much of it is already happening in today's makerspaces: project-based learning, education that is meaningful to kids, student agency, inclusive environments, new types of skills, use of technology as an expressive material. *Thus, if you redesigned schools from scratch, they would probably look like a well-functioning, inclusive makerspace.*

Since what we are doing in many makerspaces is the best form of education that we can imagine, perhaps we should stop using the terms "education" and "maker education" as we do now. We should instead use "education" to describe what goes on in makerspaces, and use another term, such as "traditional education" or "old-fashioned education," for the rest. This was something Seymour Papert suggested about conferences on "technology in education" as far back as 2004: he wanted us to simply call them "conferences on education." He recognized that, when we create a qualifier for a particular type of education, the message is that it is separate from mainstream schooling—not the "real thing" (Papert 2004). Yet, what goes on in makerspaces should not be considered a separate, special or subordinate kind of education—it should be the typical activities in which all students can take part.

This does not mean that we necessarily need to replace everything in schools completely; but this novel approach should be the backbone of the modern education system. In contrast, the old approach can be relegated to a supporting role for when traditional learning methods are perhaps better suited. Of course, we must also be wary of embracing the naïve discourse of "completely replacing schooling overnight." There is still knowledge that can be taught using more traditional methods, but it should be a much smaller percentage and should not dominate the landscape of education.

Bringing about change is possible, though, as shown by a European country with which we have been working for over seven years. Within this relatively short timespan, hundreds of the schools there have developed a maker program. They also started a regular national conference for teachers to share their experiences and conduct regular professional development programs. In fact, they have been so successful that the Minister of Education commissioned the development of a new computer science and maker education class, which will be mandatory for all students nationwide. This is a great example of how you can start building an educational infrastructure that caters to today's and tomorrow's needs. And the best part is that it is not a 50-year project: achieving great things is possible in as little as ten years with solid, research-based programs.

6 Conclusion: The *Ethos* of Our Time

We have already established that the educational system should cater to ever-evolving societal needs and the societal ethos. However, what is the ethos of our world today? We are no longer in 15th-century Venice; we are not in the industrial revolution anymore. So, where are we? Three sets of young people come to mind.

In 2018, the Dutch teenager Boyan Slat founded The Ocean Cleanup Project. He managed to fundraise millions of dollars to build enormous machines that would help remove plastics from the ocean. Also in 2018, the survivors of a high school shooting in Parkland, Florida, became activists who took it upon themselves to change the gun control legislation in the US, leading to the March for Our Lives protests. Finally, in that same year, Greta Thunberg became one of the most well-known climate change activists worldwide, starting with her modest one-person school strike. What do these stories have in common?

It seems clear that young people now cannot wait 20 years for things to happen. You cannot tell these youth that they will have to learn some disciplinary content and then wait two decades to apply it. They are connected to everything, and they have a sense of urgency because of climate change at our doorstep, the dying oceans, and social issues such as systemic racism coming to a head. All of this is happening in front of their eyes; this is a generation that feels the urgency of changing the world. They are passionate about so many things: they want to save the environment, they want to live in a better place, they want to promote gun control, they want to end systemic racism once and for all. It is essential to democratize their ability to follow their dreams and passions, because it is incredibly unfair that, while some people are allowed to do so—like those mentioned above—others are not. Boyan Slat, Greta Thunberg, and the Parkland survivors are exceptions. They also come from developed countries and more privileged backgrounds. But what if we created an education system that would make such youth the rule, not the exception?

Let us democratize the possibility of falling in love with projects and ideas, and making a difference, so that the generation growing up today can make the world a better place now—not in 20 years, when it might be too late. Schools should help children and teens become the people they want to become—not economically, but as people and citizens. To achieve that, we should make sure that the knowledge we teach students is not just about how to do arithmetic but also about climate change, social justice and equality, and any number of other current issues.

We have to make these difficult choices—arithmetic or climate change—because the school day still has a limited number of hours. Consciously choosing to cover content from five centuries ago alongside today's content will cause us to come up with a superficial curriculum that does not go anywhere and that does not fulfill the goals of today's youth. And what are those goals? What is the DNA of the society that we want? How and what do we want those children and teens to learn so that, in the future, they can become more like the inspiring young people mentioned above?

Much in the same way that, in 15th-century Venice, school served a social purpose, because of the enormous need for the skills used in commerce, the social purposes we should focus on now are social change and environmental protection. These are not superficial interests for children these days. They are what students care about and also what we as the responsible adults in their lives should foster in them. Moreover, if school does not serve this purpose of democratizing the possibility of falling in love with changing the world, then it will—once and for all—lose its purpose.

References

Autor, D.H., Levy, F., Murnane, R.J.: The skill content of recent technological change: an empirical exploration. Q. J. Econ. **118**(4), 1279–1333 (2003)

British Columbia Ministry of Education: [British Columbia Kindergarten to Grade 12 Curriculum]. British Columbia: BC's Curriculum (2016). https://curriculum.gov.bc.ca/

NGSS Lead States: Next Generation Science Standards: For States, By States. The National Academies Press, Washington, DC (2013)

Papert, S: [Keynote address]. In: 2004 Conference on 1:1 education, Sydney, Australia (2004). https://vimeo.com/9092144

Swetz, F.J.: Capitalism and Arithmetic: The New Math of the 15th Century (D. E. Smith, Trans.). Open Court Publishing (1987) with a [illegible] curriculum [illegible] does not [illegible] the [illegible] today's youth. And what are these [illegible]? What is [illegible] [illegible] want? [illegible] [illegible] those [illegible] [illegible]

Much of [illegible] 15th century [illegible] social purpose, [illegible] social purpose, we [illegible] materials for children [illegible] and lower [illegible] [illegible]

References

[illegible] (1970) [illegible]
[illegible] [illegible]
[illegible] National [illegible] Press, Washington, DC [illegible]
[illegible] [illegible]
[illegible] [illegible]

From Classroom to Learning Environment

Samuele Borri

Abstract The concept of "space as the third teacher" suggests that the learning environment is as important as the teacher in the learning process. A constructivist pedagogical paradigm requires student-centered learning processes and learners to be autonomous and active. Therefore, more and more stakeholders and policy makers interested in school innovation put school buildings and learning environments at the top of their agendas. The Organisation for Economic Co-operation and Development (OECD), the European Commission and many universities all over the world are observing case studies and promoting guidelines to implement new ways to design and furnish schools. Indire is leading a research project on educational architectures, which promotes a support framework, entitled "1 + 4 Learning Spaces for a New Generation of Schools." It is aimed at architects, municipalities, school principals and other stakeholders involved in the design, development and use of innovative learning environments.

Keywords Learning environment · Learning space · Third teacher · School architecture · Flexible · Classroom · School building

The school environments in which our parents and grandparents studied were designed to create hierarchical relationships based on criteria of control and discipline. In pedagogical terms, the classroom space, filled with rows of desks and chairs, was largely organized for "lecture-based" learning. The idea behind that model is that the knowledge possessed by the teacher needs to be transmitted to the learner. In such a context, active student participation is limited by the fact that most of the lesson is taken up with the teacher's explanations and presentations.

Many school buildings in use today were designed with this concept in mind. These schools contain separate classrooms, each for a different aged-defined group of students. Ordinary classrooms and special classrooms (such as work areas and subject labs) were connected via corridors and common spaces used only to move

S. Borri (✉)
Istituto Nazionale di Documentazione, Innovazione e Ricerca Educativa (Indire), Florence, Italy
e-mail: s.borri@indire.it

D. Scaradozzi et al. (eds.), *Makers at School, Educational Robotics and Innovative Learning Environments*, Lecture Notes in Networks and Systems 240,
https://doi.org/10.1007/978-3-030-77040-2_7

from one classroom to another. This model of school life includes an entrance, a route to get to your group-class's specific room, which no one leaves until the end of the school day, except for short breaks. Those buildings were designed and built for the criteria of that school model and no longer suit today's educational needs. The increasing complexity of the society we live in has changed students' educational needs from those of the past, when most buildings were designed for mass schooling.

Loris Malaguzzi coined the concept of the school space as a "third teacher." It is a very effective metaphor for describing the setting's role in a school system, which is not merely passive, or related only to making activities possible. The metaphor represents the way activities can be carried out and the way new generations need to get involved in student-centered tasks and initiatives.

If building design is predicated on a lecture-based schooling model, then it requires "fixed" spaces furnished with desks and chairs. A pedagogical paradigm that includes a variety of teaching methods and learner-centered strategies requires a different approach to the design of learning environments. It is no longer appropriate to determine the characteristics of the learning space without considering the changing needs and activities that will be carried out there. Since the activities are diversified and can change, the learning space should be designed for a range of different activities, featuring task-oriented zones fitted with flexible furniture and a variety of educational materials and tools. Open spaces equipped with mobile digital internet-connected devices are just one example of a new learning environment that requires a new conceptual vision. The standard classroom model loses its hegemony, and this opens new perspectives for designing and adapting innovative environments to diversified and changing needs.

We are currently witnessing more and more national school policy initiatives promoting widespread school building and reconstruction plans. These initiatives aim to revise the architectural guidelines for school buildings [1]. However, the issue is not only about updating the regulatory context. It is necessary to rethink the role of schools and modernize the educational space. There is a shift away from a paradigm based on static spaces towards a flexible and functional vision of space. In this new context, the local community is often involved in a design-oriented participatory process, or at least in a collective discussion and debate on defining needs.

Different stakeholders all over the world are interested in rethinking educational spaces:

- The OECD, with its work on the "Innovative Learning Environment" concept [2];
- The Joint Research Centre, European Commission, promoting "Creative Classrooms (CCR)" [3];
- European Schoolnet (EUN) and the group of international experts involved in the Interactive Classroom Working Group promoting international guidelines for adapting learning spaces [4];
- Ministries of Education in a number of European countries, through international projects and collaboration initiatives, such as the iTEC project and the Future Classroom Lab Initiative (FCL) [5];

- A pool of Australasian stakeholders, represented by the works of the University of Melbourne and the Innovative Learning Environments and Teacher Change project, involving organizations worldwide [6].

The need for new spaces is related to active pedagogy, which focuses the educational process on students' needs and eventually leads to a new conception of the curriculum. Problem-solving in complex situations, as well as life-long learning attitudes and learning-to-learn skills can be effectively implemented in environments that have been designed and created for active and experience-based learning, along with cooperative or collaborative activities.

In this context, Indire's research in school architecture [7] promoted a comparative analysis, developed insights and cases studies, which resulted in the publication of the manifesto *1 + 4 Learning Spaces for a New Generation of Schools* [8]. The 1 + 4 model was further developed by Biondi et al. [9] and then used as a model for case studies by Tosi [10].

The manifesto identifies five different spaces that have multiple purposes. These are: the Group Space, the Exploration Lab, the Agora, the Individual Area, the Informal Area.

The Group Space is the "1" in the 1 + 4 model. This place identifies the group-class and is the place where teachers present strategies and set up activities for students. Students can hold work groups and other activities here. The layout of this space should allow for different activities such as:

- Collaborating and working in groups, with workstations arranged in islands, equipped for interacting, planning, processing and analyzing data jointly.
- Designing in a group and creating products with instruments for dramatizing, developing, assembling, and editing multimedia content jointly (workstation arranged in islands).
- Performing individual tests, with workstations isolated to enable concentration for exams, tests, or other type of assessment.
- Presenting work, whether individual or group, with tools for collective viewing or for projecting multimedia content and sessions set up for optimal viewing.

This "1" space is the heart of daily learning activities, the home base and the place where students build their identity.

The Exploration Lab is the place where learning by doing happens: students can observe phenomena and take part in simulations here, as well as creating virtual or real artifacts to test previous hypotheses. It is where science and creativity meet in practice. It is a lab that has no specific topic, where students bring their knowledge from different subjects and apply it to individual projects.

Agora is a place where the entire school community (students, teachers, school personnel, and parents) can meet. It is the school plaza where the community can share and exchange ideas. This place is used by groups of students for creative activities, or for discussions arranged by teachers for students.

The Individual Area is where students can focus and work quietly. Students might typically withdraw to this area to read or reflect. Effective use of this space supports informal learning by enhancing individual responsibility and independent time management.

The Informal Area is for relaxation in the broader sense. It is a place for recreation, but also somewhere to hang out. This space is used for listening to music, relaxing and for informal group meetings. Different places in schools can be readapted: soft chairs or sofas in transit areas, such as corridors or under-stairs, can create spaces for students to gather.

This manifesto has several purposes. The first is to show policy and decision-makers examples of pedagogic use of space. The second is to inspire design professionals with examples of student-centered design. The third is to support local schools in the creation of innovative spaces, with a framework that is both far-reaching and inspiring.

References

1. OECD, Innovative Learning Environment, http://www.oecd.org/education/ceri/innovativelearningenvironments.htm. Last accessed 28 Sept 2020
2. Joint Research Centre European Commission, https://ec.europa.eu/jrc/en/publication/eur-scientific-and-technical-research-reports/innovating-learning-key-elements-developing-creative-classrooms-europe. Last accessed 28 Sept 2020
3. Bannister, D.: Guidelines on Exploring and Adapting Learning Spaces in Schools. European Schoolnet, Brussels (2017)
4. EUN: Future Classroom Lab homepage, http://www.eun.org/it/professional-development/future-classroom-lab. Last accessed 28 Sept 2020
5. Mahat, M., Bradbeer, C., Byers, T., Imms, W.: Innovative Learning Environments and Teacher Change: Defining Key Concepts. http://www.iletc.com.au/wp-content/uploads/2018/07/TR3_Web.pdf. Last accessed 28 Sept 2020
6. Indire Architetture Scolastiche Homepage. https://architetturescolastiche.indire.it/. Last accessed 22 July 2021
7. Borri, S., Cannella, G., Mosa E., Moscato, G., Tosi, L.: Five Learning Spaces for new generation schools in Italy. Poster presented in DGfE-Kongress 2016, Räume für Bildung. Räume der Bildung, University of Kassel (2016)
8. Biondi, G., Borri, S., Tosi, L.: Dall'aula all'ambiente di apprendimento. Altralinea Edizioni, Florence (2016)
9. Tosi, L. (ed.): Fare didattica in spazi flessibili, Giunti Scuola, Florence (2019)
10. Borri, S. (ed.): Spazi educative e architetture scolastiche: linee e indirizzi internazionali. Indire, Florence (2016)

Pedagogical Considerations for Technology-Enhanced Learning

Linda Daniela

Abstract Technology-enhanced learning is a term often used in discussions about the place and role of technology in education. Yet many gaps still need to be addressed from a pedagogical perspective to remove the veil of fascination from technology-enhanced learning and to ensure that its use is planned purposefully. Planning should take account of the intended uses of the technology, which may be multifold. This paper summarizes the pedagogical considerations for a technology-enhanced learning process. In fact it is often thought that introducing technology to the educational environment is sufficient, without specifically restructuring the pedagogical process. The paper analyzes the potential goals of using technology in the educational process. It also looks at the pedagogical factors that make technology-enhanced learning successful: learning motivation, cognitive development, cognitive load, and knowledge development in the discourse of technology-enhanced learning. The chapter is organized as follows: an introduction to the terms followed by two sections—"Technology-enhanced learning" and "Pedagogical considerations"—and a discussion.

Keywords Technology-enhanced learning · Pedagogical considerations · Smart pedagogy · Cognitive load · Cognitive development · Knowledge assessment

1 Introduction

As technologies develop, some inevitably enter the field of education, and the tasks of the education system to incorporate them successfully are multifold:

- Learning to use technology to support the knowledge acquisition process, to solve certain tasks, and to make some activities more efficient;

L. Daniela (✉)
University of Latvia, Riga 1586, LV, Latvia
e-mail: linda.daniela@lu.lv

D. Scaradozzi et al. (eds.), *Makers at School, Educational Robotics and Innovative Learning Environments*, Lecture Notes in Networks and Systems 240,
https://doi.org/10.1007/978-3-030-77040-2_8

- Learning to use technology to get access to knowledge that is only available in certain places and at certain times, owing to distance or environmental barriers, language barriers, or special needs barriers;
- Learning the principles of developing new technologies and creatively finding new solutions where technology serves as a tool for creating innovations;
- Using Smart pedagogical principles regarding the sequencing of the learning process, the interrelationships of learning motivation and learning achievements, the interrelationships of cognitive load and cognitive development, the regularities of human development, and the interrelationships of the opportunities and challenges created by technology (Daniela 2019);
- Evaluating learning achievements, taking account of different dimensions of knowledge assessment (Daniela 2020, 2021).

In this chapter, the term *technology* refers to the various digital technologies and technological solutions used in the educational environment. The advent of digital solutions marked a new turning point in computer technology, from its triumphant beginnings in 1971 in Santa Clara, California, when Intel introduced its microprocessor (Chan et al. 2006). The term *digital teaching materials* means any technological (digital) solution that provides necessary learning content. The term *educators* is used for all levels of education and can refer to a teacher in pre-school, general education, or higher education. The term *student* is used generically to describe any person who is learning something.

There is a lot of talk about the fact that technology creates many new opportunities. At the same time, numerous claims are made that technology alone has not made any new or unexpected progress in education, and that the process itself still needs to be purposeful for learning to take place. Thus, educators plan and organize technology-enhanced learning, but to do so requires a certain knowledge of the pedagogical regularities of a technology-enhanced learning process, and also certain skills in the use of technology and digital learning tools. The same applies to students and their knowledge and skills of technology and digital learning tools. In the absence of such knowledge, technology's potential is not fully exploited. This is because a negative attitude forms towards those who agree with the developers of the Technology Acceptance Model: in addition to technological factors, affective aspects, such as perceived usefulness, are important for the use of technology (Davis 1986, 1989). Moreover, it is important to bear in mind that knowledge can be assessed from (i) *knowledge growth perspectives*, (ii) *knowledge acquisition perspectives*, (iii) *perspectives on knowledge accumulation*, and (iv) *perspectives on access to knowledge* (Daniela 2020).

2 Technology-Enhanced Learning

The term technology-enhanced learning is not new or unknown in the field of education. It has its roots in terms such as computer-supported collaborative learning (Stahl 2006), computer-assisted instruction, computer-aided instruction, computer-based learning, and computer-mediated learning, which are regarded as synonyms (De Bruyckere and Kirschner 2019), and information and communication technology for education (Usluel et al. 2008). Later, in discussions on their impact on the learning process, questions were asked about the use of technologies and whether they improve learning outcomes or, conversely, whether they negatively affect learning outcomes (Kirkwood and Price 2014).

The use of technology can have several objectives, which can all be related to technology-enhanced learning and can also complement each other (see Fig. 1). In order to evaluate the successes or failures of such a process, it is important to be aware of the key objectives in each case. These learning objectives can be divided as follows:

(1) *Various technologies are used as tools for face-to-face learning* to speed up certain activities, such as preparing material, printing it, drafting presentations for display on screen, or showing students technologies and how to use them. Technology can also be used to make certain processes, such as drawing, more efficient and accurate. It can speed up the flow of information through the Internet, Bluetooth, or similar information transmission systems. For example, if a computer is only used for printing study materials, making spreadsheets, or preparing presentations, then the acquisition of certain skills should be the

Fig. 1 Objectives for the use of technologies

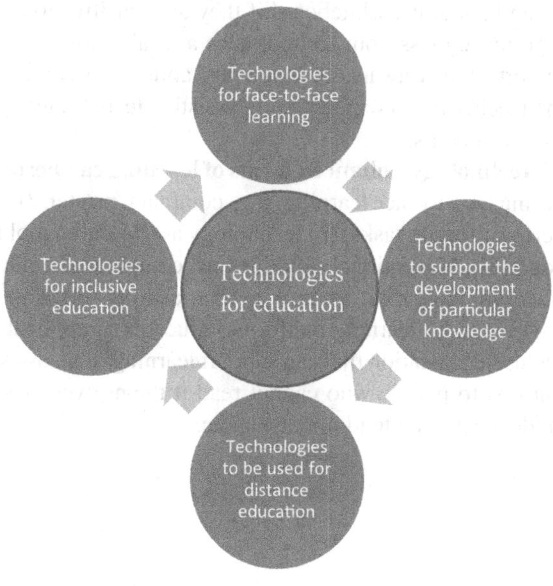

parameter used for analyzing increased knowledge. If a computer, 3D printer, or robotics kits is used to develop specific knowledge, such as the ability to program, the ability to perform accurate calculations for programming a robot, or the ability to analyze the results achieved, then increases in knowledge in a certain segment can be assessed and analyzed in terms of specific knowledge growth.

(2) We can analyze whether the activities achieved the learning objective, and this can be formulated as follows: *technologies are used for the development of knowledge* (all dimensions of knowledge).

(3) Another important goal is that *technologies ensure learning anywhere, anytime,* which can be useful for synchronous and asynchronous learning processes. In such cases, access to knowledge may also take place outside the formal education environment or it may not be available because a particular historical period is over or a certain cultural and historical value is no longer current. Other technological solutions translate material in foreign languages, enabling students to understand it. Teachers might use such solutions to deliver remote lessons to students who are unable to attend the educational institution for some reason at a given time. They might also be for a visit to a virtual museum, or access to knowledge that is not normally available, such as viewing the internal organs of humans, etc.

Despite the wide range of opportunities offered by technology and the different educational goals, one broad area that needs to be addressed is ***technologies for inclusive education***. In this area, the technologies already in use should be evaluated in terms of their usability with students with special needs; for example, can students participate in robotics activities if they have vision problems? Can they grab and connect small parts if they have muscle weakness? Or can they watch bright videos on an interactive whiteboard if they are sensitive to bright colors? Thus, if the aim is to make classroom technologies available for use with and by all students, it is important to be aware of all students' abilities to avoid social exclusion. In such cases, the teacher may consider using assistive technology to provide equal opportunities for all students.

Technology-enhanced forms of learning can be: (a) technology as a support tool during face-to-face learning (e.g., computer, printer, 3D printer, etc.) helping students perform certain tasks; (b) technology as a learning tool for acquiring specific knowledge (e.g., educational robotics kits, computer simulations for mastering computational thinking); (c) technology that can provide learning opportunities anywhere, anytime (e.g., learning platforms, educational applications); and (d) technology for inclusive education that can support learning (e.g., assistive technology that can read out text to people who cannot read it themselves or simulate a classroom when a student cannot attend in person, etc.).

3 Pedagogical Considerations

When planning and organizing a technology-enhanced learning process, the principles of Smart pedagogy should be taken into account (Daniela 2019, 2020; Borawska-Kalbarczyk et al. 2019). It is also important to analyze the digital competence of both teachers (Biezā 2020) and students (Černochová et al. 2018) so that technology-enhanced learning can take place.

Everyone involved in the learning process considers ***how cognitive development can be supported***. Promoting the development of cognitive processes—perception, sensation, attention, thinking, imagination, memory, creative thinking and problem-solving—requires knowledge of the regularities of human cognitive development, which have been discussed by various authors (Piaget and Cook 1952; Vygotsky 1978; Erikson, 1950). We can analyze learning from the perspective of Bloom's taxonomy (Bloom et al. 1956; Anderson et al. 2001), where the first stage is remembering, followed by understanding, applying, analyzing, evaluating, and creating. Thus, in a technology-enhanced learning process, depending on which of the technology use objectives (see Fig. 1) we emphasize, we need to ensure that students remember the information they have learned, understand the meaning of this information, are able to use it, can analyze what is happening, process and evaluate it, and create something new. But all this must be looked at in terms of age development, because young children are able to remember fewer items of information than older children. As they grow, their cognitive capacity increases: they are able to retain more information, create cognitive patterns, and use their thinking processes to generate new ideas. In general, this is a logical learning process, where teachers promote the step-by-step development of these cognitive processes. What changes in technology-enhanced learning, once it is clear to everyone that learning is a natural process in which a person's cognitive abilities develop gradually? One answer could be a person's ability to focus, because she has to be able to remember information, use it, analyze it, and so on. It is often assumed that technology is interesting and that students will apply this interest to acquire new knowledge and generate new ideas. In general, the concept of an 'interesting learning process' can certainly drive learning, but interest can also affect the ability to focus.

Here is an example. If a student works on a task and focuses all his attention on it in order to solve it, and he succeeds, thereby building satisfaction in his accomplishment, then his motivation to learn and achieve grows even more. If the task assigned to the student has a degree of complexity that requires a cognitive load from him, then it could turn out in one of several ways:

(a) The student will think more intensely and will look for additional information to solve the task, because he has developed learning motivation and the ability to focus, even when additional cognitive load is needed;
(b) The student will decide that it is too complicated and will look for new interesting stimuli, thus avoiding cognitive load. This could affect how much new information is learned and how much information can then be used in cognitive processes to generate new ideas. The student's previous knowledge, his

motivation to overcome difficulties and the specifics of the proposed task will determine whether he will try to connect this cognitive load to finding a solution to a more complex task.

Thus, while it is important to consider age-appropriate and goal-oriented tasks, it is also necessary to tailor the design of technology-based tasks to ensure the student directs this cognitive load towards solving them. Otherwise, he might avoid the cognitive load and instead choose to use technology for other purposes, such as scrolling through social media or playing games, which do not take him any closer to achieving the set learning goal. Technology can stimulate interest in new learning content by enabling users to see things differently, access information faster, and so on. At the same time, technology can be opaque. For example, understanding how to get access to information or the design of the technology can add a considerable cognitive burden. Sometimes design solutions are ill-considered, which again creates an additional cognitive load that affects learning processes. In a technology-enhanced environment, students can easily switch the focus of their attention.

This leads us to the ***concept of cognitive load***, which can be categorized as: intrinsic, germane, and extraneous (Sweller et al. 1998). A person's cognitive load depends on the capacity of her working memory. Intrinsic cognitive load is a person's capacity to process information. It is affected by how much a person has mastered and what information she is able to analyze. Germane cognitive load is the way a person forms thought patterns to make it easier for her to learn information, which is an essential prerequisite for learning. Extraneous cognitive load depends on the way information is presented and is directly related to information architecture and technology design.

In addition to giving students access to information, it is also important to organize it in such a way that learning can take place. For example, if we want students to learn something new that requires intrinsic cognitive load, we need to recognize that they will have to incorporate this new knowledge into their existing schemas and create new knowledge schemas (germane cognitive load will be at work here). For this to happen, the educator's task is to ensure that students are not subjected to large amounts of extraneous load during the learning process (i.e., lots of unimportant information, lots of small details, information presented chaotically, etc.). This distracts them from the essentials and creates an unnecessary cognitive load, which hinders the acquisition of new knowledge.

This means that, in a technology-enhanced learning environment, information architecture is as important as the content students have to learn. There should be no extraneous cognitive load on the student, so she can use her prior knowledge of the content both to perform a specific task and to find new information. To summarize the information analyzed in this chapter, the following are important for the organization of a technology-enhanced learning process:

(1) whether technology is perceived as being useful for acquiring knowledge or for providing access to knowledge;

(2) whether students and teachers have the appropriate digital competence to use technology, which in turn affects the extent to which the opportunities created by technology will be used in the learning process;
(3) the design of digital learning materials, which should provide an opportunity for acquiring new knowledge, while not creating an extraneous cognitive load, to ensure a context in which new knowledge is acquired;
(4) learning motivation, which depends on prior knowledge, interest, and the feedback students receive.

References

Anderson, L.W., Krathwohl, D.R., Airasian, P.W., Cruikshank, K.A., Mayer, R.E., Pintrich, P.R., Raths, J., Wittrock, M.C.: A taxonomy for learning, teaching, and assessing: a revision of Bloom's taxonomy of educational objectives. Longman, New York (2001)

Biezā, K.E.: Digital literacy: concept and definition. Int. J. Smart Educ. Urban Soc. (IJSEUS) **11**(2), 1–15 (2020)

Bloom, B.S., Engelhart, M.D., Furst, E.J., Hill, W.H., Krathwohl, D.R.: Taxonomy of Educational Objectives: The Classification of Educational Goals. Handbook I: Cognitive Domain. David McKay Company, New York (1956)

Borawska-Kalbarczyk, K., Tołwińska, B., Korzeniecka-Bondar, A.: From smart teaching to smart learning in the fast-changing digital world. In: Daniela, L. (ed.) Didactics of smart pedagogy: smart pedagogy for technology enhanced learning, pp. 23–40. Springer, Cham (2019)

Černochová, M., Voňková, H., Štípek, J., Černá, P.: How do learners perceive and evaluate their digital skills? Int. J. Smart Educ. Urban Soc. **9**(1), 37–47 (2018)

Chan, T.W., Roschelle, J., Hsi, S., Sharples, M., Brown, T., Patton, C., et al.: One-to-one technology-enhanced learning: an opportunity for global research collaboration. Res. Pract. Technol. Enhanc. Learn. **1**(1), 3–29 (2006)

Daniela, L.: Smart pedagogy for technology enhanced learning. In: Daniela, L. (ed.) Didactics of Smart Pedagogy: Smart Pedagogy for technology Enhanced Learning, pp. 3–22. Springer, Cham (2019)

Daniela, L.: The concept of smart pedagogy for learning in the digital world. In: Daniela, L. (ed.) Epistemological Approaches to Digital Learning in Educational Contexts, pp. 1–16. Routledge, Abingdon (2020)

Daniela, L.: Smart pedagogy as a driving wheel for technology-enhanced learning. Techol. Knowl. Learn. (2021). https://doi.org/10.1007/s10758-021-09536-z

Davis, F.D.: A technology acceptance model for empirically testing new end-user information systems: theory and results. Unpublished doctoral dissertation. MIT Sloan School of Management, Cambridge, MA (1986)

Davis, F.D.: Perceived usefulness, perceived ease of use, and user acceptance of information technology. MIS Q. **13**(3), 319–339 (1989)

De Bruyckere, P., Kirschner, P.A.: Computer-assisted learning. In: Tatnall, A. (ed.) Encyclopedia of Education and Information Technologies. Springer, Cham (2019)

Erikson, E.H.: Childhood and society. Norton, New York (1950)

Kirkwood, A., Price, L.: Technology-enhanced learning and teaching in higher education: what is 'enhanced' and how do we know? A critical literature review. Learn. Media Technol. **39**(1), 6–36 (2014)

Piaget, J., Cook, M.T.: The origins of intelligence in children. International University Press, New York (1952)

Stahl, G.: Group cognition: computer support for building collaborative knowledge. MIT Press, Cambridge, MA (2006)
Sweller, J., van Merrienboer, J.J.G., Paas, F.G.W.C.: Cognitive architecture and instructional design. Educ. Psychol. Rev. **10**(3), 251–296 (1998)
Usluel, Y.K., Askar, P., Bas, T.: A structural equation model for ICT usage in higher education. J. Educ. Technol. Soc. **11**, 262–273 (2008)
Vygotsky, L.S.: Mind in society: the development of higher psychological processes. Harvard University Press, Cambridge, MA (1978)

School Makerspace Manifesto

Giovanni Nulli

Abstract This contribution describes a sustainable model for makerspaces in primary and lower secondary schools. Based on Indire research on innovative school spaces, it discusses the theoretical background that schools should adopt and create before starting a makerspace lab. It also looks at which aspects of the maker culture can successfully be combined with active pedagogy. In this way, educational institutions and makers can come together to build makerspaces within schools that will be useful to both.

Keywords Makerspace · Maker culture · Active pedagogy · Sustainable model · Primary and lower secondary schools

1 Why a Makerspace Manifesto for Primary and Lower Secondary Schools

Indire has worked in several fields of study that can be considered part of the maker culture: 3D printers, robotics and coding. As part of our research into innovative spaces within schools, we also carry out observations in schools that have built makerspaces.

We consider the maker movement and the literature that it has engendered to be an important stimulus for schools. We believe the maker culture's approach to knowledge and its emphasis on practical learning are good ingredients for a new curriculum. It is important to emphasize that we consider maker culture to be related to active pedagogy. It can create an upward spiral of innovation in the curriculum and teaching methods.

Thus, active schools and the maker movement can find each other, if both parties adhere to certain conditions.

G. Nulli (✉)
Istituto Nazionale di Documentazione, Innovazione e Ricerca Educativa (Indire), Florence, Italy
e-mail: g.nulli@indire.it

D. Scaradozzi et al. (eds.), *Makers at School, Educational Robotics and Innovative Learning Environments*, Lecture Notes in Networks and Systems 240,
https://doi.org/10.1007/978-3-030-77040-2_9

2 The Potential Relationship Between Schools and Makers

We propose the manifesto from the case study we led in 2018 at Secondo Istituto Comprensivo Montessori Bilotta in Francavilla Fontana (Brindisi) and at the Istituto Comprensivo Largo Castel Seprio (now called Istituto Comprensivo Lucio Fontana) in Rome, published in Guasti and Nulli [1, p. 94–110]. Both schools have an active makerspace and both collaborate with external maker associations.

Schools and makers are two very broad, diverse and elaborate entities. Yet, as long as they can find common ground, they can create a lab where they can work together but also separately, according to their individual needs. For schools, this will be their institutional mission, which is closely related to the national curriculum; for makers, it is having a sustainable business model.

To understand why maker culture can successfully work in schools, we will look at what a maker is.

2.1 *What is a Maker?*

The Maker Movement Manifesto [2] and the Makerspace Playbook, second edition [3], describe the mindset of a maker:

1. Someone who uses their intelligence, knowledge and skills to create things;
2. Someone who learns what they need to for their projects;
3. Someone who shares their knowledge, because they believe sharing to be a fundamental facet of learning;
4. Someone who uses (and builds) tools, however simple or technological;
5. Someone who enjoys the process;
6. Someone who enjoys working with others and considers it mutually beneficial, because each may have skills that others need;
7. Someone who encourages others, because they believe everyone can get results.

We think this kind of mindset can be very useful in schools where teachers apply active pedagogy.

3 Three Principles on Which Makers and Active Schools Can Agree Before Building a Makerspace

We think schools and makers should agree on three principles before they begin collaborating. These are:

1. Recognizing the complexity of the world: creating ties and developing divergent thinking;
2. Showcasing knowledge: spreading it, and encouraging self-knowledge building;

3. Interacting with the environment and objects: intelligent artifacts and meaningful environment.

3.1 Recognizing the world's Complexity

Complexity is a key word in contemporary science. Morin [4, p. 27] defined it as an "empty word" used whenever something "is not simple, is not clear, is neither black nor white, when appearances might be deceptive, there may be doubts, we don't really know".[1] He referred to this as the "sphynx-complexity" (ibid., p. 70), indicating that it is something that should be questioned like an oracle.

In contemporary science, the emerging complex world is asking for a change in the paradigms of knowledge. For the purposes of this short work, we assume that specialist knowledge, divided into in-depth topics, does not fully grasp "the complex," and that the structure of knowledge itself[2] is no longer adequate. Therefore, an interdisciplinary approach and practical work are needed. Individual cases and points of view are more important than generalization when dealing with "the complex."

In our opinion, the maker's mindset comes from this "complex world," and the following steps represent this point quite well:

1. The objective of knowledge is to create; knowledge and creation are closely linked.
2. Team work is necessary: different points of view are needed, because there is no such thing as absolute knowledge.
3. Object creation is contingent on time and resources.
4. Having fun, because the creator's point of view is important.

If maker culture reflects the meaning of a complex world, what about schools? Are there any links to aspects of knowledge that are so theoretical?

The National Curriculum Guidelines [6, p. 11] talk about "quadri d'insieme" ("overall pictures;" from the chapter "Per un nuovo umanesimo"—For a new humanism—[3] the most "Morinian" part of the Italian Curriculum). This calls for a broader vision of learning in which schools have to create local curricular paths that bring different subjects together. Therefore, schools that already create these paths and promote the integration of students' attitudes, experiences and knowledge into their school curriculum are best placed to understand the maker's point of view, and offer students a means for questioning "sphynx-complexity."

For schools, it means that the maker way of doing things is not so distant, and there are affinities for starting a long-term shared project, such as a makerspace.

---

[1] My own translation from the Italian edition.

[2] In Morin [5, p. 17] (My own translation from the Italian edition): "We are not aware that the disjoining and fractioning of knowledge limits [...] our opportunities to understand ourselves and our world, which cause a "knowledge pathology".

[3] "For a new humanism", my own translation.

3.2 Showcasing Knowledge

Both the maker culture and schools are focused on knowledge, but they need to have more in common in order to build a connection. There are many different ways of being a maker, just as there are different ways for schools to carry out their mission, and manage knowledge. Here are some aspects that are similar for makers and schools and which give a sense of self-built knowledge that circulates.

For makers:

1. Knowledge is closely linked to personal experience and specific people.
2. Learning is fundamental and building a new object is a learning process.
3. Sharing objects and the process of building them is sharing knowledge: information about your own project can be gleaned from someone else's.
4. Open hardware and software means that the more knowledge circulates the more knowledge is available.

For schools:

1. Creating learning paths that are as personalized as possible.
2. Using competence learning, where knowledge is linked to context.
3. Promoting autonomous learning.
4. Teachers let students follow their own paths, while supporting and stimulating them, and asking them questions.

The form of knowledge emerging from the previous steps is something that is self-created by someone who explores, collaborates and builds, rather than something that already exists somewhere for the purpose of being transmitted. If we consider the learning path to be a voyage, knowledge is not the destination, but the voyage itself, where teachers are more experienced voyagers.

3.3 Interacting with the Environment and Objects

Makers use tools and create objects. For them, technology is both virtual and real. It is virtual because they share documentation, codes and projects through creative commons licenses. It is real because they use the most innovative tools for production, such as 3D printers and laser cutters.

What about schools? With the exception of laboratory work, traditional lessons involve very few tools. Active pedagogy can use the environment, tools and technological tools.

The following shows how makers and schools can meet each other in the physical world.

Makers need physical space:

1. To share beyond the virtual space.
2. To share machinery and tools.

3. To meet people, share ideas and hold courses.

Active schools (which, of course, exist in a physical space) need to give a pedagogic significance to their space and enable students to use several tools:

1. To develop active methods with objects (and to enrich interactions with technology and intelligent tools).
2. To leave students free to move as a way of fostering responsibility and autonomy.
3. To develop a pedagogic use of the environment as an area in which teachers can practice active pedagogy and share a pedagogic vision with each other.

If all of the above conditions are met, schools can start collaborating with a maker (association, group, etc.) on the design and implementation of a makerspace.

4 Starting Point and Sustainable Model

If a school wants to create a makerspace, it can be helpful to collaborate with an external group of makers. If the two can agree on the previous point, the project can go ahead. At this stage, the school principal, advised by the maker, should consider the following steps:

1. Check whether the school already has a space that is perhaps underused. Be creative.
2. It is important, almost imperative, that the chosen place has its own entrance.
3. Find/create a group of teachers the maker can train.
4. Involve people from outside the school, including public stakeholders and parents; let economic stakeholders know about the new makerspace; inform other schools; contact all important stakeholders in the local community.

The maker and the school can sign a pact and build a makerspace within school to their mutual benefit, as we can see in the following table.

The mutual benefits begin with the initial condition (first line, Table 1) and the first steps: the school finds a place, and the maker advises on machines and furniture, and creates a purchasing plan based on funds collected. Once the first step is completed, they can proceed to the second step: the school is a socially recognized stakeholder

Table 1 Mutual benefit, in order of time

School gives maker	Maker gives school
1. A place to share and create his/her activity	1. Technical and design advice
2. Advertising in the form of social recognition	2. Training courses; makerspace maintenance and development
3. Expertise on course-building for teachers/students; curricular knowledge	3. Support to teachers for designing new lessons; support to teachers during lessons in the makerspace

in relationship with other subjects. The maker can use the makerspace outside school time (especially if it has an independent entrance), for meetings or for courses for external personnel. This enables the maker to build her own business model. The maker can then hold technical training courses for teachers, while providing lab maintenance and development services, including planning new purchases.

The third step is a process of consolidation between the maker and the school. At this stage, the maker has a deeper understanding of how the school works and is able to design second-level courses, which address curricular needs. She also works for other schools as an expert, selling courses for teachers. On the other hand, the maker can be integrated into the school's activities, providing support for a new kind of lesson that bridges the makerspace activity and the school curriculum. She acts as an *atelierista*, as described by pedagogue Malaguzzi [7, p. 88], providing structured support to the teacher in the space and in the lesson design.

5 Why a Makerspace? Because It is a Disruptive Way to Make Change

Neil Gershenfeld (director of the Center for Bits and Atoms and author of the 2005 book "Fab: the coming revolution on your desktop—from personal computer to personal fabrication") identified Seymour Papert as one of the founding fathers of the maker movement [8, p. 163]: "The distinction between toys and tools for invention, [culminates] in the integration of play and work in the technology for personal fabrication. The original inspiration and instigator for bringing these worlds together was Seymour Papert, a mathematician turned computer pioneer." (own translation). We would like to conclude this work with Papert's words [9, p. 27 and 26] about active pedagogy and innovation in schools: "John Dewey's idea that children would learn better if learning were truly a part of living experience; or Freire's idea that they would learn better if they were truly in charge of their own learning processes; or Jean Piaget's idea that intelligence emerges from an evolutionary process [...]. Sadly, in practice they just wouldn't fly. When educators tried to craft an actual school based on these general principles, it was as if Leonardo had tried to make an airplane out of oak and power it with a mule [...]. In my view almost all experiments purporting to implement progressive education have been disappointing because they simply did not go far enough in making the student the subject of the process rather than the object [...]. Early designers of experiments in progressive education lacked the tools that would allow them to create new methods in a reliable and systematic fashion".[4] What Papert advocated was that active methods should be adopted along with all the repercussions and changes they bring. He also upheld the need to drive the system towards using as many tools as technology (in a wide sense) can offer, even if it means having to retrain teachers. Teachers should be prepared to come out of their comfort zones in order to invest in a different future.

[4] "For a new humanism", my own translation.

In our opinion and experience, building a makerspace is the sort of economic investment that causes the kind of disruption that leads to more stable long-term change.

References

1. Guasti G., Nulli G., Tosi, L.: Creare un makerspace: i casi dell'IC Largo Castelseprio e del Secondo IC Montessori Bilotta. In: Tosi L. (ed.) Fare didattica in spazi flessibili, Giunti Scuola, Florence (2019)
2. Hatch, M.: The Maker Movement Manifesto. S. l. McGrawHill education (2014)
3. Makespace Team: Makerspace Playbook, 2nd edn. S. l. Maker Media (2013). Retrieved 2020/09/21 https://makered.org/wp-content/uploads/2014/09/Makerspace-Playbook-Feb-2013.pdf
4. Morin E.: La sfida della complessità, Editoriale Le Lettere, Florence (2017)
5. Morin, E.: La conoscenza della conoscenza. Saggi/Feltrinelli, Milan (1989)
6. MIUR: Indicazioni Nazionali per il Curricolo, Rome, Le Monnier (2012)
7. Edwards C., Gandini L., Forman G.: I cento linguaggi dei bambini, Parma, Spaggiari Edizioni (2017)
8. Gershenfeld, N.: Fab: the coming revolution on your desktop—from personal computer to personal fabrication. Basic Books, New York (2005)
9. Papert, S.: I bambini e il computer. Rizzoli, Bologna (1994)
10. Papert, S.: The Children's Machine: Rethinking School in the Age of the Computer. Basic Books, New York (1992)

Elements of Roboethics

Fiorella Operto

Abstract Roboethics analyzes the ethical, legal and social aspects of robotics, especially with regard to advanced robotics applications. These issues are related to liability, the protection of privacy, the defense of human dignity, distributive justice and the dignity of work. Today, roboethics is becoming an important component in international standards for advanced robotics, and in various aspects of artificial intelligence. An autonomous robot endowed with deep learning capabilities shows specificities in terms of its growing autonomy and decision-making functions and, thus, gives rise to new ethical and legal issues. The learning models for a care robot assisting an elderly person or a child must be free of bias related to the selected attributes and should not be subject to any stereotypes unintentionally included in their design. As roboethics goes hand in hand with developments in robotics applications, it should be the concern of all actors in the field, from designers and manufacturers to users. There is one very important element in this—albeit one that is related indirectly—that should not be overlooked: namely, how robotics and robotic applications are represented to the general public. Of the many representations, the legacy of mythology, science fiction and the legend still play an important role. The world of robotics is often marked by icons and images from literature. Exaggerated expectations of their functions, magical descriptions of their behavior, over-anthropomorphization, insistence on their perfection and their rationality compared to that of humans are only some of the false qualities attributed to robotics.

Keywords Service Robotics · Field Robotics · Roboethics · ELS (Ethical · Legal · Societal) Issues

F. Operto (✉)
Scuola Di Robotica, Genoa, Italy
e-mail: operto@scuoladirobotica.it

D. Scaradozzi et al. (eds.), *Makers at School, Educational Robotics and Innovative Learning Environments*, Lecture Notes in Networks and Systems 240,
https://doi.org/10.1007/978-3-030-77040-2_10

1 The Birth of Roboethics

This article outlines some of the main lines of development and application in roboethics—that is, applied ethics in advanced robotics—which examines the ethical, legal and societal issues (ELS) inherent to the field. Roboethics was born—as a term and as applied ethics—in 2004, during the First International Symposium on Roboethics [2]. Roboethics analyzes the ethical, legal and social aspects of robotics, especially in relation to service and field robotics applications. These issues are related to the protection of privacy, the defense of human dignity, distributive justice and the dignity of work. Today, roboethics is the subject of hundreds of studies, applications, research, and is becoming an important component in international standards for advanced robotics, and also in various aspects of artificial intelligence. Robots are certainly formidable tools. There is no aspect of our private and social lives that cannot be improved by the introduction of robots. However, technology applied to human life always raises ethical questions. In the case of robotics, especially service robotics, these ELS issues are novel, emerging, complex and involve several disciplines.

2 A New Science?

Robotics is a field of research and application, or a new science, still in its infancy, born from the fusion of many disciplines within the humanities and natural sciences. Here, the sum is greater than the parts [2]. It is a very powerful tool for studying and increasing our knowledge, not only of the universe around us—space, oceans, our body—but also our brain/mind. This is why robotics can lead to a convergence of the so-called *two cultures*: human sciences and natural sciences. Robotics covers the following disciplines: mechanics, automation, electronics, informatics, cybernetics, physics/mathematics, artificial intelligence, and draws contributions from (and, in fact, is *invading*) logic/linguistics, neuroscience/psychology, medicine/neurology, biology/physiology, psychology, anthropology/ethology, art/industrial design.

This complexity—involving novel aspects such as ESL issues, which drive research and robot production—is giving the subject the caliber of a science, with its laws, coherent and comprehensive understanding of nature and predictive capabilities. For an overview of the state of robotics today we recommend the *Springer Handbook of Robotics* [3].

Moreover, the object of the research and development of advanced robotics, an autonomous robot endowed with learning capabilities, shows specificities in terms of its growing autonomy and decision-making functions.

3 What Ethics Should Be Applied in Roboethics?

We have different *versions* of roboethics depending on which ethical theories are adopted (utilitarianism, deontology, virtue ethics, rights ethics, Rawls' theory of justice). Yet, in all these versions, a logical and critical framing of ethics is needed, one that reveals the implicit, uncovered assumptions, and analyzes the reasons, the pros and cons, and their origin. This frame also allows us to define (the extent and limits of) human liability and machine autonomy, in cases of damage caused by a learning robot.

In addition, in the light of more complete ethical theories, the ethical framing should assess whether, according to distributive justice principles, the actors involved should be socially duty-bound to compensate for the dramatic changes caused by a heavy, rapid and unnegotiated introduction of robots to our society: job displacement and loss; privacy issues and encroachment on personal life; technological dependency; robotics divide (in terms of generations, social status, and areas of the world).

Finally, roboethics should cover a series of positive recommendations and rules that would be implemented in all contexts where robots are introduced. These should be along the same lines as the recent prescriptions being adopted for the production and use of plastics, energy, and other industrial sectors. In roboethics, analysis of ELS issues often leads to recommendations which, in many cases, have been submitted to the UN, European governments, the European Commission and the European Parliament [4].

In light of the lessons learned from the COVID-19 pandemic, we cannot afford to introduce large-scale technological applications into society without offsetting the ensuing imbalances in the environment and the disruption to social groups.

4 Emerging and Novel Roboethical Issues

Since 2004, several authors have intervened in the debate on roboethics to highlight certain ELS issues that have arisen over a very long period of time. Issues such as the rights of and the payment of and for robots and their status as moral agents can be interesting to discuss, but too far-reaching. They also do not consider the urgency of and the need for addressing ethics-related technical issues.

Given the rich and complex debate on roboethics and the sometimes unknown developments that could come over the next two decades, the author, the partners and experts of the Ethicbots European Project [5] adopted a triaging system to analyze the following issues:

Novelty: Issues that have never been looked at; the *absentia legis* and the lack of regulations is, in many cases (bionics and military robotics), evidence of a severe responsibility gap.

Emerging: Issues that are not planned for, since robotic prototypes are the result of different forces: research and business.

Complexity: Issues lying at the intersection of several disciplines (robotics, AI, moral philosophy, psychology, anthropology, law).

Social pervasiveness: Issues related to current and yet-to- be-released robotic products [6].

The sectors most directly and urgently interested in robotic applications are the military and certain areas of biomedicine (invasive prosthetics) [7, 8].

5 The Risk of Unintended Machine-Learning Bias

Issues of bias in artificial intelligence are well-known to scientists. Machine-learning models are developed to be predictive, when large datasets teach the robot learning models to predict the future, based on the past. Trained models can read and use an incredible amount of data (texts, pictures, software, other models), consuming it to identify the data patterns considered most suitable for carrying out the mission. In this way, predictions can be more accurate than with simple built-in models. The bias issue is related to the fact that machine-learning models can predict precisely what they have been trained to predict, and their predictions are as accurate as the data used to train the machine. Any errors are explained in the maxim "garbage in, garbage out." In fact, many cases of bias detection, which range from the light to the heavy involving AI products, stem from human bias intervening during the creation of data models. Either the data collected were unrepresentative of reality—as in the *portability issue*, when a model is employed out of context—or they reflect existing human prejudices—for instance, when certain attributes in the model are selected or ignored [9].

The learning models for a care robot assisting an elderly person, a child, or in a hospital must be free of bias related to the selected attributes (e.g., culture, gender, social or economic status, linguistic attributes) and should not be subject to any stereotypes unintentionally included in their design. It is easy to imagine how complex this process could be in a learning robot, especially since bias detection cannot be performed at the expense of the assisted person.

6 Ethical Guidelines for All Robots

In a review article written by Matthew Studley and Alan Winfield on the ESL aspects of industrial robots [10], the authors came to an interesting conclusion after reviewing around 84 papers on the topic: today, even robots used in industrial production are subject to similar ELS problems to those found with service robots, which means that the gap between industrial robots and other types is narrowing.

Industry is changing; converging technologies have ushered in a fourth Industrial Revolution, where new collaborative robots, or *cobots*, work alongside humans on common tasks. Unlike more common industrial robots, which largely work alone and unsupervised, collaborative robots are programmed and designed to work with humans, responding to human behaviors and actions. The authors of that review article point to the increasing importance of human–robot interaction (HRI) and the reduced differentiation between industrial robotics and other robot domains affected by the definition and range of ELS issues. In this, advanced industrial robotics may be affected by the same sorts of concerns that are faced in assistive robotics: predicting and interpreting human intentions and future actions in order to perform efficiently. Here, the interactions between humans and robots involve teaching rather than programming. The ELS issues that affect learning cobots include psychological and sociological impacts, liability, data and privacy. Cobots can be programmed for the speed, tasks and precision to which humans have to adapt. In addition, cobots can be reprogrammed rapidly for another task, forcing humans to make rapid changes with no time to adapt. Cobots can gather data about the pace of work, abilities and needs of their human co-workers. These data may be processed in cloud services and could be used by other organizations. Use of the resulting data profiles could breach European privacy legislation.

7 Representation of Robots with the General Public and Agnotology Issues

A full analysis of roboethics should not overlook another element that is indirectly related, but has a great impact on it: how robotics and robotic applications are represented to the general public. Among the many representations, the legacy of mythology, science fiction and the legend still play an important role [12]. The world of robotics is often marked by icons and images from literature. Exaggerated expectations of their functions, magical descriptions of their behavior, over-anthropomorphization, insistence on their perfection and their rationality compared to that of humans are only some of the false qualities attributed to robotics.

The word 'agnotology' was coined to refer to the study of culturally induced ignorance, and specifically its implications for individual and collective decision-making [13]. The continuous and massive dissemination of inaccurate information about robots in the media can hinder the public's understanding of the fundamental ELS aspects of robotics that already affect quality of life. The obfuscation and omission of basic, relevant, and accurate information about robotics and roboethics issues can only result in wrong assumptions which affect the public's ability to take part in the collective decision-making process.

8 Conclusions

We have briefly described only some of the elements of roboethics that should be studied by those who program, produce, market and use robots. Roboethics will have an impact on the design, programming, shape and use of robots. It should be included in engineering and architecture programs, as well as in the various disciplines of the humanities. Also, it is important to build trust between the general public and robotics laboratories through honest, concerted information campaigns.

References

1. Veruggio, G., Abney, K.: Roboethics: the applied ethics for a new science. In: Lin, P., Abney, K., Bekey, G. (eds.) Robot Ethics: The Ethical and Social Implications of Robotics. MIT Press, Cambridge, MA (2018)
2. Veruggio, G., Operto, F.: Roboethics: social and ethical implications of robotics. In: Handbook of Robotics. Springer, Heidelberg (2008). www.roboethics.org
3. Siciliano, B., Khatib, O.: Springer Handbook of Robotics. Springer, Berlin, Heidelberg (2016)
4. Resolution of the European Parliament: A comprehensive European industrial policy on artificial intelligence and robotics (2019): https://www.europarl.europa.eu/doceo/document/TA-8-2019-0081_EN.html.
5. ETHICBOTS project: Emerging Technoethics of Human Interaction with Communication, Bionic, and robOTic systems. (2005, Nov. 1–2008, Apr. 30). Coordination action, FP6—Science and society. Available: http://ethicbots.na.infn.it/ (2008).
6. Operto, F.: Ethics in advanced robotics. IEEE Robot Autom Mag Special Issue on Robot Ethics **18**(1) (2011)
7. Sharkey, N.: Automating warfare: lessons learned from the drones. J. Law Inf. Sci. **21**(2), 140 (2012)
8. Sharkey, A.: Autonomous weapons systems, killer robots and human dignity. Ethics Inf. Technol. **21**, 75–87 (2019). https://doi.org/10.1007/s10676-018-9494-0.(2019)
9. Battistuzzi, L., et al.: Embedding ethics in the design of culturally competent socially assistive robots. In: IEEE/RSJ International Conference on Intelligent Robots and Systems, 1–5 Oct 2018, Madrid, Spain (2018)
10. Studley, M., Winfield, A.: ELSA in industrial robotics. Curr Robot Rep (2020). https://doi.org/10.1007/s43154-020-00027-0.(2020)
11. Tamburrini, G.: On the ethical framing of research programs in robotics. J. AI Soc. (2020). https://doi.org/10.1007/s00146-015-0627-2
12. Operto, S.: Evaluating public opinion towards robots: a mixed-method approach. Paladyn **10**(1):286–297 (2019). https://doi.org/10.1515/pjbr-2019-0023
13. Proctor, R.N.: Londa Schiebinger. Agnotology: The Making and Unmaking of Ignorance. Stanford University Press (2008)

Making to Learn. The Pedagogical Implications of Making in a Digital Binary World

Maria Ranieri

Abstract Making has always been at the center of pedagogical reflection, as testified by the emphasis on the principle of *learning by doing*. Today, digital media and technologies provide more opportunities for expanding these concepts, given their potential to facilitate media production and creation. However, common practices in media-making in schools tend merely to emphasize the technical aspects—including the technical procedural skills, the creation of a product or the use of specific software—while overlooking the pedagogical dimensions associated with media-making processes. A reappraisal of the educational dimensions of making in the digital era may come from a reconceptualization of the relationship between manual and intellectual activities. Through the lens of Sennett's understanding of craftsmanship, this chapter first explores the ways making and thinking can be set out in a single process that characterizes the human condition. Second, it explains how the Open Source Movement's approach to software design is a sort of "digital craftsmanship" based on collaborative problem-solving, one that can inspire a renewed vision of the *learning by doing* principle for the digital world. The chapter concludes with several considerations on the implications of such an approach for the future of schools.

Keywords Learning by doing · Problem-based learning · Social learning · Open source · Digital technologies

1 Introduction

The idea that making sparks knowledge acquisition and learning is not new. Just think of the principle of *learning by doing* which is found at the root of the work of eminent psychopedagogues and education theorists such as Dewey [2] and Vygotsky [7]. Learning by doing can be defined as "the process whereby people make sense of their experiences, especially those experiences in which they actively engage in making things and exploring the world," and also indicates "a pedagogical approach

M. Ranieri (✉)
University of Florence, Florence, Italy
e-mail: maria.ranieri@unifi.it

D. Scaradozzi et al. (eds.), *Makers at School, Educational Robotics and Innovative Learning Environments*, Lecture Notes in Networks and Systems 240,
https://doi.org/10.1007/978-3-030-77040-2_11

in which teachers seek to engage learners in more hands-on, creative modes of learning" [1, p. 108]. What is new, today, is that the opportunities learners have to create or to engage in making have multiplied with the proliferation of digital media and technologies that facilitate doing and making for learners. As clearly explained by Hobbs [4, p. 7], "when we create media, we internalize knowledge deeply—we own it. *Internalization* is the process of consolidating and accepting ideas, behaviors, and attitudes into our own particular worldview. After all, if we can represent knowledge, information, and ideas in a format that makes sense to others, that's a form of mastery." Indeed, what we and our students are continually engaged in today, with the media, is a meaning-making activity through which we interpret, understand and make sense of the world around us. In short, we learn. However, as always, things are more complex than that. At school, we are surrounded by examples of media-making or making with media that merely amount to applying technical procedures, with no room for creativity, understanding or making sense [5]. To consider media production as something that goes beyond a mere technical exercise, educators need to draw attention to pedagogically significant aspects. To do this they should seek to engage children in the manipulation of symbols, problem-solving, collaboration and interaction with their peers.

2 Beyond Making as a Mere Manual Activity

One initial step we can take towards moving beyond our traditional understanding of making as a mere practical activity is to go deeper into what precisely making does entail, in terms of the socio-cognitive processes involved in the art of doing. In this vein, Richard Sennett's seminal work, *The Craftsman*, is particularly enlightening. In the opening pages, Sennett begins his argument by returning to Hannah Arendt's theoretical contribution and, at the same time, questioning its assumptions. If, in the human condition, Arendt distinguished between the three figures of the *animal laborans*, the *homo faber* and the *zoon politicon*, by contrasting them and recognizing the primacy of political action over other forms of activity, Sennett recognizes these categories, but questions their separateness, as well as the premises on which this separateness is based; namely, the dichotomy between doing and thinking, with the latter having primacy over the former. According to Sennett, integration of doing and thinking sees its concrete implementation in the figure of the craftsman, by which he meant a specific condition of humanity rather than a historically specific social category. To understand the intimate nature of this condition, he uses the concept of "craftsmanship," which he calls "an enduring, basic human impulse, the desire to do a job well for its own sake. Craftsmanship cuts a far wider swath than skilled manual labor; it serves the computer programmer, the doctor, and the artist" [6, p. 9]. Therefore, craftsmanship is not just manual labor, but art, mastery, the ability to achieve what we set out to achieve, which includes manual and intellectual activities alike. Considering it just a technical routine is a big mistake, as Sennett underlines: indeed, the craftsman achieves a synthesis between "the hand and the head," enabling concrete actions—even repetitive, habitual ones—conversing with thought and creativity. In order to grasp the mechanism that nurtures the virtuous conversation between making and thinking, one must have a deep understanding of the three fundamental abilities of craftsmanship, namely "the ability to localize, to question, and to open up. The first involves making a matter concrete, the second reflecting on its qualities, the third expanding its sense" [6, p. 277]. Finally, one further aspect highlighted by the American sociologist is linked to the social dimension of learning, which was peculiar to the transmission processes of knowledge in medieval workshops: this social dimension is inherent to mastery by way of the sociable expertise that individuals develop. "Sociable expertise doesn't create community in any self-conscious or ideological sense; it consists simply of good practices. The well-crafted organization will focus on whole human beings in time, it will encourage mentoring, and it will demand standards framed in language that any persons in the organization might understand" [6, p. 249].

3 Unlocking the Digital Box: Making to Learn

The emphasis on quality-driven work is by no means an invitation to go back to the past. On the contrary, it is a timely attempt to set off in search of that human condition represented by craftsmanship, meaning the ability to do things well for oneself, regardless of one's manual skills. According to Sennett [6], contemporary craftsmen are those who know how to use digital technologies with mastery, and consider quality, innovation and social cooperation to be fundamental values in their work. Hence, Linux operating system's developers are seen as the craftsmen of the digital age. Himanen [3] dedicated his volume *The Hacker Ethics and the Spirit of the Information Age* to the modern artisans of software, and, in particular, to Linux developers. Looking at the technological and socio-cultural practices of young developers, Himanen outlines the hacker model of learning and illustrates the ethical principles that govern the behavior of the members of the community linked to Linus Torvalds. The first feature of the model is *openness*, which is essentially based on the free circulation of ideas and access for all. Indeed, Linux's source code is public and accessible to all: anyone can use it and adapt it according to the "open source" model. Thanks to openness, more intelligences can intervene, manipulate programming languages and collaborate to solve common problems. As Himanen [3] explains, the hacker model is based on sharing the problem, the solution and the procedures that led to it. The latter two of these play a crucial role in the collaborative construction of new solutions, since the underlying information and discussions associated with the discovery of new solutions are more important than the results themselves.

The second feature of the hacker model of learning is that learning always starts with a *problem*. Here, the idea is that knowledge can be continuously improved: whenever a problem is seen, an advancement in knowledge is called for. Traces of this model can be found in the Platonic Academy, which Himanen [3] contrasts with the monastery. In the first case, the main aim of teaching is to strengthen the ability of disciples to pose problems, develop thought and voice criticisms. In the second case, teaching can be summarized in the Benedictine rule whereby «speaking and teaching belong to the teacher, while silence and listening pertain to the disciple». We can also say that both the Platonic and the hacker models are based on a critical problem-based approach, aimed at identifying problems and raising questions. Conversely, the monastery model evokes the traditional lecture-style, teacher-centered method, in which the students are the passive recipients of the training process.

Finally, in hacker ethics, sharing is not only a right but also an obligation, as is the practice of citing sources: anything can be copied and transformed, as long as the source is mentioned. Intellectual property does not entail individual ownership of an idea, since ideas belong to everyone. What matters are the credit and recognition received from the community. Indeed, since the community validates the solutions and provides support or recognition, authority resides within the community itself.

4 Conclusion

A better understanding of the making process is, today, of fundamental importance in the context of a reflection on education and digital media. As Sennett [6, p. 8] observes, "material culture matters" in the sense that "we can achieve a more humane material life, if only we better understand the making of thongs." This warning should be directed at several sectors of high-tech societies, including education and school. A school that knows and masters the technologies it uses, which develops and shapes them around its own needs, which is conversant in the technological practices of its students to promote reflexivity and distancing, and which suggests unexpected work paths to its community of learners would perhaps be able to implement an approach to technological innovation in contrast with the views that currently dominate the school arena. Instead of "viral technological injections" imposed from above and accepted misgivingly, there is participatory planning for "making well," and mastery based on openness for a more humane school life.

References

1. Bruce, B.C., Bloch, N.: Learning by doing. In: Seel, N.M. (eds) Encyclopedia of the Sciences of Learning. Springer, Boston, MA. https://doi.org/10.1007/978-1-4419-1428-6_544
2. Dewey, J.: Experience and Education. Macmillan, New York (1938)
3. Himanen, P.: The Hacker Ethics and the Spirit of the Information Age. Random House Inc., New York (2001)
4. Hobbs, R.: Create to Learn: Introduction to Digital Literacy. Wiley & Sons Inc., Hoboken, NJ (2017)

5. Parola, A., Ranieri, M.: Media Education in Action A Research Study in Six European Countries. Florence University Press, Florence (2010)
6. Sennett, R.: The Craftsman. Yale University Press, New Haven (2008)
7. Vygotsky, L.: Mind in Society. Harvard University Press, Cambridge, MA (1930)

The Game of Thinking. Interactions Between Children and Robots in Educational Environments

Luisa Zecca

Abstract Educational robotics (ER) fits into a constructivist perspective that aims to overcome the dichotomy between researchers and practitioners through collaborative research. This sparks reflection on how to develop professional training for teachers. The laboratory is the privileged setting of ER activities. It is an educational space for training and research alike, and is suitable for learning to do, whereby children can handle robots and develop scientific skills. The Laboratory of Robotics for the Cognitive and Social Sciences of the University of Milano-Bicocca, directed by Professor Edoardo Datteri, conducted a series of activities within this framework. Using an ethnographic approach informed by grounded theory, where the teacher acts as a mediator rather than an instructor, the project included: labs in primary schools involving a Lego Mindstorms robot assembled as a small vehicle; labs in lower secondary schools to study the different approaches of teachers; a robot programming activity with preschool children at "Bambini Bicocca." These research activities give children the opportunity to develop scientific and thinking skills, and show that ER can engage students in metacognitive reflection. Moreover, thanks to a well thought-out laboratory approach, robotics can cut across several educational skills, such as learning to learn, acting autonomously, solving problems. An in-depth study of the interactions between adults, children and robots also plays an important role in advancing the research with new knowledge for action: it sheds light on the problem-solving strategies of students and the behaviors of children and teachers.

Keywords Laboratory · Collaborative research · Educational robotics · Science education · Thinking Skills

L. Zecca (✉)
University of Milano-Bicocca, 20126 Milan, Italy
e-mail: luisa.zecca@unimib.it

D. Scaradozzi et al. (eds.), *Makers at School, Educational Robotics and Innovative Learning Environments*, Lecture Notes in Networks and Systems 240,
https://doi.org/10.1007/978-3-030-77040-2_12

1 Laboratory Approach and Educational Robotics

The laboratory is a specialized educational space suitable for producing critical, reflective and imaginative thinking, which is shown to be an effective theoretical and experiential device for acquiring those skills that Dewey called learning to think and learning to do [7]. Indeed, in the theory–practice nexus, different educational objectives are linked and procedural knowledge is transformed into smart expertise. This occurs through the reformulation of problems, the search for new solutions, and verification and review of thinking processes. Operating climate and collaboration are distinctive features of lab methodology and configure an educational setting of mutual help, where dialog and discussion permeate the work community [10].

Collaborative research is one of the main models of participatory research, which aims to look beyond the dichotomy between researchers and practitioners, in order to foster a dialog between theoretical knowledge and pragmatic knowledge for shared research. The aim is not merely to identify the practical implications for schools of new academic knowledge. Indeed, collaborative research also aims to spark reflection on how professional training for teachers can be developed [1, 6]. Thus, the laboratory is an educational space characterized by research and training alike. It is a privileged setting where new knowledge can be developed and, at the same time, there can be reflection on the activities and experiences, in a continuous process of destructuring and restructuring.

Educational robotics fits into this constructivist perspective, because it involves a form of electronics that can enhance learning through new ways of constructing meaning. Educational robots have come a long way since Papert, especially thanks to the approach of learning by doing through multi-level interaction. First, they are relatively safe and children can handle and experiment with them themselves, without the need for close teacher supervision. Also, although robotic behaviors depend in part on environmental and internal factors, educational robots are relatively predictable. This makes analyzing robots less frustrating than studying animals or plants. Their use helps children to develop research skills, such as observation, explanatory hypothesis formulation, hypothesis testing, and review of hypotheses in light of the results observed. Furthermore, they can encourage them to reflect metacognitively on the fundamentals of scientific research methodology. Indeed, as they carry out these activities, children have to think actively about what they are doing. This causes them to adopt a metacognitive perspective involving epistemological and methodological reflection [5].

The "Riccardo Massa" Department of Human Sciences for Education of the University of Milano-Bicocca has pursued these goals for years in a research project that aims to observe the development of scientific skills in children and devise methodological guidelines to plan and assess ER activities.

2 Towards the Game of Thinking in Primary Schools

Since 2011, our department has held several ER laboratories in primary schools in Milan, involving a Lego Mindstorms robot assembled as a small vehicle, programmed in advance with the NXT-G software to function as a Braitenberg-like vehicle. During these activities, instead of programming a robot, children have to find out how the robot has been programmed. It is in this that these learning activities differ from conventional approaches to educational robotics: instead of identifying sequences of motor commands, they are asked to explain the robot's behavior. To encourage deep learning in various scientific disciplines and develop abstract thinking and problem-solving skills, children are asked to describe what the robot was doing and to explain why it was doing it. They are free to interact with the robot, to get close to it and to put their hands near the sensors [2–4].

These robotics activities were labeled the Game of Science by the first group of eight-year-olds to do this activity with us. It involves studying the robot's behavior in ways similar to how ethologists study living animals, which is why it can be defined as "roboethology." We audio-recorded the verbal interactions among children and between teachers and children and adopted an ethnographic approach informed by grounded theory to identify and categorize the scientific and abstract reasoning skills displayed by children. The teacher acts as a mediator rather than as an instructor: he or she should avoid answering the children's questions or correcting their views, and instead should reformulate the questions raised by the children and ask them for explanations and justifications. This dialogical approach stimulates independent reasoning in children and encourages them to experiment.

2.1 Considerations on Experimental Adequacy and Refining the Setting

Among the other hypotheses formulated by children at one of the sessions was the idea that the robot was moving randomly. To test this theory, the children devised an experiment. It consisted of making a door in an experimental arena with several obstacles, and seeing whether the robot was able to go through it to leave the arena (see Fig. 1). The children reasoned that a randomly moving robot would collide with the obstacles, therefore it would fail the objective of finding the way out.

However, as we can see from the exchange below, formulating an experiment means evaluating the suitability of the experiment and, potentially, refining it:

T[eacher]: C1, should the door be wide or narrow?

C[hild]1: Narrow…

T: Like this?

C1: No…

T: Should it be narrower?

Fig. 1 The experimental arena with a door and some obstacles. Is the robot moving randomly?

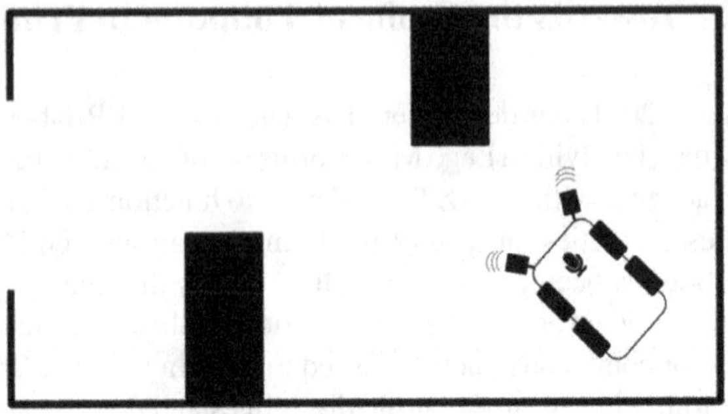

C1: We need to add a brick [to make the door narrower].

C2: Yes, because if you put the robot there, it will go out straightaway.

The point is that if the door is too wide the robot may leave the arena by chance. Therefore, an arena with a wide door is a bad experimental setting for testing whether or not the robot moves randomly. For this reason, children suggest refining the experimental setting in order to rule out alternative explanations.

2.2 *Drawing Theoretical Conclusions and Identifying Alternative Explanations*

At this point, one child imagines that if the robot is not moving randomly, it will go out. In short, he is predicting what the robot will do, if a given theory is true. But not everyone agrees, as we can see from the excerpt here:

T: If it goes outside the arena, that means it is not moving randomly. Do you all agree?

C: I don't agree. If it leaves immediately, it isn't moving randomly. But if it hits a box and then goes out, then it's moving randomly.

The child identifies two possible experimental results and he concludes that, if the robot exits after bumping into obstacles, it means it cannot see and therefore is moving randomly, even if it is able to exit. He is showing the ability to draw theoretical conclusions from possible experimental results. But we also notice that different interpretations can be made from the same behavior, as is evident in the following excerpt:

C: […] but maybe the robot hits the obstacles because it's got a map inside its mind, and it can't see the obstacles.

Here the child identifies a potential alternative (as yet unseen) explanation for the robot's behavior. So, we can conclude that the door experiment is not adequate for deciding whether or not the robot's movements are random.

Children were asked to observe a target system, describe it, identify the phenomena, propose explanatory hypotheses, make predictions, design experiments, compare the experimental results with their predictions and revise their hypotheses in line with the results [11]. Thus, educational robots appear to be particularly suitable tools for organizing scientific laboratories in schools, where we can also examine, among other things, the different direction styles of teachers, which have a strong impact on the knowledge environment.

3 Robotic Labs and Different ER Approaches of Teachers

In the 2013/2014 academic year, we conducted a case study in three lower secondary schools that had joined Amicorobot, a network of schools that promotes research, training, and experimentation in ER. These three schools held robotics labs for 20 years. This enabled us to study in-depth the different ER direction styles of teachers leading the labs, especially related to the verbal interactions between teachers and the class as a group. The analysis of their interactions was compared with the objectives declared by the teachers themselves and with the organizational environment, with the idea that there is strong reciprocity between objectives, interactions and organization [9].

During this case study, the teachers were the first to want to start a metacognitive analysis of their practices. With the aid of video recordings, we identified three different ER approaches among the teachers:

1. content-based learning, which views the laboratory as a space where the acquisition of knowledge can be tested along with the capacity to apply it;
2. cooperative learning, which aims to solve problems creatively and works toward developing team-building skills;
3. problem-based learning, which stimulates dialog between equals and an autonomous and shared design, as well as testing the skills acquired during previous educational robotics laboratories.

In addition to the construction of educational robots, understanding different lab approaches helps advance educational research with new knowledge for practice. This would be helpful to other teachers taking similar paths. Specifically, studying the interaction between children, adults and robots in order to understand problem-solving attitudes could help structure targeted education interventions, as in the following example.

3.1 Programming a Robot with Preschool Children at "Bambini Bicocca" Infant School

In 2018, our department held a robot programming workshop at the "Bambini Bicocca" Infant School. We conducted seven separate sessions lasting around 40 min each, with the participation of seven children aged between 4 and 5. Educational robots are a useful focus of scientific inquiry, even for preschool children. Cubetto, a robot produced in 2013, is particularly suitable for children up to the age of six. After explaining the operating procedures with the robot, the teacher-researcher posed some programming problems to children. The activities were recorded using three video cameras and the data collected were examined in a specific software program. The main purpose was to examine the interaction between children, adults and robots, to explain the problems and solving strategies faced by the children and to understand what types of tutoring the teachers implemented [8].

The children faced various problems, such as:

1. issues with defining the objectives: children were given the freedom to suggest how to structure the objective; all the problems that arose originated from their spontaneous suggestions;
2. programming problems, related to building the code;
3. problems related to Cubetto's interlocking blocks;
4. verification issues, that is, all the problems arising from solving the objective, from building the code to checking its effectiveness, once it was built;
5. directional issues related to how the robot moved;
6. issues with defining all the movements that Cubetto should make to achieve the assigned goal (these are related to programming issues);
7. issues with detecting errors: all the problems that enable children to locate the errors they have made.

The children employed strategies that were specific to each problem they were confronted with, such as:

1. observation and intuition, whereby they figured out what to do to achieve an objective by looking at the problem;
2. decomposition of a problem or an objective into several parts;
3. mental simulation, i.e., mental representation of the movements the robot should make to achieve its objective;
4. body simulation, where the child represents the itinerary physically with his own body;
5. strategy of direct verification during programming;
6. verbal simulation, meaning, all the heuristic procedures guided by a verbal explanation of what movements the robot should make;
7. indifference to problems, which enables children to choose other strategies to reach the objective;
8. distress call and listening strategies;
9. trial and error.

As mentioned above, it may be useful to have an understanding of what children see as problems and their strategies for addressing them, to enable teachers to structure educational interventions. First, this classification can predict the behavior of children involved in robotics activities. Second, it can be used as a guide for observing their behavior. Lastly, identifying problems and strategies can provide effective support tool for teachers or experts who want to work in the field of educational robotics.

4 Conclusions

In recent years, our RobotiCSS Lab has conducted other robotics activities in schools—in addition to the research already mentioned—involving other types of robots, including Lego WeDo 2.0 and Blue-Bot. These activities provide children with an opportunity to develop science skills and competences and a stimulus to reflect metacognitively on the fundamentals of scientific research methodology. These experiences also showed that robotics extends beyond the didactic, through technology laboratories and active methods, to encompass cross-cutting educational features. A well-thought-out laboratory approach helps children to learn to learn, plan, communicate and collaborate, act independently and responsibly, solve problems, identify connections and interpret information.

It is a fact that the educational use of robots stimulates productive, creative and divergent thinking and is useful for understanding how students learn. However, in order to achieve these goals and continue to make progress, we must not forget that it is extremely important to strengthen the professionalism of teachers, through research-training and participatory action research.

References

1. Asquini, G. (ed.): La Ricerca-Formazione. Temi, esperienze e prospettive. Franco Angeli, Milan (2018)
2. Datteri, E., Zecca, L.: Il "gioco dello scienziato": un robot per imparare a spiegare. Mondo Digitale **14**(58), 978–988 (2015)
3. Datteri, E., Zecca, L.: The game of science: an experiment in synthetic roboethology with primary school children. IEEE Robot. Autom. Mag. **23**(2), 24–29 (2016)
4. Datteri, E., Bozzi, G., Zecca, L.: Il "Gioco dello Scienziato" per l'apprendimento del metodo scientifico nella scuola primaria. TD Tecnologie Didattiche **23**(3), 172–175 (2015)
5. Datteri, E., Zecca, L., Laudisa, F., Castiglioni, M.: Learning to explain: the role of educational robots in science education. Themes Sci. Technol. Educ. **6**(1), 29–38 (2013)
6. Desgagné, S.: Le concept de recherche collaborative: l'idée d'un rapprochement entre chercheurs universitaires et praticiens enseignants. Revue des sciences de l'éducation **23**(2), 371–393 (1997)
7. Dewey, J.: Democracy and Education: An Introduction to the Philosophy of Education. Macmillan, New York (1916)
8. Scannapieco, S.: A robot in kindergarten: problems and strategies for problem solving, Thesis in Educational Sciences, University of Milano-Bicocca, Supervisor: Prof. Edoardo Datteri, Co-Supervisor: Prof. Luisa Zecca (2019)
9. Zecca, L.: Conversazioni con i bambini e stili educativi. In: Demetrio, D., Zecca, L. (eds.) Appunti per una ricerca sugli stili educativi, pp. 145–168. CUEM, Milan (2000)

10. Zecca, L.: Didattica laboratoriale e formazione Bambini e insegnanti in ricerca. Franco Angeli, Milan (2016)
11. Zecca, L.: I pensieri del fare. Verso una didattica metariflessiva. Junior, Parma (2012)

Maker Spaces and Fablabs at School: A Maker Approach to Teaching and Learning

Furniture Design Education with 3D Printing Technology

Meltem Eti Proto and Ceren Koç Sağlam

Abstract Three-dimensional printing technology has an important place in furniture and interior design, a strong global sector that responds rapidly to the changing needs and expectations of the individual and society. The main objective of design education should be to equip us to imagine new models of life. Among the most attractive benefits of 3D printing technology that make it a boon to designers working in the building and furniture sector are that it enables them to seek original forms that cannot be produced in molds, it generates less waste, and is accessible to all. Today, innovation in the profession, innovative materials, and knowledge of innovative production technologies that feed creative thinking have become ever important features of design education. This knowledge will allow us to imagine, discuss and pioneer design production ideas for new life models. This paper discusses 3D printing technology, the furniture design studio method and its contribution to design education in the Production Techniques courses of the Interior Architecture Department of Marmara University's Faculty of Fine Arts led by Professor Meltem Eti Proto, Instructor Can Onart, Lecturer T. Emre Eke, and Research Assistant Ceren Koç Sağlam.

Keywords 3D printer · Furniture design education · Innovation · Production · Design training · Prototype

1 Introduction

Three-dimensional printing technology has an important place in furniture and interior design, a strong global sector that responds rapidly to the changing needs and expectations of the individual and society. The main objectives of design education should be to equip us to imagine new models of life and to respond to the rapidly changing needs and expectations of the individual and society. Today, innovation in the profession, knowledge of innovative materials, and the ability to recognize innovative production technologies that feed creative thinking have become ever

M. Eti Proto (✉) · C. Koç Sağlam
Department of Interior Architecture, Faculty of Fine Arts, Marmara University, Istanbul, Turkey

D. Scaradozzi et al. (eds.), *Makers at School, Educational Robotics and Innovative Learning Environments*, Lecture Notes in Networks and Systems 240,
https://doi.org/10.1007/978-3-030-77040-2_13

important features of design education. Indeed, as Achille Castiglioni [1] would say, "If you are not curious, forget it." This knowledge will enable us to imagine new ideas for life models, and will encourage discussion and inspire design production.

"In every design or engineering exercise, we as creatives are looking to innovate, to do more with less, and to always do something better than what came before us. Parallel to this we are expected to produce faster, more efficiently, and cheaper, with less time than ever. The demand from consumers, as well as investors has created a world where design exploration is under immense pressure." These words by Paul Sohi [2], leader of Autodesk's iconic projects program, describe the dilemmas of today's design production. Among the most attractive benefits of 3D printing technology that make it a boon to designers working in the building and furniture sector are that it enables them to seek original forms that cannot be produced in molds, it generates less waste, and is accessible to all. As the technology has developed and different forms of it have emerged, the applications of 3D printing have multiplied to cover fields as diverse as medicine, the automotive industry, consumer products, architecture, construction, textiles, mold applications and the food sector.

1.1 Design with 3D Printing Technology

In recent years, 3D printing technology has shaken up many of the world's industries, from the food sector to the construction industry, the automotive industry and the health sector; in fact, few sectors have been left untouched by it. This technology is also changing the world of furniture. Made from melted parts of an old refrigerator, designer Dirk Vander Kooij's "Endless Pulse Chair" is a good example of the possibilities of 3D printing [3]. The result is a durable, strong and lightweight design. The chair is environmentally friendly and recyclable. Another environmentally friendly project is "Print Your City" by The New Raw design team, who recycle household plastic waste into raw materials for 3D-printed street furniture [4] (Figs. 1 and 2).

3D printing simplifies and reduces the cost of furniture design. The ability to quickly and cheaply produce furniture prototypes through 3D printing provides designers with more scope to test their creations and maximize the useful properties of the finished product. This technology enables furniture companies to design forms that cannot be produced by conventional methods, while consuming less energy and fewer resources in the manufacturing process. 3D printing is not only more cost-effective for companies than traditional furniture creation processes, but is also more environmentally friendly and sustainable.

The benefits of 3D printing for the consumer. While custom-design furniture is traditionally not cost-effective, because of the related operating expenses, 3D printing reduces the cost of customization. Changing the color, design, and other customized options adds no costs to the printing process. Although 3D printing will drastically change the furniture industry, it will not completely replace traditional manufacturing. The designer and manufacturer will continue to combine 3D printing with traditional production capabilities. As shown in the diagram in Fig. 3, the speed

Fig. 1 Endless pulse chair, by Dirk Vander Kooij (credit: Dirk Vander Kooij, 2011) [3]

Fig. 2 Print your city, XXX Bench in Amsterdam by The New Raw (credit: The New Raw, 2017) [4]

of production will have a big impact on the habits of designers, manufacturers and consumers, and will also change the way we shop for furniture. It will not be long before consumers will be able to have chairs, tables and fittings printed at their local 3D printing shop (see Fig. 3).

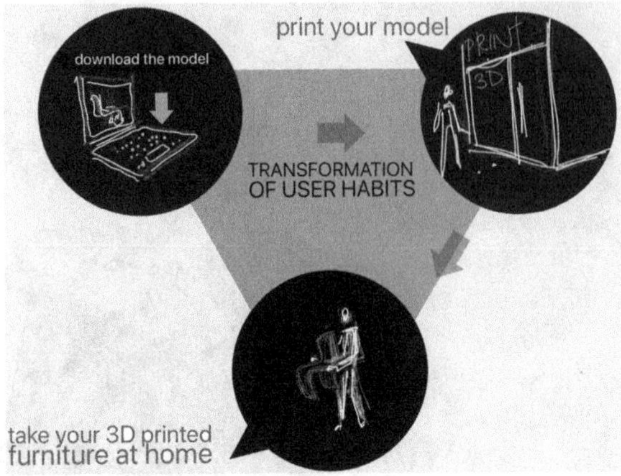

Fig. 3 Design process of 3D-printed furniture and changes in consumer habits (Eti Proto, 2019)

2 Furniture Design Studio with 3D Printing Technology

The aim of the course is to bring awareness to students regarding the global economic, social and political factors that give rise to change in furniture manufacturing methods and the introduction of new technologies. This will help students become part of the solution. The furniture designed during the course offers solutions to the transportation, assembly, production, raw materials, sustainability and cost issues common to many manufacturing industries. These designs are inspired by the democratic design philosophy, which emphasizes the need to provide functional and quality design for all [6]. Democratic design means affordability, environmental protection, sustainability and manufacturing efficiency.

The study was conducted with the participation of third-year students within the framework of the Free Furniture Design course of the Department of Interior Architecture, Marmara University. The Free Furniture Design—Production Techniques courses are led by Professor Meltem Eti Proto, Instructor Can Onart, Lecturer T. Emre Eke, and Research Assistant Ceren Koç Sağlam. The project also emphasizes that such values as originality, flexibility, communication/interaction, and interdisciplinarity, which inspire art and design, should be featured in furniture design courses.

In the first phase of this study, a modular furniture project was developed according to democratic design principles. The interlocking parts that connect and lock together the functional plywood parts of the furniture are produced using FDM 3D printing technology (see Fig. 4). The furniture designed and produced by the students has a variety of functions. For example, a table that can be used for educational purposes with children, or a modular shelf for display, storage, a hanger, a home-play area for pets, or a seating element.

An entirely different method was employed in the second phase of the study to develop experimental techniques that lead to creativity. The idea was to establish

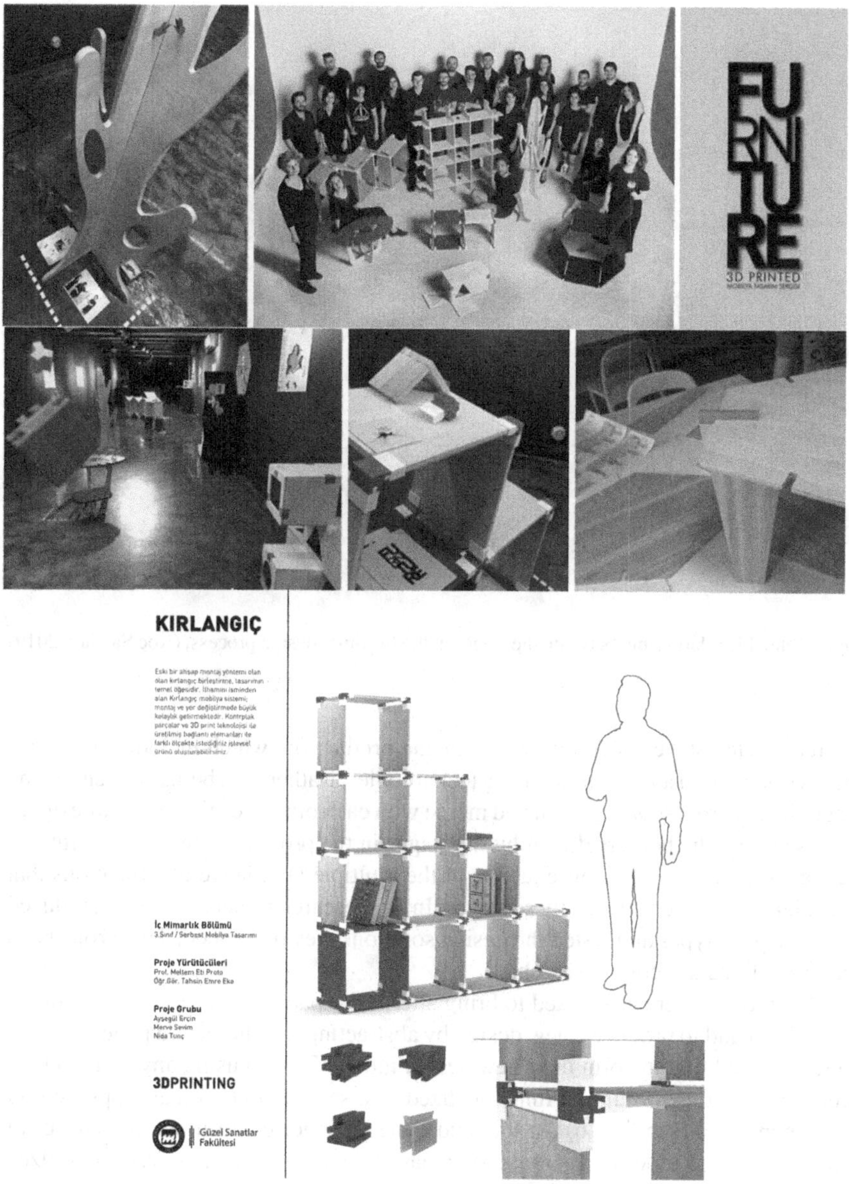

Fig. 4 Exhibition of 3D-printed furniture, Kadıköy youth and art center, İstanbul, 2018

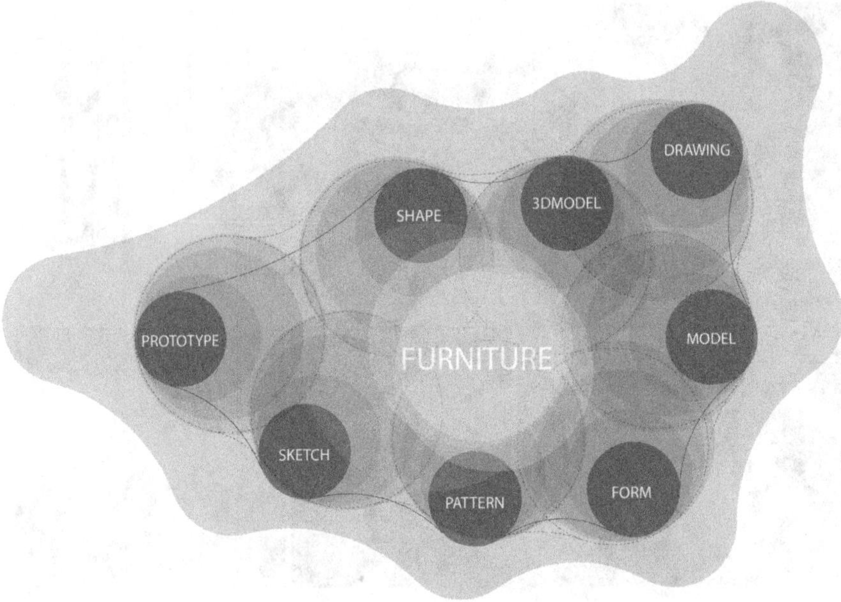

Fig. 5 Flexible relationship between the stages in the furniture design process, (Koc Saglam, 2019)

interconnected stages of furniture design and production, which included 3D drawings, visualization and manufacturing technologies. Rather than being sequential, the stages in this process are cyclical and merge with each other (see Fig. 5). The designer moves around the borders drawn by the stages in the process. However, furniture is defined as an alternative consequence of the multiple alternative combinations that come into existence between these stages. In this research project, students produced 1:1 scale prototypes and tested the design solutions they developed. This project can thus be defined as action research.

The students were also asked to bring alternative patterns they could draw inspiration from and to produce a new design by abstracting the alternative patterns. They were expected to transform their new design into a 3D form using any method they preferred; the form would be functionalized as a seating unit with appropriate size and ergonomics (see Fig. 6). In the study, the designer could move in and out of the stages in the process any way she pleased, and the outcomes were optimized in terms of function, ergonomics and durability. Although the students were free to move between the stages, priority was given to developing the products through drawing and then 3D modeling. The students experienced the production stages and the design stages simultaneously, including technical drawing, modeling, visualization, and presentation of 1:5 and 1:1 scale prototypes which they printed on FDM 3D printers. They were able to perform unique actions within a free framework that highlighted the design process and 3D printing technologies (see Fig. 6).

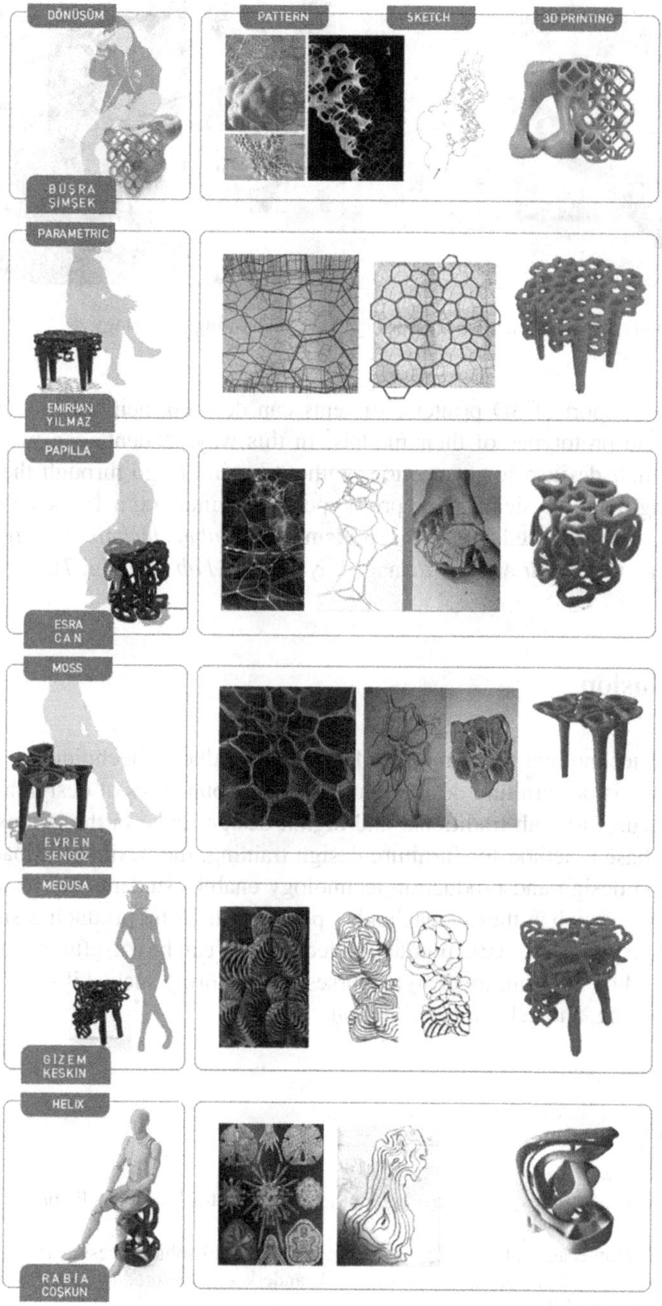

Fig. 6 Images of the stages, creating the design process, and 3D prototypes (Eti Proto, Koc Saglam, 2019)

Fig. 7 A prototype printed on a 1:1 scale using FDM 3D printing technology

With the support of 3D printers, students can develop their designs and build different-scale prototypes of their models. In this way, students can produce 1:1 models of their designs for inspection, without having to go through the process of producing them physically. The prototypes, 3D-printed on a 1:1 scale by FDM technology, are produced under the trademark *Istanbul Technical University* in cooperation with "*Tiridi Atölye*" managed by *Magnet Fab* (see Fig. 7).

3 Conclusion

Advances in technology-based modern design and production techniques have made positive impacts on furniture design practices. The boundaries of design have been expanded thanks to both traditional and digital design tools. In the context of this experience-based method for furniture design training, the flexible feedback introduced by 3D design and production technology enables students to create unique ideas, rapidly visualize these ideas and experience them by producing simultaneously. Such experiences in 3D production technology can be a useful method to use in design studios on furniture design courses for exploring such skills as creativity, abstract thinking, and 1:1 scale production.

References

1. FLOS Homepage, https://flos.com/news/not-curious-forget-achille-castiglioni/. Accessed 10 Sep 2019
2. Develop3D Homepage, https://d3dliveusa.com/speakers/paul-sohi/. Accessed 10 Sep 2019
3. Dirk Vander Kooij Homepage, https://www.dirkvanderkooij.com/products/endless-chair-2010. Accessed 10 Sep 2019
4. The New Raw Webpage, https://thenewraw.org/Print-Your-City-Amsterdam. Accessed 15 Nov 2019
5. Bell, M.L.: (2018), http://unyq.com/unyq-partners-up-with-ikea-and-area-academy-for-the-gaming-community/. Accessed 15 Nov 2019

6. Sanoff, H.: Multiple views of participatory design. Int J Archit Res ArchNet-IJAR **2**(1), 57–69 (2008)

Makerspaces for Innovation in Teaching Practices

Giuseppe Alberghina

Abstract "MakIN Teach—MAKerspaces for INnovation in TEACHing practices" is a two-year project recently funded by the Erasmus + program. It aims to promote the exchange of good practices in the use of rapid prototyping techniques, tools and spaces in the fields of education and training. The project supports teachers/educators working with students/learners showing poor educational results in theoretical subjects (e.g., mathematics, biology, geography, history, language, communication etc.). The project will conduct three short-term joint staff training activities and one student learning event, which will provide teachers/educators and students/learners with practical experience in makerspaces/fab labs. The materials and ideas developed during these transnational activities will be collected into an interactive e-book containing information about rapid prototyping technology, as well as tools for innovating the educational system and integrating fab labs and makerspaces into traditional teaching practices. All the partners involved in this project will have the opportunity to be part of a fruitful and durable network of educational institutions and makerspaces/fab labs at local and European level.

Keywords Digital fabrication · Makerspaces · Fab labs · Education

1 Introduction

"MakIN Teach—MAKerspaces for INnovation in TEACHing practices" is a two-year project recently funded by the European Union's Erasmus + program, part of Key Action 2: "Cooperation for innovation and the exchange of good practices" [1]. The project partners are AFP Patronato San Vincenzo (IT), I.E.S. Juan Ciudad Duarte (ES), Tallinn Polytechnic School (EE), Transit Projectes (ES) and Center for Creative Training Association (BG). All these organizations have the same mission of supporting students/learners on their path of growth, education and inclusion in

G. Alberghina (✉)
Associazione Formazione Professionale del Patronato San Vincenzo, Via Mauro Gavazzeni, 3, 24125 Bergamo, Italy
e-mail: giuseppe.alberghina@afppatronatosv.org

D. Scaradozzi et al. (eds.), *Makers at School, Educational Robotics and Innovative Learning Environments*, Lecture Notes in Networks and Systems 240,
https://doi.org/10.1007/978-3-030-77040-2_14

today's society. Educational underachievement (especially in theoretical disciplines) and early school leaving are common problems for all project partners.

The objectives of this project stem from inquiries that teachers/educators in each organization have made at local level. The teachers/educators involved have found concerning rates of educational underachievement and early school leaving (the respective figures by country for these are: IT 22 and 27%, ES 20 and 19%, BG 40 and 13%, EE 12 and 11%). In addition, they have found this worrying situation to be linked to the teaching methods currently in use, which are often too static and not engaging enough for students/learners with educational difficulties. The teachers/educators realized that it is crucial to develop more captivating teaching methods, as well as learning environments that are closer to real life [2].

These results are consistent with the Digital Education Action Plan of the European Commission, which states that "education can benefit from opening classrooms, real-life experiences and projects, and from new learning tools, materials and open educational resources." It claims that innovation in education systems can help improve learning outcomes and that "it is most effective and sustainable when embraced by well-trained teachers and embedded in clear teaching goals" [3]. Given this, the project partners believe that makerspaces and fab labs can support their teaching staff in this change [4].

2 Methodology

The project is inspired by the maker movement and fab lab ideas, and the closely related concepts of "experience learning" and "active learning." As stated by Rivoltella [5]: "Experience learning is traditionally related to active learning. This means to make possible that students could be actors of their own learning. As Activism demonstrated, there is no chance for learning if teacher speaks all the time, thinking that education is only information giving. Learning is fostered if it is experienced, that is related with emotions and real-life situations. Classroom has to be re-designed as a lab: lessons become workshops into which problem-solving and collaborative learning are the main students' activities. Digital technologies and mobile devices can empower these activities making possible that every student could be able to produce its contents and share them with his/her colleagues."

The methodology applied in "MakIN Teach" combines do-it-yourself (DIY) principles [6], typical of the maker movement and fab lab programs, with Activism theory and the related ideas of "experience learning" and "active learning." By combining and implementing these concepts, the consortium aims to exchange good practices in the use of rapid prototyping techniques, tools and spaces in the fields of education and training. The project supports teachers/educators working with students/learners showing poor educational results in theoretical subjects (e.g., mathematics, biology, geography, history, language, communication etc.).

To optimize the exchange of expertise and experiences, the partners have planned three short-term joint staff training activities and one student learning event, which will provide teachers/educators and students/learners with practical experience in makerspaces and fab labs. Each activity is hosted by a project partner and is organized in collaboration with a local makerspace/fab lab, either on or off the partner's premises. Makers will hold training sessions during part of the event, whereas the rest of the day will be used to develop innovative educational resources and define changes in teaching practices related to these resources.

The activities are thematically interrelated to create a coherent path of innovation in teaching practices, and provide an opportunity to develop educational resources that will be published in an interactive e-book. The planned activities are:

1. "FabLab/Makerspace 101—modelling of teaching artefacts" held at Fablab Bergamo (IT);
2. "How to create teaching artefacts by using rapid prototyping tools and machines" held at MakerConvent in Barcelona (ES);
3. "Turning the FabLab and Makerspace into the new classroom" held at Tallinn Polytechnic Institute (EE) and the first student learning activity.

However, these transnational training/learning activities are not an end in themselves. They are the starting point for building a fruitful and durable network of educational institutions and makerspaces/fab labs at local and European level. Indeed, in just a few years, rapid prototyping technology will be affordable for most schools, and vocational education and training (VET) centers and institutions [7]. By implementing the project activities, collaborating with the associated partners (fab labs/makerspaces or educational institutions) and using the project results, the consortium's goal is to plan ahead proactively and to acquire the elements, information, expertise and tools to exploit this technology and promote change in traditional teaching practices and spaces.

3 Objectives

Even though rapid prototyping technology is becoming more and more popular, educational institutions are still in the early stage of adopting it, and teachers and educators need training to use it. "MakIN Teach" aims to foster innovation in teaching practices by training teachers/educators (from VET centers and general schools) to use and adapt rapid prototyping techniques and tools to develop learner-centered approaches, problem-solving learning, participative learning and learning-by-doing.

For this, the project "MakIN Teach" has the following operational objectives:

- to gather best practices and develop a set of modern teaching insights for publication in an interactive e-book containing tutorial video-clips, pictures, texts, exercises, artifact designs and information about rapid prototyping machines and tools, with a view to innovating the educational system and integrating fab labs/makerspaces into traditional teaching practices;

- to provide elements, information and expertise for re-evaluating current and traditional educational environments, especially classrooms used for general subjects, and adapting them into a more engaging, unstructured, laboratory-style, less static setting;
- to open up classrooms by building networks between educational institutions and fab labs/makerspaces at European and local level.

4 Expected Results and Impact

Organizations, teachers, educators and makers working with students/learners who underachieve in predominantly theoretical subjects (mathematics, biology, geography, history, language, communication etc.) will be involved with the aim of improving their teaching practices. The training and learning activities and the interactive e-book developed by the participants and the staff of the partner organizations will have an impact on the teaching practices, tools and spaces of the project's target groups, that is:

- education and VET providers;
- education and training professionals, particularly teachers, trainers, guidance counselors;
- experts, specialists, and professionals involved in the maker movement and fab labs;
- associations of schools/VET centers.

A positive impact is predicted for the final beneficiaries, who are learners with poor school results in predominantly theoretical subjects, and early school leavers. First, they will benefit from being involved in an inspiring, transnational experience. They will also get the indirect benefit of the training their teachers/educators will receive, which should result in improved, more experiential methods. Lastly, they will benefit from new teaching material (the interactive e-book and its parts) validated in a European context, as well as having access to the network of education/training institutions and fab labs/makerspaces (locally and transnationally) created during the project. These innovations will help them strengthen their competences and employability. The long-term social impact will be greater opportunities for social inclusion.

From a European perspective, the project will enable education/training providers in other countries to benefit from its results and materials, which include open educational resources and deliverables licensed Creative Commons 3.0. Finally, the networks built during the project will be able to implement new materials, offer new services (e.g., seminars, CPD courses etc.) and reach new users. In professional terms, the project will provide important continuing education and professional development opportunities for teachers/educators.

At least 35 teachers and educators and at least 12 students/learners will take part in the transnational training/learning activities. Local team activities will involve at least 100 teachers and educators and at least 90 students/learners with poor school results.

5 Monitoring and Evaluation

The target groups and activities we have discussed will be involved in a multi-level assessment process, in which the training/learning activities are evaluated according to the following criteria (Kirkpatrick Model):

- Reaction: on the last day of each training/learning activity, the participants complete a satisfaction questionnaire;
- Learning: after each training/learning activity, the participants attend a co-evaluation of the acquired knowledge and skills;
- Behavior: the coordinators perform a pre- and post-event evaluation of the participants' behavior;
- Results: the coordinators check the teachers/educators' ability to use the fab lab/makerspace tools and spaces in their teaching practices.

At the same time, the educational resources and the interactive e-book features will be evaluated according to these criteria:

- Functionality: the quality of being suited to serve a purpose and meet stated or implied needs.
- Usability: a set of attributes that have a bearing on the effort needed for use (learnability, operability, attractiveness, satisfaction).
- Efficiency: the relationship between the tool's performance when it is used and the amount of time and other resources needed to use it.
- Portability: the ability of the tool to be adopted in different learning environments or with users with different needs/capabilities.

References

1. Erasmus+ Project Results Platform, https://ec.europa.eu/programmes/erasmus-plus/projects/eplus-project-details/#project/2019-1-IT02-KA201-062448
2. Blikstein, P., Krannich, D.: The makers' movement and FabLabs in education: experiences, technologies, and research. ACM Int. Conf. Proc. Ser. 613–616. https://doi.org/10.1145/2485760.2485884
3. COM/2018/022 final: Communication from the Commission to the European Parliament, the Council, the European Economic and Social Committee and the Committee of the Regions on the Digital Education Action Plan

4. González, C., Arias, A., Gonzalo, L.: Maker movement in education: maker mindset and makerspaces (2018)
5. Rivoltella, P.C.: Episodes of situated learning. A new way to teaching and learning. Res. Educ. Media VI **2**, 79–87 (2014)
6. Watson, M., Shove, E.: Doing it yourself? Products, competence and meaning in the practices of DIY (2005)
7. Jiang, R., Kleer, R., Piller, F.T.: Predicting the future of additive manufacturing: a Delphi study on economic and societal implications of 3D printing for 2030. In: Technological Forecasting and Social Change, vol. 117(C), pp. 84–97. Elsevier (2017)

Montessori Creativity Space: Making a Space for Creativity

Tiziano Fattizzo and Pierfrancesco Vania

Abstract We describe the creation of the "Montessori creativity space," the maker space of the "Secondo Istituto Comprensivo" primary and middle school of Francavilla Fontana, Italy. The space was inspired by the concept of Malaguzzi's atelier as a learning environment for students, teachers, local associations and artisans. In it, the imagination and action come together to ensure cross-disciplinary learning around robotics, educational electronics, logic, computational thinking, and digital and hand-built artifacts.

Keywords Malaguzzi · Maker space · Creative atelier · Fab lab · PNSD · Creativity

1 Introduction

Until the late 1960s, the term "atelier" was used in Italy with the meaning of art studio, that is, a place for painting, sculpture, dance and haute couture. Then, the national education system adopted it to mean a setting in which the "aesthetic dimension of learning" can become an educational and training strategy [1].

In early 2016, about 60 years after the initial experiences in the preschools of Reggio Emilia, and more than 30 years after the theorization of Loris Malaguzzi's "Hundred Languages of Children" method [2], the Ministry of Education, University and Research invited primary and lower secondary schools to set up creative ateliers, as part of the National Plan for Digital Education (PNSD).

The objective was to "*reinstate the importance of workshop teaching, as a meeting point between knowledge and know-how*" and to transform workshops into "*places of innovation and creativity.*" Malaguzzi's Atelier, "*a place of research, invention,*

T. Fattizzo (✉)
Secondo Istituto Comprensivo, Francavilla Fontana, Brindisi, Italy
e-mail: tfattizzo@libero.it

P. Vania
Via Enrico Toti, 49, Oria, Brindisi, Italy

D. Scaradozzi et al. (eds.), *Makers at School, Educational Robotics and Innovative Learning Environments*, Lecture Notes in Networks and Systems 240,
https://doi.org/10.1007/978-3-030-77040-2_15

and empathy expressed through 100 languages," becomes "meeting point for manual skills, crafts, creativity and technology" [3].

These are the cultural foundations of the "Montessori Creativity space" of the "Secondo Istituto Comprensivo" primary and middle school of Francavilla Fontana, whose mission is to provide space for the "creative potential" of students, to improve their basic and soft skills. "*We all have creative potential,*" Mark Runco says; "*our commitment as parents and teachers is to help children achieve it*" [4].

A maker space is a place for students, teachers, local associations and artisans, where action and the imagination come together to encourage cross-disciplinary learning around robotics, educational electronics, logic, computational thinking, and digital and hand-built artifacts [5].

2 The Context

The "Secondo Istituto Comprensivo" primary and middle school of Francavilla Fontana has about 1400 students on different sites, including four preschools, one middle school and two elementary schools [6]. The maker space is located in the largest of these, "M. Montessori" elementary school. Opening in 2017 as a place dedicated to non-formal and informal learning, where all students can cultivate the wonder of discovery, as well as their manual skills and creativity.

Within about a year, all the funding needed had come from contributions from the European Regional Development Fund (ERDF) and the Ministry of Education fund for "Creative Ateliers," [7] as well as donations from parents and local businesses, and crowdfunding. We had around €50,000 to achieve what at the beginning was only a "dream": creating a learning space for educational activities during curricular and non-curricular hours, which would include collaborations between students of different ages, and between teachers. It would be a place where we could work with other educational institutions to offer work experience schemes for secondary school students, and shared "learning by doing" projects in which older students could pass their knowledge, emotions and experiences onto their younger counterparts.

The "Montessori Creativity Space" has become an activity hub for a variety of accredited public and private institutions in and around Francavilla, with the shared intent of collaborating together, including by providing adult education courses. It is a bridging space between the school and the local area.

3 Work Method

The "Montessori Creativity Space" is now a reality, and many of the ideas presented above have come to fruition. Covering an area of approximately 180 m^2, the maker space can accommodate small groups of students, but also several classes at once, if necessary. It is a place where students can tinker, code, experiment with robotics and educational electronics, make prototypes, do 3D printing, and digital and traditional crafts.

It is a multifunctional space that can be arranged into different configurations. An area of about 40 m^2occupied by the school's fab lab, is separated from the rest of the space with movable colored panels, and contains 3D printers and scanners that even work with chocolate and clay, a filament extruder with a plastic shredder for recycling PLA and other plastic waste, a laser cutter and chiller, three cutting plotters, a thermoformer, work tables and various tools (compressor, orbital saw, welding equipment, drills, hammers, screwdrivers, etc.).

An area of about 20 m^2, covered in rubber flooring and scattered with colorful cushions, is available for reading, relaxing, debating and metacognition. The remaining 120 m^2 feature a large table for robotics and several brightly colored modular tables of different shapes and sizes, arranged into islands, which are meant for team work activities, training, and peer collaboration. Colorful cabinets lining the walls contain various types of robot for younger preschool children [8], and educational electronics kits and various accessories for elementary and middle school students. Finally, graffiti and colorful artwork provide a warm and cheerful atmosphere [9].

A teacher manages the inventory and the schedules of the different spaces and the tools. A voluntary association of teachers, professionals and parents with experience in digital fabrication and networking provides free equipment maintenance, as well as training courses and educational experiments, in which they have involved higher education establishments and craft associations, in order to develop and promote youth entrepreneurship schemes and other innovative projects.

An "innovation department" called Codingkids has been set up in which the digital animator, the innovation team and all the teachers compare good teaching practices.

4 Relationship Between Space, Technologies, Teaching and Learning Practices

Studies in recent years have shown that the types of activities conducted at the Montessori Creativity Space help develop lateral thinking, which in turn enhances children's acquisition of skills that are crucial in the 21st century, such as critical thinking, problem-solving, creativity, curiosity, imagination, communication, and collaboration.

Having a maker on hand has inspired teachers to adapt their teaching methods, and stimulated them to experiment with new ideas, and new ways of thinking and solving problems. The maker plays several roles in our project: he provides technical training to teachers, helps them re-evaluate their teaching methods, and does maintenance on the machines [10]. He also acts as a technical teacher, a direct interface between teacher and students, disrupting their traditional relationship, and favoring an atmosphere of informality that benefits students' creativity and autonomy [11]. The maker's role is similar to that of Malaguzzi's *atelierista*, someone with specific skills, who supports the pedagogical practices of teachers. The maker provides support as the teacher holds the rudder that steers curricular learning, enabling students to express their creativity through multiple idioms.

Regarding student behavior, we have seen that interactions in the mostly group activities are mainly between peers, but also involve teachers and makers. Irrespective of when group activities are set up by the teacher, students express themselves openly and without fear of being judged.

5 Conclusion

The maker space was set up with the active participation of all actors in the school's community: head teacher, digital animator, teachers, parents and external technical staff. The equipment purchased meets the teachers' main requirement of revising their teaching practices. All teachers have access to the equipment, and have the assistance of external technical staff who provide courses to help them acquire the skills to use these tools, and practical support during teaching activities.

Finally, the action-research experiences within the maker space in coding, 3D printing, educational robotics and electronics, and tinkering are the basis for sharing good teaching practices [8], even beyond our school community, through seminars in other schools and participation in national training programs for teachers [12].

References

1. Atelier creativi e laboratori per le competenze chiave—MIUR, http://www.istruzione.it/allegati/2016/Allegato_1.pdf
2. Malaguzzi L.: L'educazione dei cento linguaggi dei bambini. In Zerosei n. 4–5 (1983)
3. Atelier|Reggio Children, http://www.reggiochildren.it/attivita/atelier/
4. Runco, M.A.: Creativity: Theories and Themes: Research, Development, and Practice, Elsevier Science Publishing Co Inc (2006)
5. Feeney, C.: Makerspaces: An Introduction to Innovation Learning Spaces. Think, San Francisco (https://issuu.com/mkthink/docs/makerbook_final_aw)
6. Secondo Istituto Comprensivo "Montessori-Bilotta" di Francavilla Fontana (Brindisi), https://www.secondocomprensivo.edu.it
7. Atelier creativi—La scuola digitale—MIUR, http://www.istruzione.it/scuola_digitale/prog-atelier.shtml
8. Guasti, L., Newint-Gori, J.: Maker@Scuola. Stampanti 3D nella scuola dell'infanzia. Florence, Asso+ Ed. (2017)
9. Guasti, L.: Fare didattica in spazi flessibili: Progettare, organizzare e utilizzare gli ambienti di apprendimento a scuola, Tosi, L., Scuola, G. (eds.), pp. 74–110 (2019)
10. Blackley, S., Sheffield, R., Maynard, N., Koul, R., Walker, R.: Makerspace and reflective practice: advancing pre-service teachers in STEM education. Australian J. Teacher Educ. **42**(3), 22–37 (2017)

11. Blackley, S., Rahmawati, Y., Fitriani, E., Sheffield, R., Koul, R.: Using a 'makerspace' approach to engage Indonesian primary students with STEM. Issues in Educ. Res. **28**(1), 18–42 (2018)
12. Guasti, L., Rosa, A.: Maker@Scuola. Florence, Asso+ Ed (2017)

Fab the Knowledge

Sofia Scataglini and Daniele Busciantella-Ricci

Abstract This paper draws a link between what happens in maker spaces and how these processes can be simulated in the mathematical collaborative model (co-model) of the research through collaborative design (co-design) process (RTC). The result is the ability to identify the main variables for simulating the "making" dynamics of the RTC model. This outcome is discussed with an emphasis on the "intangible" role of "making," alongside the proposed concept of "fab the knowledge." Speculative thinking is used here to link the innovative and theoretical aspects of design research to their application in and for innovative learning contexts. The RTC co-model can be used to compute, simulate and train a co-design process in intangible spaces, such as fab labs. In these spaces, multiple actors with different skills and backgrounds, who may or may not be experts in design, collaborate on setting a design question and identifying a shared design answer, in a process of RTC. A "network" of neural mechanisms operating and communicating between design experts and non-experts, like a computing system of a biological mechanism, can be used to train and simulate a research answer, thereby "fabricating" knowledge.

Keywords Research through co-design · Design research · Co-design · Neural network

1 Introduction

This paper focuses on the research activities of the FabLearn network for "project-based, inquiry-driven education" (https://fablearn.org/research/). These concepts have in common their relationship with the field of design research [1, 2] and the

S. Scataglini (✉)
Department of Product Development, Faculty of Design Sciences, University of Antwerp, Antwerp, Belgium
e-mail: sofia.scataglini@uantwerpen.be

D. Busciantella-Ricci
Design Research Lab, Department of Humanities, University of Trento, Trento, Italy
e-mail: d.busciantellaricci@unitn.it

D. Scaradozzi et al. (eds.), *Makers at School, Educational Robotics and Innovative Learning Environments*, Lecture Notes in Networks and Systems 240,
https://doi.org/10.1007/978-3-030-77040-2_16

research through design (RTD) approach [3, 4]. Adopting a speculative approach, this paper draws a link between what happens in maker spaces and how these processes can be simulated in the linear mathematical co-model of the research through co-design process [5]. The result is a model designed to simulate the "making" dynamics of the RTC model. This is discussed with an emphasis on the "intangible" role of "making," alongside the proposed concept of "fab the knowledge."

1.1 Making and Prototyping in Contemporary Design Domains

Making and prototyping are widely explored in maker spaces. They are traditionally recognized as supportive actions with tangible content (Fig. 1, right). They are also activities within design domains (Fig. 1, left), where design is seen as "making" (traditional design practice), or as "integrating," and for value creation (see [6, 7]). At the same time, there is a growing interest among practitioners and researchers in understanding the relationships between design culture and complex social systems. As this awareness increases, "making" and "prototyping" take on a decisive role in complex societal and policy-making contexts, where design processes demand a good understanding of the concept of intangibility. Given this premise, this work is driven by the question of how we can model the "fabrication" of intangible things like knowledge.

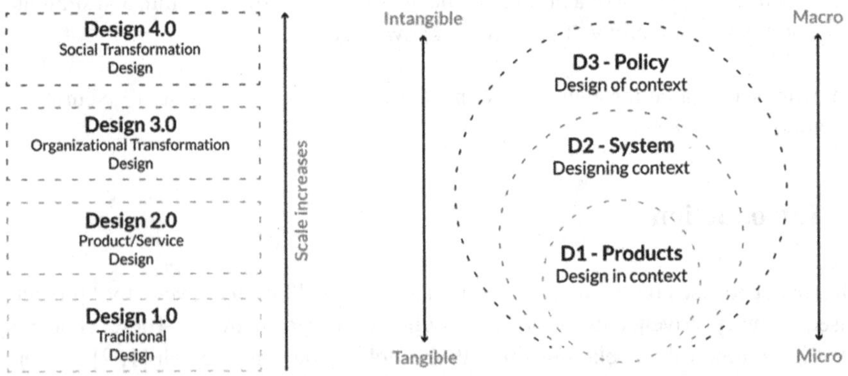

Fig. 1 Mapping design domains (left) (revised from [7]) and levels of design content (revised from [8])

1.2 The Research Through Co-design Co-model

The RTC theory describes how the RTD process works within collaborative design. Previous work by the authors of this paper [5, 9] explain this theory, which is based on the control system theory [10]. The result was the design of a collaborative model (co-model) in the form of a "mathematical model of cognitive control that describes the process in doing research by an RTC process" [5]. This model can be used to understand how to gain knowledge from co-design within a wider process of research. As a form of collective creativity in a design process (see [11]) in learning contexts such as those found in maker spaces, co-design can also be "understood as a creative process developed collaboratively by teachers, students and researchers to design inquiry-based and technology-enhanced and networked learning scenarios" [12]. Participatory design is a process in which people learn from each other [13]. In this sense, the action of co-designing, as a specific instance of co-creation [11], can be related "to actions of collective creativity and co-creation of knowledge" [12].

2 Methodological Approach

This research uses a speculative approach [14]. Specifically, the authors used the RTC theory co-model as a speculative design proposal [15], linking the co-model variables to the learning activities of maker spaces (Fig. 2).

For this reason, this kind of space is shown as the variable G(s), which represents the design process within an RTC co-model (collaborative model). When we input a research question R(s) through a co-design process G(s) and test it H(s), the co-model uses a closed-loop system [16–18] to calculate the error between the research answer obtained and the research answer set C(s) in Fig. 3 [5].

In the co-designing process, the co-designers' team is a cross-disciplinary team made up of the product coi and coj:

$$Co = \sum_{i,j=1}^{n} co_i * co_j \tag{1}$$

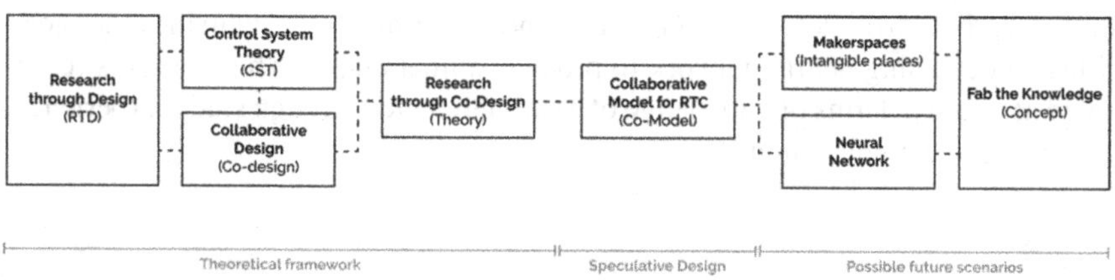

Fig. 2 Diagram summarizing the speculative approach adopted by the authors

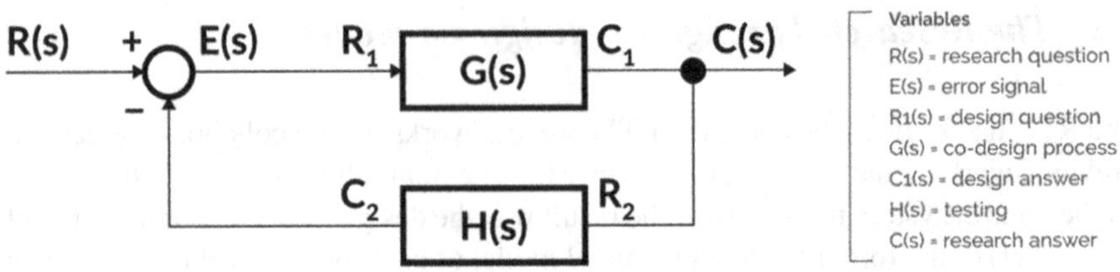

Fig. 3 A co-model based on a closed-loop system in research through co-design [5]

where coi and $co_j \geq 2$. Consequently, with a design question R1(s), the co-designer should use the design tools T in the co-design process G(s) to solve the design answer C1(s): G(s) = Co * T/R1(s). A design tool is a context-related set of actions, thoughts, or objects that makes actions, thoughts, or objects possible for design-related tasks (cf. [19]).

3 Results and Discussion

If a maker space and/or a fab lab is linked to the G(s) variables, then they can be used to simulate, calculate and model research through co-design processes. Accordingly, the maker space can be considered an intangible (non-physical) "space." This means that it is a simulated space that is mathematically linked to the cognition process that leads to a new kind of knowledge.

Physical or tangible spaces can be perceived by the senses, whereas non-physical or intangible spaces are not available to the senses [20], only to conscious perception [21]. Fab labs and maker spaces are intangible spaces where co-designers can convert the intangible values, factors and constraints of "conscious making" into tangible and measurable actions within RTC ensuring sustainability. Indeed, when we define G(s) as a design process in this intangible space called the fab lab, we can represent collaboration in an RTC as a "network" of neural mechanisms that acts and communicates like a computing system of a biological mechanism [22]. The neural network mechanism [23] works by computing, training and simulating the best criteria possible for maximizing the accuracy of the research answer through an RTC process. This is achieved by associating a node in the co-designing neural network with each co-designer in G(s). Each node communicates with the other node in this co-designing neural network to produce a final output. This output is represented as a gain in terms of knowledge of a skill fabricated through study, experience, or teaching in a co-design process.

4 Conclusions

The RTC co-model can be used to compute, simulate and train a co-design process in an intangible space like a maker space. In these spaces, multiple actors with different skills and backgrounds, who may or may not be experts in design, collaborate on setting a design question and identifying a shared design answer, in a process of RTC. A "network" of neural mechanisms operating and communicating between design experts and non-experts, like a computing system of a biological mechanism, can be used to train and simulate a research answer, thereby "fabricating" knowledge. The RTC co-model is therefore a support model for collaborative learning processes and for understanding how to fabricate intangibility through participatory design processes. Accordingly, the authors argue that the RTC co-model can be used in educational research for experimentation in project-based and inquiry-driven education programs. Finally, the authors argue the co-model can play a crucial role in the FabLearn network to increase the knowledge of its various stakeholders in innovation and educational research processes.

References

1. Archer, B.: A view of the nature of design research. In: Jacques, R., Powell, J. (eds.) Design Science Method, pp. 30–47. Westbury House, Guildford (1981)
2. Jonas, W.: Research through DESIGN through research: a cybernetic model of designing design foundations. Kybernetes **36**(9/10), 1362–1380 (2007)
3. Frayling, C.: Research in art and design. Royal. Coll. Art. Res. Pap **1**(1), 1–5 (1993)
4. Jonas, W.: A cybernetic model of design research. In: Rodgers, P.A., Yee, J. (eds.) The Routledge Companion to Design Research, pp. 23–37. Routledge (2014)
5. Busciantella Ricci, D., Scataglini, S.: A Co-model for research through co-design. In: Di Nicolantonio, M., Rossi, E., Alexander, T. (eds.) Advances in Additive Manufacturing, Modeling Systems and 3D Prototyping. AHFE 2019. Advances in Intelligent Systems and Computing, vol. 975, pp. 595–602. Springer, Cham (2019)
6. Jones, P.H., Van Patter, G.K.: Design 1.0, 2.0, 3.0, 4.0: the rise of visual sense making. Next. Des. Leadersh. Insts. (article), New York (2009)
7. Jones, P.H.: Systemic design principles for complex social systems. In: Metcalf, G.S. (eds.) Social Systems and Design, pp. 91–128. Springer, Tokyo (2014)
8. Young, R.A.: An integrated model of designing to aid understanding of the complexity paradigm in design practice. Futures **40**(6), 562–576 (2008)
9. Scataglini, S., Busciantella Ricci, D.: Toward a co-logical aid for research through co-design. In: M. Di Nicolantonio, M., Rossi, E., Alexander, T., (eds.) Advances in Additive Manufacturing, Modeling Systems and 3D Prototyping. AHFE 2019. Advances in Intelligent Systems and Computing, vol. 975, pp. 623–634. Springer, Cham (2019)
10. Levine, W.S.: The Control Handbook: Control System Fundamentals, 2nd edn. CRC Press, Boca Raton (2011)
11. Sanders, E.B.N., Stappers, P.J.: Co-creation and the new landscapes of design. Co-design **4**(1), 5–18 (2008)
12. Garcia, I., Barberà, E., Gros, B., Escofet, A., Fuertes, M., Noguera, I., López, M., Meritxell Cortada, M., Marimón, M.: Analysing and supporting the process of co-designing inquiry-based and technology-enhanced learning scenarios in higher education. In: Networked Learning Conference, pp. 493–501 (2014)

13. Ehn, P.: Scandinavian design: On participation and skill. In: Schuler, D., Namioka, A. (eds.) Participatory design, pp. 41–77. L. Erlbaum Associates Inc. (1993).
14. Wilkie, A., Savransky, M., Rosengarten, M.: Speculative research: the lure of possible futures. Routledge (2017).
15. Raby, F.: Critical design. In: Erlhoff, M., Marshall T. (eds.) Design Dictionary: Perspectives on Design Terminology, pp. 94–96. Birkhäuser, Basel (2008)
16. Golnaraghi, F., Kuo, B.C.: Automatic control systems, 10th edn. McGraw-Hill Education (2017)
17. Isidori, A.: Sistemi di controllo, 2nd edn. Siderea (1993)
18. Bolzern, P., Scattolini, R., Schiavoni, N.: Fondamenti di controlli automatici. McGraw-Hill (2004)
19. Whitfield, T.: Tools. In: Erlhoff, M., Marshall, T. (eds.) Design Dictionary: Perspectives on Design Terminology, pp. 404. Birkhäuser, Basel (2008)
20. Agustina, I.H.: "From tangible space to intangible space" Kanoman palace and Kacirebonan palace. J. Sampurasun: Interdiscip. Stud. Cult. Heritage. **3**(2), 105–111 (2017)
21. Gamez, D.: Conscious sensation, conscious perception and sensorimotor theories of consciousness. In: Bishop, J., Martin, A. (eds.) Contemporary Sensorimotor Theory. Studies in Applied Philosophy, Epistemology and Rational. Ethics., vol. 15, pp. 159–174. Springer, Cham (2014)
22. Barrett, D.G., Morcos, A.S., Macke, J.H.: Analyzing biological and artificial neural networks: challenges with opportunities for synergy? Curr. Opin. Neurobiol. **55**, 55–64 (2019)
23. Tang, H., Tan, K.C., Yi, Z.: Neural networks: computational models and applications. In: Studies in Computational Intelligence. Springer (2007)

Teaching Environmental Education Using an Augmented Reality World Map

Anastasia Nancy Pyrini

Abstract The aim of this short paper is to provide readers with a comprehensive lesson plan for elementary schools, which seeks to improve digital citizenship and competency in students and teachers, by fostering digital literacy skills through an augmented reality (AR) application. The lesson plan was developed within the framework of the "Digital, Responsible Citizenship in a Connected World (DRC)" project funded under Key Action 2: Cooperation for innovation and the exchange of good practices, part of the European Union's Erasmus+program. The DRC project aims to infuse contemporary pedagogical practices into quality lifelong learning for students and teaching professionals, including teachers, school leaders and teacher educators, across Europe. Specifically, the project aims to improve digital citizenship and competency in students and teachers by fostering digital literacy skills through education.

Keywords Environmental education · Habitats · Augmented reality · World map

1 Introduction

This short paper presents how the material and resources of the Digital Citizenship program and curricula on digital literacy [1] enhanced by the Clever Books Augmented Reality App for Geography [2] were introduced and evaluated in an elementary class of the 1st Primary School of Rafina in Greece, for teaching environmental education and, more specifically, the topic "habitats."

The aim of the intervention was to test and assess the usability of the lesson plan entitled "Habitats" within the "Communication and Collaboration" competence area of the European Digital Competence Framework for Citizens (DigComp), which offers a tool to improve citizens' digital competence [3].

A. N. Pyrini (✉)
Hellenic Ministry of Education, Research and Religious Affairs, East Attica, Greece
e-mail: nancypyrini@icicte.org

D. Scaradozzi et al. (eds.), *Makers at School, Educational Robotics and Innovative Learning Environments*, Lecture Notes in Networks and Systems 240,
https://doi.org/10.1007/978-3-030-77040-2_17

1.1 Profile of School and Students

The 1st Primary School of Rafina is a public elementary school overseen by the Ministry of Education, Research and Religious Affairs.

Number of teachers: 28.

Gender: Male 7, Female 21.

Number of students: 276.

Besides DRC, there is no other program running currently that promotes digital citizenship.

Demographics of class where workshop took place:

Number of students: 23.

Gender: Male 7, Female 16.

Age: 9 years old.

The group included students who have been diagnosed with autism, attention deficit hyperactivity disorder (ADHD) and dyslexia.

1.2 Description of the Workshop With Students

The teacher selected the "Communication and Collaboration" competence area on: interacting through digital technologies; sharing through digital technologies; engaging in citizenship through digital technologies; collaborating through digital technologies; netiquette; managing digital identity. The lesson plan entitled "Habitats" was based on the following rationale.

1.3 Grade Level—Age of Students

The workshop was for third-grade students (Class C2), aged 9–10.

1.4 Material/Resources

Although the teacher is an expert in the use of new technologies in education and the students are digital natives, the only technology available in the classroom is a chalkboard. The school's computer lab is located in another building and the process for borrowing laptops and projectors and setting them up is time-consuming. The teacher therefore decided to use mobile devices, which students could easily bring from home, along with an offline mobile app that students could run, regardless of connectivity, to exploit the possibilities offered by augmented reality. The "Digital competences—Self-assessment grid" [4] was used to evaluate the activities.

1.5 Interdisciplinary and Constructivist Approach

Multiple disciplines, including environmental studies and language, were integrated, for an interdisciplinary curriculum approach. This was to maximize the students' ability to understand the major themes and ideas associated with Habitats.

Students would need to activate their previous knowledge from any of the disciplines, especially environmental education, in order to achieve a constructivist approach to learning and an understanding of habitats. This workshop not only integrated multiple disciplines, but challenged students constructively to strengthen their social, abstract-thinking, and creative skills. The worksheet in English can be retrieved at: http://bit.ly/2RVvsSI.

1.6 Parental Involvement

Throughout the workshop, parental involvement was widely encouraged to support and enhance student performance. Parents had the opportunity to work with their child at the annual Students' Festival to exploit the learning opportunities offered by augmented reality.

1.7 Active Citizenship

Many of the ideas, concepts and discussions associated with the workshop, such as habitat conservation, would motivate students to analyze the community they live in and hopefully in still a need for active citizenship.

1.8 Data Collection

The teacher translated the "Parent Consent Form" into Greek and distributed it to parents. Parent testimonials were recorded during the Students' Festival held at Karamanlis Park on 11 May 2019.

An "Observation Form" and a "Teacher Evaluation for Digital Literacy and Citizenship Resources for Workshops with Students" form were completed by the teacher/workshop facilitator after the workshop.

The teacher distributed a testimonial template to students and informed them that completing it was not obligatory.

The teacher took photographs of all workshop activities.

2 Findings

2.1 Use of Digital Literacy and Citizenship Resources

AR app Demonstration at the Students' Festival. The group of students led by their teacher participated in the Students' Festival held at Karamanlis Park on 11 May 2019. The students demonstrated the augmented reality application to their parents and to visitors.

Workshop Based on the Lesson Plan Entitled "Habitats". The student workshop took place on May 15 and was based on the lesson plan developed by the 1st Primary School of Rafina entitled "Habitats." The worksheets in Greek were completed by each group of students, then presented by each work group and discussed in a session that included feedback.

Workshop Based on a Lesson Plan that the Students Had Developed. The students developed their own lesson plan based on a book powered by an augmented reality app. The lesson was implemented by the students on May 30.

2.2 Course: Study of the Environment

Lesson: The little explorers and monuments of the world.

Digital Competence Area: 2. Communication and Collaboration.

Grade Level: C Time frame: 2 h.

Lesson Overview.

This lesson plan summarizes the significance and importance for the class of famous monuments. The augmented reality app helps students to learn about specific monuments.

Learning Objectives. On completion of this lesson, students will be able to:

- list and describe examples of famous monuments

- discuss the importance of conservation in relation to our national monuments.

Material/Resources. The book *Οι μικροί εξερευνητές και τα μνημεία του κόσμου* [The Little Explorers and Monuments of the World]. Publisher: iWrite. ISBN: 978–618-5218–24-9.

Lesson Activities. The students read the book and explore the monuments using the augmented reality application that they have downloaded. They can even take photos in front of their favorite monuments. Discussion follows about the importance and the preservation of world monuments.

The digital literacy and citizenship resources and, more specifically, the subject of "Environmental Education" were integrated into the curriculum.

2.3 Successes

The greatest success of the intervention was that the students were inspired to create their own lesson plan and implemented it successfully in the classroom. A community of self-directed elementary school learners with enhanced homonomy was created.

2.4 Challenges

The challenge the teacher faced was designing a five-week intervention with no technological resources available in the classroom.

This was overcome by using free resources that did not require internet access, along with the kind cooperation of parents, who gave their children permission to bring their mobile devices to school for the workshops.

The students faced the following challenges:

- Poor battery life
- Tablet/phone or app crashes.

Students were advised to bring extra devices in case they encountered technical problems, therefore there was no real inconvenience, just frustration whenever the issue arose.

2.5 Comments and Feedback

Teacher's Views/feedback. Based on the teacher's evaluation form:

- The workshop was worth completing, met her expectations and was beneficial for the students in her class.
- The resources used were of excellent quality.
- Unfortunately, the Digital Citizenship app was not available at the time the intervention was designed.
- It was very easy for the teacher to integrate the resources into her teaching practice.
- The workshop had a positive impact on the teacher. It raised her awareness of digital literacy and citizenship and of how they can be taught to young students. It also helped her to support her students to further develop their digital literacy.

Students' Views/Feedback. Students enjoyed working with the AR apps and, particularly, collaborating in groups. There were a few challenges, which they overcame easily. They also liked sharing their experience with their parents and visitors at the Students' Festival.

References

1. Digital, Responsible Citizenship in a Connected World (DRC): Project website (2017) https://digital-citizenship.org/
2. CleverBooks. Geography with Augmented Reality: Lesson Plan. Primary/Elementary School (2019) https://www.cleverbooks.eu/product/augmented-reality-treasure-hunt-e-book/
3. EU Science Hub. DigComp 2.0: The Digital Competence Framework for Citizens. Update Phase 1: the Conceptual Reference Model (2019) https://ec.europa.eu/jrc/en/publication/eur-scientific-and-technical-research-reports/digcomp-20-digital-competence-framework-citizens-update-phase-1-conceptual-reference-model
4. European Union.: Digital competences—self-assessment grid (2015) https://europass.cedefop.europa.eu/sites/default/files/dc-en.pdf

Laboratory Teaching with the Makers Approach: Models, Methods and Instruments

The Maker Movement: From the Development of a Theoretical Reference Framework to the Experience of DENSA Coop. Soc

Valentina Costa

Abstract The Maker Movement, which has for years received much attention, still presents many economic, social and educational implications that are ripe for investigation. The movement's community of practice can be defined as "*a knowledge-building community*" (Scardamalia and Bereiter, The Cambridge handbook of the learning sciences, pp 97–115, 2006) as cited in Martin (Martin, J Pre-Coll Eng Educ Res (J-PEER) 5(1):4, 36 2015). This apt definition, which refers to the hyper complex, connected society that engendered it, opens up new possibilities in the field of education. The main goal of this reflective paper is linked to the creation of a theoretical framework that could explain and support the movement's background. This will lead to an analysis of three different pedagogical models (Célestin Freinet, Loris Malaguzzi, Bruno Munari) that have much in common with the Maker Movement. We focused our study on the most positive traits of makers: social inclusion, democracy and the failure-positive/collaborative approach. Considering the importance of the European Union's Key Competences, our aim was to create a bridge between the "maker mindset" (Dougherty, Design, make, play, pp 9, 25–29, 2013) and these competences, in order to consider the possibility of introducing the movement into the national curriculum. The point of contact between the two can be the basis for promoting *active citizenship*, grounded, naturally, in the Key Competences. Subsequently, to test our analysis in the first part of our reflection, we look at the experience of DENSA (Developing Edutainment for New Skills and Attitudes) Coop. Soc.

Keywords Maker movement · Making · Key competences · Failure-positive · Social inclusion · Democracy · Collaborative learning

---

V. Costa (✉)
University of Perugia, 06121 Perugia, Italy

D. Scaradozzi et al. (eds.), *Makers at School, Educational Robotics and Innovative Learning Environments*, Lecture Notes in Networks and Systems 240,
https://doi.org/10.1007/978-3-030-77040-2_18

1 Introduction. Children, Makers, Key Competences

This paper has its roots in the work of DENSA Coop. Soc., a young social cooperative in Italy, working within the extracurricular system, whose mission has much in common with the principal features of the Maker Movement. This reflective paper considers the role of maker didactics (culture) in the development of active citizenship through the Key Competences framework.

We first looked at the social function that the maker space—a place where people create things—can fulfil. The maker space appeared not only to be a place where interpersonal relationships, social ties, business connections can develop, it also facilitates the learning of new skills (Sturges 2013).

The maker space can also stimulate the creative process, and help the individual cultivate a passion, thanks to the "playful" (Tanenbaum et al. 2013, p. 2603) approach that characterizes the movement and can be found in these labs. In addition, the paradigm of the maker space as a place that can foster social inclusion, tolerance and peer collaboration can give definition to the concept of active citizenship (another pivotal theme of democratic citizenship).

2 Community and Participation: Makerspace and Social Inclusion

Anderson (2012), a leading contributor to the literature on the Maker Movement, has said that society's current cultural (and fertile) structure is what has influenced and helped spread the democratic maker revolution. At the same time, it has assisted the spread of democratizing tools, experiences and customs, including: the Open Source model ("Open Source is not just an efficient innovation model—it's a belief system as powerful as democracy or capitalism for its adherents") (Anderson 2012, p. 93); the dissemination of maker spaces and Maker Faires; the launch of low-cost 3D printers (e.g., MakerBot); and the availability of tools. It should be stressed that relationships can develop within maker communities (both online and offline) based on mutual exchange, assistance and collaboration. In these settings, a maker has the opportunity to use tools (e.g., a laser cutter, a vinyl cutter, a CNC router, a CNC milling machine) to create, collaborate, form partnerships, and facilitate interoperability.

According to Tanenbaum et al. (2013), the materials needed for production already seem to suit collaborative making. Again, standardized structures and materials are functional for creating and building a *knowledge-sharing* system. This means that every participant is involved in the same system of values and meaning, and this helps build community, collaboration and a collective learning process (Tanenbaum et al. 2013). This is why, according to Martin (2015), the Maker Movement can be considered "*a knowledge-building community*," a kind of community that "works collectively to build and share new knowledge" (Martin 2015, p. 36) (like the scientific community), in a non-competitive manner (Scardamalia and Bereiter 2006).

As a further matter, we analyzed how the maker approach can be related to social issues; and for this, Jeff Sturges and his story are an excellent case in point. At the 2013 TEDx Midwest conference, Sturges, generally considered a community idealist and a *maker*, presented a project taking place in the basement of the Church of The Messiah in Detroit. There, the idealist and his group had created a maker space for the local community. When describing his idea, Sturges emphasized that he was strongly motivated by social issues when starting this project. His aim was to create a space where people could connect and have an opportunity to work and spend time together. He stressed that the young people attending the space, born and raised in under protected areas of the city, found a "safe haven" in the maker space. They were aware that this new environment could open a door to their future working lives, thanks to the opportunity to learn new skills. Everyone in the film he presented talked about his or her own experience of social inclusion. One of the adolescents said that when he is focusing on designing or creating something at the maker space, time flies and he is not drawn to the usual "temptations" (Sturges 2013). These stories portray the happiness and pleasure of getting together, within a community, getting involved in creative activities in a safe and secure place. As reported by Taylor et al. (2016) the maker space can be defined as a "third place," a social space (away from the home or the workplace) that can play a pivotal role in the individual's public life.

To sum up, the maker space is a place that can offer an array of positive interactions and stimulating activities. It fosters the development of social ties and relationships, the learning of essential skills for the future, and the opportunity to create things and cultivate a hobby. The maker space should be considered a *social leveller*, a place where there can be no discrimination (Taylor et al. 2016).

3 Key Competences and Active Citizenship

The community framework and the collaborative mechanisms that characterize maker communities (in the way of Tim Berners-Lee's theorization of *interoperability* and Pierre Lévy's *collective intelligence*) seem to be compatible with the development of active citizenship. Making and the maker mind set (fostering social inclusion, democracy tolerance and collaboration) can play a crucial role in mastering the key competences. First, we examined four (of the eight) competences: on the one hand *mathematical competence and competence in science, technology and engineering, digital competence* (disciplinary skills), and on the other *citizenship competence, cultural awareness and expression competence* (cross competencies). When we consider the two disciplinary skills mentioned earlier, their encounter with the maker min set can yield surprising results. If we analyze the text of the EU Council Recommendation of 22 May 2018 on Key Competences for Lifelong Learning, many points in common emerge between disciplinary competences and the maker mind set, sparking a reflection on the potential work that can be achieved.

(Mathematical competence and competence in science, technology and engineering). "... It includes the ability to use and handle technological tools and machines as well as scientific data to achieve a goal or to reach an evidence-based decision or conclusion." For this reason, the European Council's guidelines clarify the importance of becoming familiar with the use of "technological tools and machines." Thus, it is natural to think of the Maker Movement and the role that tools play in the maker community; introducing the maker mind set can promote the learning of these technical practices. It is also important to remember that "experimental play" (Martin 2015, p. 35) is the makers' usual way of doing things: *a recreational perspective on technical learning*. At the same time, *digital competence* also fits perfectly into the movement's structure. In fact, again according to the European Council's declaration (2018), attaining good *digital awareness* means having an interest in digital technologies and in related topics, and, hence, dealing responsibly with digital literacy and all related practical aspects (such as programming, coding and so on … cf. Jeanette Wing and her viewpoint).

In short, the other two cross competencies, namely, citizenship competence, cultural awareness and expression competence, can be linked to the Maker Movement through its fondness for cooperative learning (team work) and the failure-positive approach. "Competence in cultural awareness and expression involves having an understanding of and respect for how ideas and meaning are creatively expressed. […] It involves being engaged in understanding, developing and expressing one's own ideas and sense of place or role in society in a variety of ways and contexts." Thus, a positive and active outlook on the development of personal ideas is related to the maker's tendency to work on certain theoretical ideas in order to bring them into reality.

"Citizenship competence is the ability to act as responsible citizens and to fully participate in civic and social life, based on understanding of social, economic, legal and political concepts and structures, as well as global developments and sustainability." The usual dynamics of a maker space can be seen as an explanatory example of active citizenship. The very structure of the community requires a form of active citizenship that can help build shared knowledge. To sum up, these analogies can lead us to imagine certain beneficial results. For example, the maker mind set can help children learn key competences and, as a result, become active citizens.

4 The Experience of DENSA Coop. Soc

While reflecting on this subject, we decided to relate the Maker Movement to three different pedagogical models of the last century—Célestin Freinet, Bruno Munari, Loris Malaguzzi—to show the maker mind set's compatibility with the educational system. In this paper we will only consider the French pedagogue, Freinet. As a pragmatic teacher, Freinet always promoted a concrete, tangible form of education, and considered work to be an important ally of learning. In his view, education required experience; *a technique for life* (Legrand 1993). This direct (conceptual) connection is linked to Dougherty's (2012) description of a maker mind set. According to the latter author, a maker is a creator who plays an active role *in the creative process*, and exploits manual techniques.

Moreover, according to Freinet (2002), school subjects should not forgo the collaborative approach, which should be part of an internal system of sharing and interacting; each learner should be an active participant in the network. This is an early theorization of active citizenship.

Retracing the three authors' steps and rediscovering their mind sets in the light of the Maker Movement were crucial and preparatory to the study of the DENSA experience. The cooperative is profoundly connected to the Maker Movement, as is evident in its mission of inclusion, and its aim of providing learning opportunities to all. It applies the maker mind set to enhance skills and capabilities. We had an opportunity to study DENSA (and its activities) for four months (November 2018 to February 2019), analyzing the links between making skills and the learning process. In doing so, we looked at two workshop activities that aimed to introduce the maker culture to an extracurricular project: *Making Culture—Our Heritage in 3D* (which involved a 3D printing system) and *Coding Unplugged* (involving educational robotics).

The aim of *Our Heritage in 3D* was to reflect on artistic and cultural heritage (tangible and intangible) and on territorial identity (past and present). The project appeared to be a good opportunity for participants to practice all four of the competences we considered. First, it is noteworthy that the workshop was linked to a collective reflection on the recent earthquake in the region (2016). In fact, the event caused serious damage to Perugia's Carducci-Purgotti secondary school, which later had to be demolished. The idea behind *Our Heritage in 3D* was to invite students to think about the aftermath of the disaster, by creating 3D prototypes. In addition, the project had a specific intercultural theme (inspired by the multicultural class). Students analyzed the experience of *Superkilen* in Copenhagen as an example of social integration through urban design. These important topics allowed us to consider that all four of the competences co-existed harmoniously within the project. First, using Sugar CAD and other tools, participants were able to put their digital skills into practice, through 3D prototyping designs to reconstruct their local area, and also the competences of mathematics, science, technology and engineering. At the same time, the social and cultural themes of the project facilitated artistic expression and cooperative learning. Moreover, *Our Heritage in 3D* was related to INDIRE's *Maker@Scuola* project (2014), mainly because DENSA's activities were built on the same tools and software: In3dire, Sugar CAD, Raspberry Pi, OctoPrint, Cura. We also looked at 3D printing as an "emotional technology" (Guasti et al. 2017, p. 129), which allowed us to examine the project in all its complexity. According to this definition, 3D printing takes time, reflection and critical thinking. From here, we structured this complex project into four phases to allow students to master and develop the competences. First, students had an opportunity to get an idea about the subject and then analyze it in more depth. Next, they did some research on the theme and on icons, which was followed by a design phase using Sugar CAD. The final phase involved the TMI (Think, Make, Improve) approach.

5 Conclusions

To sum up, the aim of our reflection on the Maker Movement was to underline how the maker mind set can foster learning, not only in individual disciplinary subjects, but also of cross-disciplinary competencies. It is interesting to note that each approach that we considered (Freinet, Munari, Malaguzzi) focuses on the fundamental importance of techniques, which can be door-openers for learners. In cross-disciplinary terms, it is essential to acquire the know-how to handle all forms of learning: "Rhetoric in the Maker Movement often focuses on skills rather than abilities" (Martin 2015). As Célestin Freinet used to say, the only educational system that can help learners to cultivate their culture and prepare them for life, is one that is based on techniques (Freinet 2002).

References

Anderson, C.: Makers. In: The New Industrial Revolution. Crown, New York (2012)
Dougherty, D.: The maker mindset. In: Design, Make, Play, pp. 25–29. Routledge (2013)
Freinet, C., Eynard, R. (eds.): La scuola del fare. In: Quaderni di Cooperazione Educativa, Junior 2002
Guasti L., Rosa A., Maker@Scuola. Stampanti 3D nella scuola dell'infanzia. sFlorence, Assopiù Editore, (2017)
LEGRAND, L.: Célestin Freinet, Prospects: the quarterly review of comparative education, vol. XXIII, no. 1/2, pp. 403–418 (1993)
Martin, L.: The promise of the maker movement for education. J. Pre-Coll. Eng. Educ. Res. (J-PEER) **5**(1), 4 (2015)
OJ C 189, 4.6.2018, pp. 1–13
Scardamalia, M., Bereiter, C.: Knowledge building: theory, pedagogy, and technology. In: Sawyer, R.K. (ed.), The Cambridge handbook of the learning sciences, pp. 97–115. Cambridge University Press, New York, NY (2006)
Sturges, J.: TEDx Talks. The Maker Movement: Jeff Sturges at TEDx Midwest Video file. https://www.youtube.com/watch?v=uIXJclJE2Y&t=633s. Accessed 6 Jan 2014
Tanenbaum, J.G., Williams, A.M., Desjardins, A., Tanenbaum, K.: Democratizing technology: pleasure, utility and expressiveness in DIY and maker practice. In: Proceedings of the SIGCHI Conference on Human Factors in Computing Systems, pp. 2603–2612, April 2013
Taylor, N., Hurley, U., Connolly, P.: Making community: the wider role of maker spaces in public life. In: Proceedings of the 2016 CHI Conference on Human Factors in Computing Systems, 1415–1425 May 2016

Chesscards: Making a Paper Chess Game with Primary School Students, a Cooperative Approach

Agnese Addone and Luigi De Bernardis

Abstract The game of chess can be too theoretical for children and can even be quite a challenge for teachers and chess masters. It is hard to make it approachable and, at the same time, technically correct. The *Chesscards* educational project arose from these observations, and is intended to be a way to translate chess theory actively, by tinkering with paper and colors. This delightful experience was conducted from 2015 to 2019 with 10-year-olds in a primary school in Rome, Italy, and enabled children to develop good chess skills by cooperating in making. Small groups of children aged 7–10 created playing cards and a paper chessboard along the lines of some of the most famous games. The initiative's huge success, and the reason it was repeated in these last years, lies in its strictly constructionist approach to making: *Chesscards* became an original way to learn, and an easy social game that any child can play.

Keywords Chess · Making · Constructionism · Constructivism · Cooperative learning · Primary school · Tinkering · Pattern recognition · Social learning

1 Introduction

The idea behind this project is to get primary school students to play chess playfully and cooperatively.

They are introduced to the difficulties of the game, along with the rules and strategies. From the first experience in 2015 until 2019, the number of classes involved increased, and the project officially became one of our school's main educational focuses. A book [1] and a paper [2] have been published on the workshop.

A. Addone (✉)
Department of Computer Science, University of Salerno, Fisciano, Italy
e-mail: addone@unisa.it

L. De Bernardis
Luiss Guido Carli University, Rome, Italy

D. Scaradozzi et al. (eds.), *Makers at School, Educational Robotics and Innovative Learning Environments*, Lecture Notes in Networks and Systems 240,
https://doi.org/10.1007/978-3-030-77040-2_19

Both girls and boys achieved a good level of proficiency in chess, which was seen in their total autonomy in playing and in their participation in contests and tournaments at school and externally.

Chesscards is a making workshop where children design, color, cut and use their own cards, similar to the more famous Pokémon and Yu-Gi-Oh! games.

During school hours, the project always involves a class teacher and a chess master, one class at a time. This is an important setup, which maintains the link between theory and practice in the game of chess.

The children work in small groups to boost collaboration, cooperation [3] and participation. They enter a process of peer learning [4], which is central to the activity: the group's observation is essential to the success of the learning process [5].

The children achieve numerous skills in this experience: individual responsibility to the other members of the group, interdependence and a sense of having a common goal; review of each other's work; logical and algorithmic reasoning, social networking and team building; controlling negative or aggressive emotions.

The project was developed over four years and is still in progress. The number of children involved by year is as follows: 61 (2015/16), 251 (2016/17), 193 (2017/18) 200 (2018/19), 220 (2019/2020). We briefly describe a concrete example of application of the *Chesscards* methodology in a course organized in classes (10-year-olds) in a primary school in Rome, Italy. Our sample consisted of children aged 7–10 years; the ratio of males to females was more or less equal.

2 Making *Chesscards*

The first stage in the workshop is an introduction to the history of the game of chess and its development through the centuries in different countries. This is an essential step which focuses the children's attention on the game and explains its extraordinary position at the meeting point of games, sports, science and art.

During this phase, the chess master also introduces all the pieces, their different roles and values, and the basic strategies that are important to know to start playing. There is also great emphasis on the objective: conquering the center of the board. The teachers put the children into small groups and assigned a paper template to each child. The template is a sheet with several blank fields, an image of the piece, the name of the piece, a brief description of its value in points (3 for the bishop or knight, 5 for the rook and 9 for the queen) and how it moves, arrows indicating the direction in which it moves. At this stage the template is quite large (A4 format), to make it easier for the children to draw. Students are free to represent the piece in the central frame of the template as realistically as they want or using a metaphor or a symbol.

This is a very important stage, because it allows each child to express him or herself individually, and it is a kind of passport for obtaining the group's respect (Figs. 1 and 2).

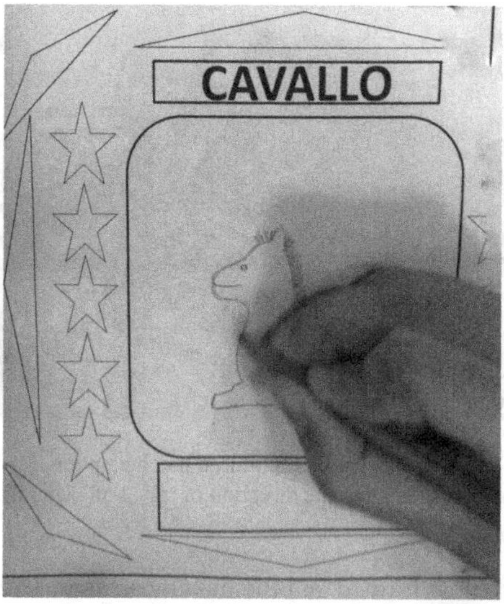

Fig. 1 *Chesscards* template: name of the piece, caption/description, stars indicating its value, arrows indicating the direction of travel. The children can unleash their creativity within the rules of the game

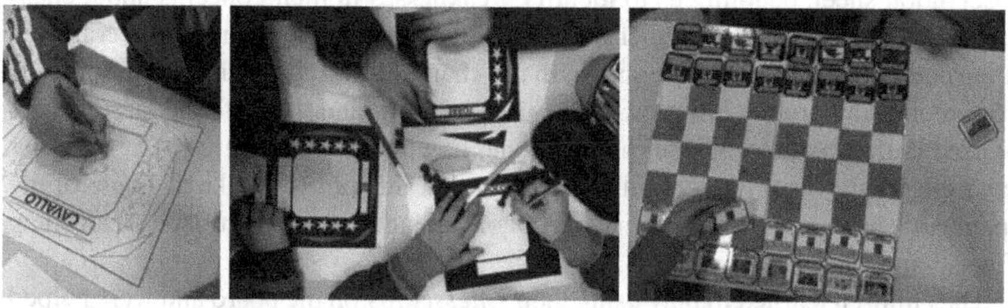

Fig. 2 The three stages of making *Chesscards*: designing, cooperating, testing. First drawing, then the final version of the cards, when the game can begin on the chessboard

After a group discussion of the results and a peer review, the second step is to color the drawings. Much attention is given to how the stars and arrows are colored in, which has to be consistent with the rules of the game.

In the last step, the *Chesscards* are reduced and mounted onto white or black backgrounds, then cropped and laminated to protect them during play. It is essential for the background of the cards to be black or white: each player has a full set in one of the two colors.

Separately, children prepare a large paper chessboard, with squares the same size as the cards, and then place the pieces onto it.

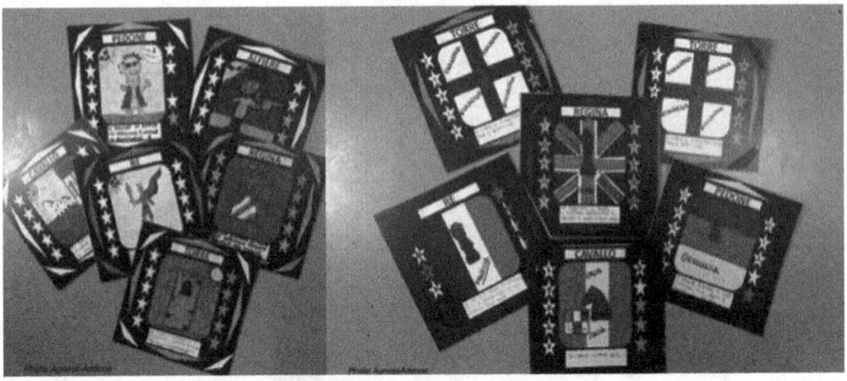

Fig. 3 Two different versions of *Chesscards*: children use metaphors for their pieces. The stars indicating its value and arrows indicating the direction of travel of each piece are colored

Pairs of players are now formed and the match can start. Every child knows the rules and, if in doubt, can refer to the rules on the Chesscards so they can be sure they are moving correctly on the board.

To measure learning effectiveness at the end of the course, we ask children in the first few lessons to draw their own map, on a blank form. This also acts as our observation sheet, enabling us to identify weaknesses in their concepts about chess and which ones can be strengthened. The results are compared with those collected at the end of the course and are shown in the chart below. It is clear that the concepts that are missing are the different stages in the match (and the different strategies to be used in them), the center of the board and how the pieces progress. On the other hand, there is a very frequent misconception that the aim is to take as many of the opponent's pawns as possible. The concept of stalemate is not mentioned on any map (Figs. 3 and 4).

The children learned to play using their *Chesscards* on the board and were helped by the information written on the cards and the board. This information does not appear on a traditional board, but makes the process of learning the game easier for the children.

3 Outputs

Creativity is actively encouraged in this project. The children use their imagination and choose how to portray the subject, and can represent their pieces through symbol or metaphor. The most interesting results include characters in a castle, people in the street, country flags, fantasy characters and symbols. Each child has his or her own set of *Chesscards* and a board. Many exchange pieces or create a new set to exchange.

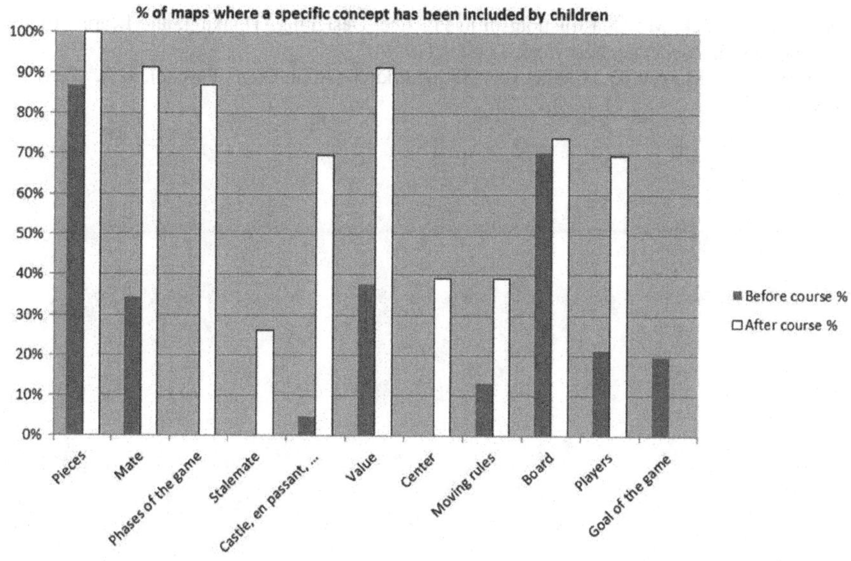

Fig. 4 Survey of learning effectiveness based on the problems raised in cognitive maps; each bar represents a node on the map, and is a key concept in the game

The constructionist approach [6] employed in the *Chesscards* project shows that learning by doing is vital. The maker approach may seem minimal here, but the children's active involvement and the results in educational terms are effective. By making their own cards, children become actors in their learning process and boost their motivation, engagement, accountability, and participation.

The social context plays an enormous role in the process [4]: the small group size is fundamental for achieving the best environment for constructing the game [3].

Playing with *Chesscards* is not so different from playing with real pieces, but the game is much more approachable for all children and teaches them the fundamentals of chess [7] in a fun and playful way.

References

1. De Bernardis, L., Corvi, M.: How a cooperative approach can improve children chess learning, 1st edn. Lambert Academic Publishing, Saarbruecken (2016)
2. Addone, A., et al.: Co-costruire le *Chesscard*: un'esperienza di apprendimento cooperativo degli scacchi in una scuola primaria. Psicologia e Scuola **47**(9–10), 58–61 (2016)
3. Johnson, D.W., Johnson, R.T., Holubec, E.J.: Cooperation in the Classroom, 8th ed., Edina, MN, USA (2008)
4. Vygotsky, L.S.: Mind in Society: The Development of Higher Psychological Processes. Harvard University Press, Cambridge (1980)
5. Darnis, F., Lafont, L.: Cooperative learning and dyadic interactions. Phys. Educ. Sport Pedagog. **20**(5), 459–473 (2015)

6. Kafai, J., Resnick, M.: Constructionism in Practice. Designing, Thinking, and Learning in a Digital World. Routledge, New York, London (1996)
7. Horgan, D.D., Morgan, D.: Chess expertise in children. Appl. Cogn. Psychol. **4**(2), 109–128 (1990)

A New Graphic User Interface Design for 3D Modeling Software for Children

Laura Giraldi, Mirko Burberi, Francesca Morelli, Marta Maini, and Lorenzo Guasti

Abstract Digital technologies have become a central part of the everyday experience the younger generation children and young people, born and raised in the ITC era, are known as digital natives. At school, this generation prefers active learning instead of being passive recipients of knowledge. This research proposes a new graphic user interface concept for a 3D modeling software program, designed for preschool and early primary school children. The interface of SugarCad Kids was improved to be more understandable, friendly, intuitive and enjoyable to use, in keeping with the principles of human-centered design.

Keywords Children · Preschool design · 3D software · UX/UI · GUI

1 Context

1.1 Digital Natives and ITC

From the early 1990s, new definitions began to appear defining the generation after Generation X (born between 1961 and 1981 in the era of the video game console and

L. Giraldi (✉) · M. Burberi · F. Morelli · M. Maini
University of Florence, Florence, Italy
e-mail: laura.giraldi@unifi.it

M. Burberi
e-mail: mirko.burberi@stud.unifi.it

F. Morelli
e-mail: francesca.morelli@unifi.it

M. Maini
e-mail: marta.maini@unifi.it

L. Guasti
Istituto Nazionale di Documentazione, Innovazione e Ricerca Educativa (Indire), Florence, Italy
e-mail: l.guasti.tecnologo@indire.it

D. Scaradozzi et al. (eds.), *Makers at School, Educational Robotics and Innovative Learning Environments*, Lecture Notes in Networks and Systems 240,
https://doi.org/10.1007/978-3-030-77040-2_20

the personal computer). This new generation of people was dubbed millennials [1], the net generation, Generation Y, digital natives, and digital immigrants [2, 3].

These two generation of people have in common a world full of technology (ITC) and real-time information, which are also both useful factors for the school system, where learning should be more active than passive. According to cyberpsychology, new technologies (especially digital technologies) affect cognitive processing in digital natives, particularly their intuition, and the way they organize and implement their actions [4].

1.2 School Education and Learning for Digital Natives

According to Papert, knowledge construction is much more significant if it takes place in a context where the learner is engaged in producing something concrete and shareable. Errors are not negative, they are simply part of the learning process.

Making a mistake means exploring, learning from one's mistakes and looking for alternative solutions to the problem. The teacher's role is to guide children towards understanding their mistakes (debugging).

This form of mental construction and *concrete thinking* [5] can be summarized in this way: not learning to make, but making to learn [6].

Constructionism assigns a particularly important role to real constructions, as they support constructions in the mind. According to these theories, school should become a place where construction takes place, rather than being only a place for learning passively. Furthermore, it should feature dedicated learning environments with technology, where children are the main users (student-centered) and practical activities are the source of learning.

In February 2018, the Italian Ministry of Education, University and Research published new national guidelines highlighting the importance of improving scientific, mathematic and computational learning in three- to seven-year-olds. This is necessary to help them developing rational, logical, and critical thinking skills with which to solve problems.

The National Digital School Plan (Piano Nazionale Scuola Digitale—PNSD) calls for the production of creative ateliers in primary schools (Law no. 107, 13 July 2015 known as the "Good School" law). These are physical workshop spaces where maker learning can take place.

1.3 A New Teaching Methodology: Maker Pedagogy

The Maker Movement is a new frontier in artisanship; called 2.0, it combines the spirit of traditional crafts with new technologies.

In pedagogical terms, this culture has its roots in the principles of pedagogical activism and new schools, and also the Montessori and Munari methods.

Maker Movement pedagogy has combined the spirit of crafts with experimental play; it connects digital objects to material ones, through specific solutions that generate design models that can be produced using different fabrication technologies, including 3D printing.

In *maker pedagogy*, children are at the center of learning and their interests guide doing and learning activities through play, fabrication, designing and exploring (learning by making). According to Piaget [7], until the age of six, children learn mainly by working. This is why it is fundamental for them to work with all five of their senses in the first years of life.

An international study [8] shows that children three- to four-year-old have the skills to distinguish between geometrical forms, and they can understand their more basic properties like sides and corners. Children's interest in geometry could be enhanced and their knowledge expanded through targeted programs and early experiences [9]. These exercises are easy to achieve using 3D printing, offering children the opportunity to see their product being made, and increasing their metacognitive competence.

2 The Aim of the Research

This research *refers to Maker@Scuola*: "*Nuove Tecnologie per la Didattica*" [10], a project running since 2014 and developed by INDIRE (a government agency for research in educational innovation). The aim of the research project is to experiment new forms of laboratory teaching that are enhanced by new technologies (such as the 3D printer and the hydroponic greenhouse) at all school levels beginning with preschool and primary school.

In3Dire—a system that gives schools the ability to design and print 3D artifacts, even without an Internet connection—and a 3D modeling software called *SugarCad*, which is integrated into the in3Dire System, are part of the Maker@Scuola project. SugarCad was developed with the technological setting in mind, to resolve issues found in school environments. It was also designed to be used by students (and teachers) at different school levels and offers multiple interfaces for different user levels of experience.

The aim of this research paper is to define a concept for a GUI (graphical user interface) for a new level in SugarCad called *SugarCad Kids*, dedicated to children, three- to seven-year-old. This new level will join the two already available (basic and advanced). The aim of SugarCad Kids is to adapt the SugarCad GUI so that it is as intelligible, intuitive and enjoyable as possible to use, thereby improving its educational purpose.

Another goal of this research is to propagate a new way of teaching young users based on *learning by doing* and *learning by making*. This method can bolster the natural development of their skills and attitudes, starting from pre-primary level.

3 Research Method

3.1 Child-Centered Design

A good *user experience* from a product, a system or a service comes when the design is functional, and is easy to understand and use.

The research considers the user-centered-design (UCD) approach as the main method for child-centered design (CCD) [11]. The work also concerns human-centered design (HCD), a problem-solving process that focuses on understanding children's needs and behaviors in relation to the context they live in and their psychological traits and features. The HCD approach considers the needs and behaviours of people during the *design thinking* process and later when testing and validating the results [12, 13].

Furthermore, the CCD that goes into a product, a service or a GUI communicates clearly and interacts not only with children but also with teachers and parents. In fact, GUIs that have been designed taking into consideration the user experience (UX), not only meet the satisfaction of children and their friends, but are also more effective for learning. To summarize, a GUI should be playful, easy to use and suit the user's age [14].

A digital interaction for children should:

1. *Be efficient* (when the action time is too long, children think they are wasting time and stop playing);
2. *Be coherent* (when children recognize actions and interactions they feel at ease);
3. *Be responsive* (children need *feedback* for their actions so they can understand what is happening; they need confirmation and gratification);
4. *Enable them to test it and/or make errors* (this helps guide children around the GUI, because they sometimes lose focus and forget what they were doing, because they have found something interesting to click on);
5. *Be able to* give *visual clues* (preschoolers cannot read and even those who can, usually avoid doing so, particularly if they are using a digital device) [15].

3.2 Analysis

Users: children aged 3–7. It is a common mistake to think of children as a single category, and not relate their cognitive and mobility skills to their specific ages [16–21].

A four-year-old child's reading and mobility skills need different design solutions. Therefore, sub-categories of user/child should be taken into consideration.

Some researchers and professionals in the field of new digital technologies use a *3–6–9–12* age grouping model [22], which contains three age groups: 3–6, 6–9 and 9–12 years. Others suggest a *2–4–6–8–12* format with four different groups. This one is more appropriate, because it reduces the range within each group from three to two years, [23] and defines the user more accurately. As the target for SugarCad is children in preschool and the first year of primary school, users have been analyzed according to the 2–4–6–8–12 model, mainly considering those aged three to seven.

2–4 years. This is an important age range because during this period children become autonomous (from babies to small children).

The GUI should have the following features:

1. A clear visual hierarchy (children usually focus on details, which is why it is necessary to create a clear visual difference between interactive and non-interactive elements);
2. Few bright colors (too many colors confuse them during activities);
3. Elements on the screen should correspond to a single behavior (2–4-year-olds can only link one function to an element or object they are interacting with);
4. The background should be clearly distinguishable from the foreground;
5. Images and icons with no text.

4–6 years. A GUI for this age group should be more than a modified version of the one for two- to four-year-olds based on their different skill levels. Their interaction with the interface should be stimulating and should take into consideration what they already know and their collective imagination. As they approach the age of six, students are able to solve relatively complex problems and make mental classifications quite efficiently. They are also quite loud and have become technologically adept.

4 The Project: "SugarCad Kids"

4.1 Wireframe and Logo

The practical activities on the SugarCad software were focused on the design concept for the new GUI. Since it had features that were closer to those needed for the target group, the existing platform's basic level interface was used as a starting point for analyzing how to improve usability, interaction and UX in general.

This analysis identified the actions users would perform when working from this new level (SugarCad Kids). This led to a new task analysis, a new task flow and, subsequently, the layout of the prototype for the new GUI.

The wireframe design also tooks account of factors like content (text, images, icons, etc.), functionality (commands, buttons, sliders, feedback from commands, etc.) and navigation (the way the content and the various functions are located and used). The project proposes four wireframes for the *welcome page*, the *switch page*, the *freehand drawing page* and the *3D geometric shapes page*. The idea of giving SugarCad Kids its own identity led to the creation of a dedicated logo, which kept a sense of continuity with the other levels of the software. At the same time, it had its own recognizable identity more suited to younger users and their collective imagination.

4.2 *Graphic User Interface for Children (3–7-Year-Old)*

Neutral colors were used for the background to create greater contrast and the features become more readable. It also helps the visual separation of different interactive elements.

The action icons were enlarged, using homogeneous colors, depending on their position on the screen. Where possible, these positions were kept the same on both main pages (*freehand drawing* and *3D geometric shapes*), to ensure continuity, avoid visual confusion and help users find their way around the GUI. In some cases it was seen that icons were difficult to interpret, even for children who can read, so a text reference was added in the web version of an easy-to-read, dyslexia-friendly font. The icons of the 3D geometric shapes were given anthropomorphic characterization, making them as close as possible to comic or cartoon characters that children are familiar with, and more playful. It may be possible to add a simple animation that can be activated when a finger hovers over or touches the character.

Message boxes have simpler, clearer, child-friendly graphics to ensure instant feedback and further improve UX.

5 Conclusion

This paper has considered children's abilities in relation to the maker learning approach. Unlike traditional learning, this method gives a child more scope to express his/her creativity actively. The distinctive aspect of this project is the attention given to the graphic design, with the purpose of making the interface more appealing and intuitive for children. It also underlines the significance of adapting the interface to suit different users (children).

The study of children's characteristics made it possible to define project rules that could be applied to the graphic user interface for younger users, to make it as child-friendly as possible. This method could be used to design other CDC GUIs for digital products, and not only in learning environments.

The following points summarize the proposed elements that can be used to design this kind of "kid-sized" interface.

1. *Functionality* in relation to children's abilities. All the interactive commands and feedback must be easy to understand and easy to use.
2. The *visual design* of the GUI. All icons and shapes must be fun-looking, inspired by children's collective imagination (e.g., cartoons). The meaning of

actions should be immediate to avoid confusion and distraction; they should be accompanied by animation and video, where possible.

3. *Cognitive aspects*. The cognitive skills of young users change rapidly.

The GUI's visual language should take account of children's imaginations; they need fun, playful elements to enable them to approach maker learning. This helps children's experiences become a new way of learning that is more active and less passive, and leads to better results. We hope designers can use this approach to GUI design for future CCD digital products for teaching and learning in line with *maker pedagogy*. Our aim for the next year is to finalize implementation of the interface and begin the first experiment with students at school. This will enable us to analyze the strengths and weaknesses of the interface and improve it.

References

1. Howe, N., Strauss, W.: Millennials Rising: The Next Great Generation. Paperback, USA (2000)
2. Prensky, M.: On the horizon (NCB University Press). Digital Natives, Digital Immigrants **9**(5), 1–6 (2001)
3. Riva, G.: Nativi Digitali. Crescere e Apprendere nel Mondo dei Nuovi Media. Bologna, Il Mulino
4. Riva, G.: Psicologia dei Nuovi Media. Bologna, Il Mulino (2008)
5. Papert, S.: Constructionism: A New Opportunity for Elementary Science Education. MIT, Media Laboratory, Epistemology and Learning Group (1986)
6. Papert, S.: Mindstorms. Children, Computers, And Powerful Ideas. Basic Books, New York (1993)
7. Piaget, J.: Lo sviluppo mentale del bambino e altri studi di psicologia. Einaudi, Turin (2000)
8. Clements, H.D., Sarama, J.: Teaching children mathematics, young children's ideas about geometric shapes **6**, 482–488 (2000)
9. Clements, H.D.: Teaching children mathematics, mathematics in the preschool **7**(5), 270–275 (2001)
10. Guasti, L., Rosa, A. (eds.): Maker@Scuola. Stampanti 3D nella scuola dell'infanzia. Florence, Assopiù Editore (2017)
11. Idler, S.: Child-centered design is user-centered design, but then different. http://uxkids.com/blog/child-centered-design-is-user-centered-design-but-then-different/. 16 Aug 2013
12. Dam, R., Siang, T.: What is design thinking and why is it so popular*?* (2018). https://www.interaction-design.org/literature/article/what-is-design-thinking-and-why-is-it-so-popular
13. Dam, R., Siang, T.: Design thinking: getting started with empathy (2018). https://www.interaction-design.org/literature/article/design-thinking-getting-started-with-empathy
14. Idler, S.: Child-centered design is a mindset, not rocket science (2013). http://uxkids.com/blog/child-centered-design-is-a-mindset-not-rocket-science/. 04 Oct 2013
15. Idler, S.: 5 key rules for designing interactions for kids (2017). http://uxkids.com/blog/interaction-design-for-kids. 23 Nov 2017
16. Liu, F.: Design for kids based on their stage of physical development (2018). www.nngroup.com/articles/children-ux-physical-development/. 08 June 2018
17. Liu, F.: Designing for kids: cognitive considerations (2018). https://www.nngroup.com/articles/kids-cognition/. 16 Dec 2018
18. Sherwin, K., Nielsen, J.: Children's UX: usability issues in designing for young people (2019). https://www.nngroup.com/articles/childrens-websites-usability-issues/. 13 Jan 2019

19. Gossen, T., Nitsche, M., Nürnberger, A.: Search user interface design for children: Challenges and solutions. In: CEUR Workshop Proceedings, 909 (2012)
20. Kientz, J., Anthony, L., Hiniker, A.: Playful interfaces: designing interactive experiences for children. User Experience Mag. **18**(1) (2018). https://uxpamagazine.org/playful-interfaces/
21. Pugnali, A., Sullivan, A., Bers, M.U.: The impact of user interface on young Children's computational thinking. J. Inf. Technol. Educ. Innovations Pract. **v16** (2017)
22. Tisseron, S.: 3–6–9–12. Diventare grandi all'epoca degli schermi digitali. In: Rivoltella, P.C (ed.) s.l., Ed. La Scuola (2016)
23. Levin Gelman, D.: Design for kids. Digital products for playing and learning. NYC, Rosenfeld Media (2014)

Museum Education Between Digital Technologies and Unplugged Processes. Two Case Studies

Alessandra Carlini

Abstract This document presents the results of architectural design and prototyping of educational kits within the museum context, two case studies featuring a combination of digital technologies and unplugged processes. The field of application is cultural heritage and the topics are part of school curricula. The first case study is a museum display of digital video installations and educational kits that reproduce mechanisms of symmetry from patterned flooring ("www.formulas.it" laboratory, Department of Architecture, Roma Tre University and Liceo Scientifico Cavour" high school). The second case concerns the setting up of a school fab lab in which 3D-printed prototype educational kits are made for schools and museums in Rome, in partnership with the Municipality of Rome and the Ministry of Cultural Heritage and Activities (General Directorate for Education and Research). The cases involve professional, research and didactic experiences which led to funding-supported projects. The experiences showcase good practices in informal and cooperative learning, and highlight the relationship between education and popularization that draws on our architectural heritage.

Keywords Museum education · Architectural design · Educational kit · 3D printing · Digital manufacturing · Fab lab

1 Introduction

Digital technologies are transforming traditional learning methods. What are the implications of this for the museum's role in lifelong learning and in today's knowledge-based society? In some of the experiences we describe, unplugged activities and digital technologies are used to increase motivation and spark interest and curiosity in learning . Although targeting people of different ages and backgrounds,

A. Carlini (✉)
DARCH, Department of Architecture, Roma Tre University, Largo Giovanni Battista Marzi, 10, 00154 Rome, RM, Italy
e-mail: alessandra.carlini@archiworldpec.it; alessandra.carlini@uniroma3.it

D. Scaradozzi et al. (eds.), *Makers at School, Educational Robotics and Innovative Learning Environments*, Lecture Notes in Networks and Systems 240,
https://doi.org/10.1007/978-3-030-77040-2_21

the projects are particularly effective in informal learning contexts in public schools. The field of application is cultural heritage. The teaching of STEAM (Science, Technology, Engineering, Art, Mathematics) subjects is at the heart of the creative process, calling attention to soft skills and a more conscious digital citizenship, as required by the National Plan for Digital Education (PNSD) [1]. All prototype kits were used in a hands-on, peer-to-peer workshop. Finally, the experiences feature best practice examples of "authentic tasks" [2]. In a context of "laboratory teaching" [3] and "competency-based learning" [4], they pose an open-ended question by simulating real contexts.

2 Museum Display for Science Popularization

This section presents the experience of designing for science popularization in a museum context. The goal is the study of symmetry trough the visual stimuli provided by historic artifacts (geometric pattern paving). In the first case, the communicative potential of digital tools is exploited when an exhibit is created that centers around the relationship between the museum space and the visitor. Digital tools provide an experiential stimulus for mathematical content. In the second case, the museum is an "authentic" setting for developing scientific teaching skills, and for designing educational kits on the subject of symmetry, for use in activities taking place at the museum.

2.1 Video Floor Installation Showing Symmetries in Motion

The first design opportunity took place at the Genoa Science Festival, 2007. The "www.formulas.it" laboratory (Department of Architecture, Roma Tre University) presented the exhibition entitled "Mathematics and Archeology" (scientific supervisor Laura Tedeschini Lalli, Roma Tre University). The original appearance of the floor is reconstructed with virtual animations starting from the archaeological fragments found on site (Trajan's Markets, Rome, imperial period), by applying the mathematical theory of periodic tessellation [5]. In the exhibit a video installation is projected onto the floor, rather than a screen, because the object is mathematical and related to architecture (Fig. 1).

The floors appear from above and visitors can walk on them. Perception and conceptual content are linked through physical experience.

This method seeks constant interaction between the Italian experience on science popularization linked to the school of Enrico Giusti ("Garden of Archimedes", first mathematics museum, currently in Florence, Italy), and 20th-century museography.

Fig. 1 The video projected onto the floor and, at the top, a "Velarium" help recreate the original spatial dimension of the Roman *tabernae* and convey the correct proportions to the floor. The projected animations were created by Elisa Conversano

Herbert Bayer's experiments (1900–1985) at the Bauhaus Institute in Dessau (1919–1933) used the relationship between the mechanics of the eye and the physical dimension of perception as design tools [6] (Fig. 2).

After the Second World War, Italian architects like Carlo Scarpa (1906–1978) and Franco Albini (1905–1977) created museum installations in which they reflected on the relationship between content and container, by making a connection between the object and the physical dimension of perception. Both the experiences cited viewed

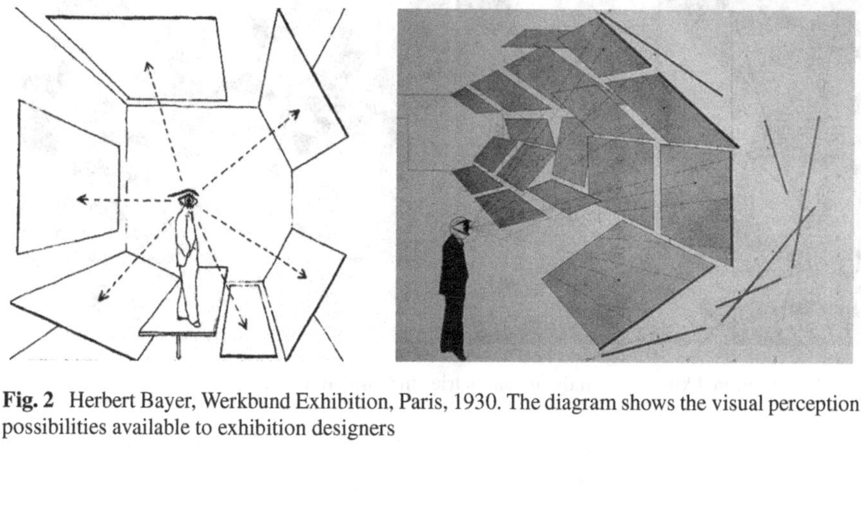

Fig. 2 Herbert Bayer, Werkbund Exhibition, Paris, 1930. The diagram shows the visual perception possibilities available to exhibition designers

museography as a system of relationships. The relationship between architecture and the popularization of science was discussed in Carlini and Tedeschini Lalli [7] and Carlini et al. [8].

2.2 *Extended Museum of Cosmati Floors. Educational Kit*

The experience described in the previous paragraph inspired the 2016 project "Educational kit for an extended museum [9] of Cosmati floors", a PCTO (training course in transversal skills and orientation) for the "Liceo Scientifico Cavour" high school and the National Museum of Palazzo Venezia (Rome, Italy; Director Sonia Martone). Carried out with architect Teresita d'Agostino, the project is mentioned as one of the Ministry of Education's best practices [10] and was selected by CNR (National Research Council of Italy) for the conference "Officina 2018" [11] (Fig. 3).

In this case, the study of symmetries concerns the geometric pattern of Cosmati floors in Rome and Tuscania. The symmetry educational kit was made using digital and unplugged techniques: mirror rooms [12], inventory of magnetized pieces, smart tourism with immersive Google Cardboard, interactive map for smartphone [13]. Since 2017, the Ministry of Education and Research (MIUR) has included the project in the "Visiting Teacher" workshop activities for peer-to-peer teacher training

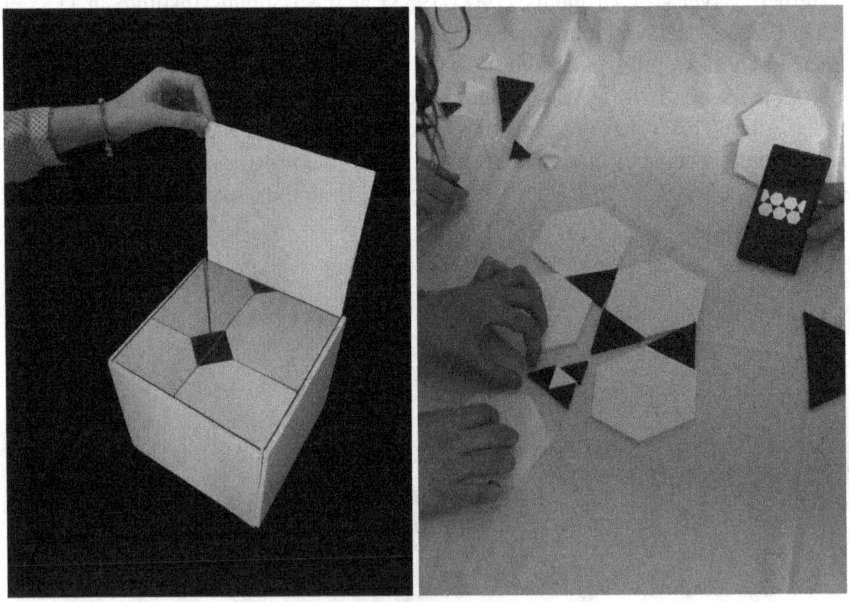

Fig. 3 Educational kit for the study of symmetries in Cosmati floors

in secondary schools (project managers at "Liceo Scientifico C. Cavour" school: Alessandra Carlini and Teresita d'Agostino).

3 Museum Education. Prototyping Educational Kits with 3D Printing in the School Fab Lab

The educational potential of a school fab lab is illustrated below. The structure was built at "Liceo Scientifico Cavour"school in Rome (headteachers: Ester Rizzi 2017; Antonella Corea 2019; Claudia Sabatano 2020) together with Antonina Amadei, the school's digital coordinator. The goal was the creation of a learning environment open to local institutions, a prototyping center for educational kits with a 3D printer, applying the living lab approach. This strategy uses digital fabrication to stimulate scientific inquiry and inclusion, by strengthening the role of science as an economic lever in the knowledge society. Museum education also means using the museum as a learning environment, not merely for the passive enjoyment of cultural heritage.

The activities of a school fab lab provide an experiential approach to knowledge that uses the body as a cognitive tool through the tactile exploration of objects made with a 3D printer, and direct engagement with historical and cultural heritage. The experiment fits into the cultural context of the "Tactile Laboratories" developed by Bruno Munari (1907–1998) after his "*Le mani guardano*" exhibition (Palazzo Reale, Milan, Italy, 1979) [14].

Students aged 16 to 18 in the third or fourth year of *Liceo Scientifico* engage with younger students, aged nine upwards, at hands-on workshops conducted with prototype models printed in the school fab lab. The goals are: inclusion through museum education; cultural heritage education through an experiential approach; informal education and cooperative learning; dissemination of science culture through visual and experiential input; development of digital manufacturing skills; new technologies applied to cultural heritage in the knowledge society. Skills were evaluated using: *Ex ante* evaluation (analysis of starting levels, *ex ante* observation section of MIUR protocol for PON-FSE 2014–2020 projects); formative evaluation (presentation of intermediate steps and review of partial results, intermediate observation section); and *ex post* evaluation (presentation of the results to a commission made up of teachers and external tutors, interviews, dissemination workshops, *ex post* observation section of MIUR protocol). The kit can be used in different contexts: educational workshops in museums; hands-on peer-to-peer workshops with younger students; curricular activities in the classroom.

3.1 Creative Geometry Kits: Detachable 3D-Printed Apollonius's Cone

The school fab lab was used to prototype a 3D-printed creative geometry kit for use in schools in the Municipality of Rome (2017) [15]. Students chose Apollonius's cone for project development [16]. This object allows the visualization of conic sections—circle, ellipse, parabola, hyperbola. The project uses magnetic surfaces to enable disassembly and reassembly. The cone, of which no other examples of "hourglass" or double-branch models are known, was prototyped using CAD software (Tinkercad, GeoGebra) and 3D printing (Cura) in the school fab lab (Fig. 4).

The project won the international prize at the 2018 edition of Maker Faire, Rome. Here, too, the didactic results were achieved through years of work, starting from the experience of the Ministry of Education's PLS (Scientific Degree Program) entitled "*Coniche e macchine da disegno*" (Conics and drawing machines) [17].

The activity takes place in four steps:

1st step—analysis of market attractiveness and identification of the topic through brainstorming;

2nd step—designing the prototype, including identifying parts, deciding on printing procedures and choosing materials and colors. Storyboard for the video tutorial on conic sections;

3rd step—development of the 3D model in Tinkercad and GeoGebra software; thematic insights, word and image processing for the video tutorial;

4th step—3D-printed (Cura software) prototype of the kit, surface finishing and gluing magnetized surfaces.

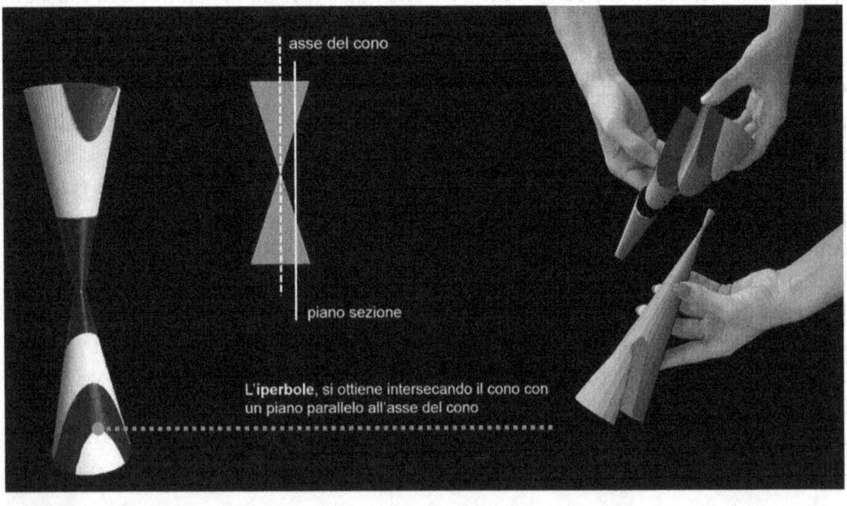

Fig. 4 Detachable apollonius's cone made with a 3D printer

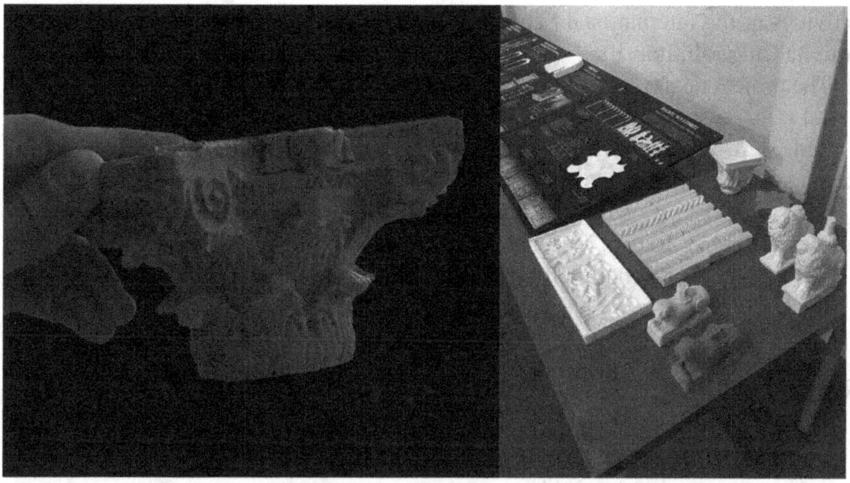

Fig. 5 Tactile kit of Romanesque architecture in the National Roman Museum (Rome, Italy)

Inclusivity is one of the goals achieved with this project, with the teaching kit being used during curricular activities of recovery training gaps.

3.2 ART-TOUCH-LAB. Tactile Kits Made with a 3D Printer

The educational kit illustrated in Fig. 5 is a tactile device to improve museum access and inclusion. It was prototyped in the school fab lab and commissioned in 2019 by the Directorate-General for Education and Research of the Ministry of Cultural Heritage and Activities and Tourism (MiBACT). Autodesk software (Zephyr, Recap Photo, Meshmixer) and hardware (smartphone and 3D printer) were used to make a photogrammetric survey of Romanesque architecture and decorations [18].

The printed kit allows tactile exploration for the blind and the sighted. More generally, it is of interest to everyone as it stimulates cognition and is a different approach to art. The 1921 "Manifesto on tactilism" by Filippo Tommaso Marinetti (1876–1944) is a fundamental premise for any cognitive experience of art that uses sensory channels other than sight. The "Omero" tactile museum (Italy) is an important venue for social inclusion through art. The project was carried out with architect Teresita d'Agostino and won the PON-FSE "Alternanza scuola-lavoro" competition and was implemented with European structural funds (Partners: Centro S. Alessio per ciechi, Myosotis, Scuola Innovativa start-up; A-Sapiens. Evaluation: Daniela Liuzzi). In the context of STEAM, cultural heritage and museum inclusion is meant to encourage active citizenship: a more conscious digital citizenship, with inclusion as an instrument of democracy and popularization of scientific culture to reduce social inequalities. For these reasons, the program is one of the case studies in the training activities of the International School of Cultural Heritage (Fondazione Scuola Beni Attività Culturali, Italy).

The project takes place in four steps:

1st step—historical and artistic documentation;

2nd step—BYOD (Bring Your Own Device) photogrammetric survey of Romanesque architecture and art;

3rd step—data processing in a school multimedia laboratory using Autodesk software (Zephyr, Recap, Meshmixer);

4th step—3D printing of prototype tactile devices in the school fab lab.

References

1. MIUR: http://www.istruzione.it/scuola_digitale/allegati/Materiali/pnsd-layout-30.10-WEB.pdf. Last accessed 20 Aug 2018
2. Glatthorn, A.A.: Performance Standards and Authentic Learning. Eye on Education, Larchmont (1999)
3. Dewey, J.: Esperienza e educazione. Cortina Raffaello, Milan (2014)
4. Recommendation of the European Parliament: https://eur-lex.europa.eu/legal-content/IT/TXT/?uri=celex%3A32006H0962. Last accessed 20 Aug 2018
5. Carlini, A., Conversano, E., Tedeschini Lalli, L.: Mathematics and archaeology. In: APLIMAT—Journal of Applied Mathematics, vol. 1(2), pp. 61–68. Bratislava (2008)
6. Rudofsky, B.: Notes on exhibition design: herbert Bayer's pioneer work. Interiors **12**, 60–77 (1947)
7. Carlini, A., Tedeschini, L.L.: Interrogare lo spazio. Gangemi editore, Rome (2012)
8. Carlini, A., Millán Gasca, A., Tedeschini Lalli, L.: La matematica in mostra. In: Poce, A. (ed.) Memory, inclusion and cultural heritage first results from the Roma Tre inclusive memory project, pp. 205–223. Edizioni Scientifiche Italiane, Naples
9. Drugman, F.: "Il museo diffuso". In: Hinterland n. pp. 21–22 (1982)
10. MIUR: http://www.alternanza.miur.gov.it/_RMPS060005.html. Last accessed 10 Aug 2019
11. CNR, Officina 2018: https://www.cnr.it/it/eventi/allegato/11038. Last accessed 08 Apr 2018
12. Bertolini, M., Cazzola, M., Dedò, M., Di Sieno, S., Frigerio, E., Luminati, D., Poldi, G., Rampichini, M., Tamanini, I., Todesco, G.M., Turrini, C.: mateMilano. Percorsi matematici in città. Springer, Milan (2004)
13. YouTube: Museo diffuso dei pavimenti cosmateschi. https://www.youtube.com/watch?v=6vEc018Hl-M&t=29s. Last accessed 20 Aug 2018
14. Munari, B.: I laboratori tattili. Corraini, Mantua (2014)
15. YouTube, Coniche S.T.E.A.M.: https://www.youtube.com/watch?v=wIyHp2uHLI8. Last accessed 20 Aug 2018
16. Giusti, E., Conti, F.: Oltre il compasso. Diagonale, Palermo (1999)
17. Farroni L., Magrone P.: A multidisciplinary approach to teaching mathematics and architectural representation: historical drawing machines. In: Proceedings of History and Pedagogy of Mathematics, pp. 641–651. Montpellier (2016)
18. YouTube: "ART-TOUCH-LAB". https://www.youtube.com/watch?v=yuqBtTEzWqc. Last accessed 20 Aug 2018

Officina Degli Errori: An Extended Experiment to Bring Constructionist Approaches to Public Schools in Bologna

Sara Ricciardi, Stefano Rini, and Fabrizio Villa

Abstract In this contribution we describe an extended experiment to bring constructionist approaches to public schools in Bologna. Specifically, we focus on our latest project called *Officina degli Errori*, which is an extended teacher training program for primary school teachers based on tinkering. We highlight our motivation, the structure of *Officina degli Errori* and the lessons learned co-designing the activities and implementing them in the reality of public schools in Bologna. We also interviewed teachers to understand the critical issues affecting implementation of constructionist approaches in public primary schools.

Keywords Constructionism · Tinkering · Public primary school · Teacher training

1 Introduction

The basic goal of *Officina degli Errori* is to provide teacher training and strong support to constructionist practices in the classroom through tinkering. Through this project, we are also interested in exploring and understanding what changes are needed in the school's organization to enable proper implementation of these practices. In Sect. 2, we describe the origins of the idea, its associated values, and the close collaboration between the researchers (science) and teachers (education) involved in co-designing the activities for schools. Section 3 describes the activity carried out from September 2018 to June 2019, lasting the entire school year. Conclusions and future prospects are discussed in Sect. 4.

S. Ricciardi (✉) · F. Villa
INAF-OAS, via Gobetti 101, 40129 Bologna, Italy
e-mail: sara.ricciardi@inaf.it

S. Rini
IC12 Bologna, Via Bartolini 2, 40139 Bologna, Italy

D. Scaradozzi et al. (eds.), *Makers at School, Educational Robotics and Innovative Learning Environments*, Lecture Notes in Networks and Systems 240,
https://doi.org/10.1007/978-3-030-77040-2_22

2 Values, Aims and First Round of Co-design

The authors launched *Officina degli Errori* after more than five years' involvement in science education and outreach activities. We begin our description of our approach by briefly mentioning how we came to be involved in constructionist practices. The starting point was a typical lecture given at "Marella" primary school within Istituto Comprensivo 12 (IC12) in Bologna, which had been requested by S. Rini, a teacher at the school at the time. The lecture was on cosmology and was delivered by S. Ricciardi. As a researcher at INAF (National Institute of Astrophysics) in Bologna, S. Ricciardi was actively involved in analyzing data and publishing papers on the European Space Agency (ESA) Planck satellite [1]. She prepared a lesson for seven-year-olds on the CMB (Cosmic Microwave Background), the first light in our universe, and about the Planck satellite and her personal involvement in the research. The lesson went very well, the children were amazed, and everybody was satisfied. It is important to say that this was the expected outcome, given the known appeal of the subjects, which include dark matter and dark energy, and the origin and fate of the universe. As F. Villa, the third author of this paper and fellow INAF researcher, had had similar experiences, we discussed whether the students truly learned something from such lectures. We were particularly interested in the "wow factor" of experimental cosmology and, more generally, astronomy, and whether the experience was truly positive for everyone in the science education classroom.

We discussed it with teachers and concluded that children who were already interested in STEM subjects (science, technology, engineering and mathematics) were certainly inspired. We noticed that this was not the case for all children. We believe self-stereotypes (socio-economic conditions, gender issues, family culture) are already strong in elementary schoolchildren [2], and, hence, that some children might not feel they are smart enough to be involved in STEM.

This reflection made us rethink our work in the classroom and inspired us to find a better way to express our values, including the basic concepts of trial and error in science research, and the sharing of knowledge and skills.

We identified what, in our view, are the top characteristics that should be met by all educational activities in the sphere of astrophysics and STEM in general:

(i) **Democratic**: activities should be designed to be truly inclusive.
(ii) **True/real/honest**: we have to tell students the truth about our research work, including the failed attempts and errors, especially when we do frontier research. We also have to let them know that scientists cannot know everything; they might be unable to explain natural phenomena; research is about constructing knowledge.
(iii) **Meaningful**: activities should be relevant to students, so they can relate to them.
(iv) **Empowering**: the learning process must be designed so that students feel they belong to the STEM environment, and feel empowered by the process itself.

Frontier research is based on new ideas, and exploring new techniques and technologies. Obviously, there is no set road map to follow to get to the final destination, since most things are under development, and researchers use their expertise to build and develop things from scratch. In more than 15 years' experience developing the ESA Planck satellite and more than 20 years developing new instrumentation and codes for astrophysics, we have come to see that creativity is one of the main drivers, especially at the start of a research project.

Although there is no doubt that (frontier) research requires creativity, children generally have a different perception of research (and researchers). Hence, another characteristic should be added to the list:

(xxii) **Creative**: activities should be designed to stimulate creativity and invention.

We had an opportunity to study and improve our understanding, thanks to the online material and MOOC (massive online open courses) offered by the Tinkering Studio at the San Francisco Exploratorium [3–5]. We also drew inspiration from the work of Mitch Resnick and the Lifelong Kindergarten group, and their vision of a STEM pedagogy that aims to be inclusive (low floor), democratic and, at the same time, allows projects to evolve (high ceiling), allowing children to build knowledge that is personal and meaningful to them (wide walls) [6, 7].

Tinkering is a holistic way to engage people in STEM subjects, by mixing them with art and combining hi-tech materials with low-tech, recycled ones. Knowledge is not merely transmitted from teacher to learner, but is actively constructed in the mind (and the hands) of the learner. Constructionism [8] suggests that learners are more likely to develop new insights and understandings while they are actively engaged in making an external artifact. This method supports the construction of knowledge within the context of building personally meaningful artifacts, and the more self-directed the work is, the more meaningful the learning becomes.

From 2014, we offered several workshops to students in our local community based on the activities originally developed by the Tinkering Studio. We have been collaborating with teachers to design, promote and deliver hands-on, self-directed and playful activities to engage children in STEM, focusing particularly on gender inclusion [2].

By 2017 our labs had matured and were ready to be taken to a larger arena. Hence, in October to December 2017, we took an informal program of four tinkering activities to Bologna's Museum of Industrial Heritage and named it *Officina degli Errori*. We worked with a group of 20 children aged 6–12, over four workshops held on Saturday afternoons in the museum's conference area [9–12]. In the 2018/2019 school year, we offered a training course at the museum for primary school teachers in Emilia Romagna, to help them to become more autonomous tinkerers in their classrooms, and also to develop new insights into how these practices can work in the ecosystem of Italian public schools. These activities were attended by around 16 teachers and around 400 students. We also hosted a shorter version of this training course at Istituto Comprensivo 12 in Bologna, involving 20 teachers and their students (about 500).

3 *Officina Degli Errori*: Tinkering Goes to School

Officina degli Errori was an extended teacher training program lasting from September 2018 to June 2019. We organized this experience in three blocks:

Experiencing: We organized three intensive sessions in early September, where we developed the building blocks. We presented the constructionist framework in pedagogical terms, drawing parallels with the way the scientific research community works, and we discussed the ideas of constructivist epistemology, following the path of Piaget [13] and Khun [14].

We hosted three hands-on workshops where teachers experienced the constructionist approach through tinkering. We had a group of 15 teachers and three facilitators and we proposed several classic workshops originally developed by the Tinkering Studio, which we tailored to our audience (e.g., scribbling machines, paper automata, marble machines);

Engaging: we provided the teachers with a kit containing 9 motors, 20 cables, and 20 battery holders so they could try the scribbling machine with their classes, and hopefully continue using the materials in other creative ways. We invited the teachers to come to the museum with their students and fellow teachers to participate in a workshop, conducted by experienced facilitators. We also invited another teacher, who had already participated in the class workshop, to be a facilitator. This gave us a good student/facilitator ratio of about 6 to 1, and we created a relaxed environment where teachers could facilitate a challenging workshop, perhaps for the first time, without feeling overwhelmed. We also provided a space where teachers could just observe their students and reflect, without the urge to intervene;

Reflecting: we gave teachers more than four months to try different things in their classrooms, providing them with feedback and assistance. We also asked them authentic questions to help them and ourselves reflect. We asked them to describe their feelings when they first tried tinkering as "students." Specifically, we discussed group composition, how they prepared the groups and how things went, focusing particularly on the participation of girls. At the end, we asked them to think about whether tinkering could be organized in their schools, in terms of space, time, and human resources. We also asked them to report their activities to the other teachers in a final session in late June. The final step from our point of view was a recent interview with some of the teachers.

Our goal was mainly to bring tinkering and constructionist approaches to public schools in Bologna, especially in areas where there is a higher risk of student dropout or difficulties integrating hands-on STEM workshops in formal school activities. This means supporting teachers far beyond the workshops, and helping them with all they need to get started. For this reason, our program was offered free of charge and hosted at the Museum of Industrial Heritage in Bologna. We gave teachers a kit they could use in their classrooms, and the opportunity to facilitate a couple of tinkering sessions in relaxed conditions. We also gave them some guidance on collecting materials and designing new settings. The main differences with earlier training courses we had organized were the time scale (very extended), the ability to offer materials free of charge, and the commitment to each teacher we trained to facilitate at least one workshop with their students.

4 Conclusions and Future Prospects

From a preliminary analysis of the reports and interviews with teachers, it is clear that this experience was very successful in terms of the teachers' engagement, the quality of the materials and the support provided. Constructionist practices are difficult to implement in schools, and are contingent on teachers making a special effort. One common issue for schools is the availability of physical space. In the schools that have an *atelier creativo* (or a similar environment) or access to one, teachers do tinkering more often (every day/three times a week), and the practice is easier to incorporate into their everyday learning environments. When such settings are not available, sessions are less frequent and more intensive (e.g., one full week of tinkering twice a year). Adapting a classroom for tinkering can be demanding even for a teacher who already uses cooperative learning and innovative teaching styles (e.g., *scuola senza zaino*), simply because there is not enough space to store materials and tools. Teachers have to re-configure spaces every time they use them. Another important issue is the number of facilitators available. We suggested working as a pair with the other class teacher (all the classes have a *tempo pieno* schedule—full-time, including afternoons—and two main teachers, or are module-based with at least three teachers) but this was not always possible. It is extremely hard to facilitate a workshop in these conditions, with only one teacher to about 28 students, especially if there is no dedicated environment. Conversely, workshop facilitation was more relaxed and productive in schools with a better student to teacher ratio or in schools where co-teaching could be arranged. Despite the teachers' motivation and their readiness to overcome difficulties, they were often deterred by the general constraints of how their schools were organized.

On the positive side, some teachers reported that they intend to include constructionist approaches in their teaching practices, not limiting them to STEM, but also applying them to other subjects (such as language). The focus of training for teachers in STEM and educational technology is often on the devices and the literacy associated with the proposed activities. Instead, our extended training offered a more general "pedagogy," which, hopefully, every teacher can tweak to their interests and their class's needs, building genuinely new knowledge not restricted to a specific discipline.

After this experience, we need to continue analyzing the interviews with teachers and highlight what can really help to establish these practices in public schools, without overloading teachers. We intend to continue designing and offering *Officina degli Errori*, working with any partner interested in spreading the constructionist approach and tinkering in Italian public schools. In today's knowledge society [15], the ability to develop 21st-century skills (e.g., [16]) is a crucial question of democracy.

For us teachers and researchers employed by the Ministry of Education, University and Research, it is a firm moral commitment to help schools innovate and offer the best practices in education.

During these years of practice, we developed a particular interest in tinkering as an effective way to engage girls in STEM. To help schools understand the value of these practices, we have launched a preparatory program with the Department of Psychology at the University of Bologna on students' perception of science and technology, with a particular focus on gender differences (190 school students). INAF and IC12 have also co-designed a STEAM (science, technology, engineering, art and mathematics) learning space, which includes tinkering with technology (coding and robotics), storytelling and interacting in pairs, with areas dedicated to experimenting, showing and presenting, and sharing. We were also interested in opening up those approaches to secondary schools, so in the 2018/2019 school year we collaborated with IC12 in the afternoon program "Girls code it better." Eighteen 18 girls aged 11 to 13 took part in the ESA challenge "MOON BASE CAMP," which saw them work with INAF experts, a maker from Bologna FabLab and the teacher who coordinated the overall project on 3D modeling and printing.

We want to offer *Officina degli Errori* to teachers every year, but to be effective we have to engage closely with them and involve them in designing the activities. From our interviews we found that external conditions (mainly the way schools are organized and the availability of space) can be a huge deterrent to these practices. This is why, in addition to tinkering tools and strategies, we also need to provide an "organizational framework" to help them bring those activities to Italian public schools.

References

1. Planck Satellite: https://sci.esa.int/web/planck
2. Bian, L., Leslie, S.J., Cimpian, A.: Science **355**(6323), 389–391 (2017)
3. Wilkinson, K., Petrich, M.: The Art of Tinkering, Weldon Owen (2014)
4. http://tinkering.exploratorium.edu/
5. Petrich, M., Wilkinson, K., Bevan, B.: It looks like fun, but are they learning? In: Design, Make, Play: Growing the Next Generation of STEM Innovators. Routledge (2013)
6. Resnick, M.: Lifelong Kindergarten: Cultivating Creativity through Projects. Peers, and Play. MIT Press, Passions (2017)
7. Presicce, C.: Thesis master of science in media arts and science (2017)
8. Papert, S.: Mindstorms: Children, Computers, and Powerful Ideas. Basic Books (1980)
9. Ricciardi, S., Villa, F., Rini, S.: Tinkering with the Universe: a primary school project. In: Proceeding of the Communicating Astronomy with the Public Fukuoka, Japan (2018)
10. Ricciardi, S., Villa, F., Rini, S.: Officina degli Errori: a Tinkering Experience in an Informal Environment. In: Proceeding of International Conference New Perspective in Science Education Firenze (2018)
11. Ricciardi, S., Villa, F., Rini, S.: "Il tinkering va al museo" Scuola Officina numero 2 (2018)
12. Ricciardi, S., Villa, F., Rini, S.: Tinkering la coraggiosa arte di sbagliare incontra un museo civico. In: Città come Cultura, Fondazione MAXXI (2019). ISBN 978-88-942824-6-7
13. Piaget, J.: Genetic Epistemology (E. Duckworth, Trans.). Columbia University Press, NY (1970)

14. Kuhn, T.S.: The Structure of Scientific Revolutions. University of Chicago Press, Chicago (1962)
15. Lisbon's strategy: European Council (2000)
16. OECD Report: New Vision for Education: Unlocking the Potential of Technology (2015)

Service Learning: A Proposal for the Maker Approach

Irene Frazzarin and Danila Leonori

Abstract Education plays a vital role in promoting the acquisition of 21st-century skills. This is why it is important to implement a social intervention project in which makers and technology are key to the learning process, and students can come together to deal with issues within their community.

Keywords Service learning · Learning by doing · Project-based learning · Maker approach · Coding · Digital storytelling

Education plays a vital role in promoting the acquisition of 21st-century skills. However the "pedagogical methods many educational programmes use today primarily serve a standardised, industrial, content-delivery-and-reproduction model, with little or no room for learners to acquire or demonstrate skills beyond content knowledge. This has meant that such systems have increasingly diverged from the needs of those being educated, today's society, and the current global economic system" [1]. This is the reason why a school's physical position in the community is crucial. The reality is that there is often no connection between it and the classroom, the very place where knowledge is taught: access is now far more widespread than ever before, thanks to the extensive availability of technological resources. We teachers and the children in our schools are witnessing a significant rise in social inequalities, while values such as acceptance, solidarity, and civil cooperation are on the decline.

In light of this, we have taken this opportunity to implement a social intervention project to enable students to come together to deal with issues within their community. We decided that makers and technology are key to the learning process in this project.

We consider service learning to be the ideal way to breathe new life into school. The aim is to view it as a key establishment within its community, where local

I. Frazzarin (✉)
1 IC Zanella, Padua, Italy

D. Leonori
IIS Mattei, Recanati, Italy

D. Scaradozzi et al. (eds.), *Makers at School, Educational Robotics and Innovative Learning Environments*, Lecture Notes in Networks and Systems 240,
https://doi.org/10.1007/978-3-030-77040-2_23

citizens and associations can return to the social dimension of learning, improve their relational skills and foster a sense of community.

Just as important as disciplinary skills are 21st-century skills, such as creativity critical thinking, collaboration, cooperation, and communication, which can all be strengthened. As Fiorin argues "[…] Strongly rooted in the experience of both the students and the social context, service learning enhances the value of the students' protagonism, and employs the best active and socio-constructive methodologies"[1] [2].

Added to this is the impact of the maker approach. As Donaldson argues "learning happens best when learners construct their understanding through a process of constructing things to share with others" [3].

1 Service Learning, Coding and Digital Storytelling: A Methodological Proposal

Service learning began as a teaching and learning strategy that connects academic curricula to community problem-solving. It was originally intended for university students, but today's participants are usually elementary, middle, high, and post-secondary schools, with the backing of the Italian Ministry of Education, University and Research and the "Avanguardie educative—INDIRE" movement [4].

It is an approach that teaches students how to think rather than what to think [5, 6].

Service learning projects are problem-based. Students or participants learn and develop through active participation, which eventually leads to civic engagement, thereby meeting needs within a community. It is also part of a school's curriculum, enhancing and building upon the students' knowledge and key skills.

Project-based learning is a comprehensive perspective whose goal is to engage students in investigation. Within this framework, students seek solutions to nontrivial problems by "asking and refining questions, debating ideas, making predictions, designing plans and/or experiments, collecting and analyzing data, drawing conclusions, communicating their ideas and findings to others, asking new questions, and creating artifacts" [7].

2 The Maker Movement Approach and Coding

Our project was introduced at Istituto Mattei, a secondary school in Recanati, at the beginning of 2019, and will probably continue for the next two years.

During this first year, 100 students ranging from 16 to 18 years will take part in the project. Our students are active participants, in keeping with the maker approach, and they will develop team working skills that will enable them to come up with

[1] Own translation.

creative solutions to real-life problems affecting the entire school initially, and then the entire community.

They will create an app, drawing on a range of traditional construction materials, as is usual in maker pedagogy, but also embracing the unique properties of digital tools [8].

The project is in two phases:

- Phase 1: the "Welcome" app prototype
- Phase 2: the "Welcome" app.

2.1 *Phase 1: "Welcome" App Prototype*

The prototype will be developed by first-year students to deal with the main problem in their school community: finding your way around the building. The prototype can include functions specific to the school, which has a maze-like structure that is easy to get lost in, and it can be hard to find the right classroom. It also has many labs; and it would be useful for "school consumers" or new students to have information about the school's amenities, places and rules. The year 11 students will use free tools like Unity 3d and C# coding language, whereas older students will be introduced to Android coding.

The experiment focuses on the Think, Make, Improve (TMI) cycle, an adaptive decision-making process. This method helps teachers to design their lesson structure and students to understand the process and organize their work. Students will apply the principles of adapting, designing and creating [9], to get into a maker mindset, and increase their learning activities and learning outcomes.

By "adapting" we mean the freedom to use a technology for new purposes. For example, the Unity 3d software is normally used for video games, whereas in our project, students create a virtual school with it to help newcomers find their way around the school.

As students designing and creating software, we believe they will apply their skills in cooperation with others, and come up with different solutions, which they can also combine.

Throughout the process, students will be exposed to the "learning by doing" method, coding, and project- based and problem-based learning.

2.2 *Phase 2: The "Welcome" App*

In the second year, the same students will manage the Welcome app for the local community, and will coach new students during its development. They will be expected to give their new colleagues useful advice about the coding process and also about the learning path they themselves have already been on.

Students are divided into heterogeneous groups based on the needs of the community and will create a geolocation-based app that operates as a virtual tour guide to help Italian and foreign members of the community. Functions will include identifying available resources in their area, historical events, nearby institutions and the customs of foreign citizens.

Users have access to a lot of information, thanks to AR (augmented reality) technology. By scanning a retailer's door, they might see a video interview pop up or the rules of a public place, or information about future events.

The app will need to be multilingual (right from the start screen), to encourage maximum inclusion. Italian will always be the language spoken in videos, but there will be subtitles in several other languages.

The students will interview foreign and Italian retailers in their neighborhoods, as well as local elderly residents, and will record their stories about migration and life in general. The video clips from these interviews will be geo-localized and shared through a child-friendly app called Storymap. The aim of sharing these videos on the neighborhood's social media pages and the school website is to increase local awareness and gather useful feedback.

Maintaining an open communication channel with the community is fundamental to understanding and bringing all those involved closer together on a deeper personal level. According to Bruner, in "the analysis of the stories we tell about our lives: our "autobiographies [...] In the end, we *become* the autobiographical narratives by which we 'tell about' our lives" [10].

Thus, by working on the prototype first and then on the Welcome app itself, students will take part in peer-to-peer teaching and will be able to deal with small failures by identifying and correcting errors.

3 Objectives

The learning objectives we have set will be adapted for the different classes involved in the project. These are: service learning objectives and curricular objectives.

3.1 Service Learning Objectives for Students

These objectives will be differentiated depending on what grade the students are in. Overall we can summarize them as follows:

- A better understanding of the circumstances of immigrants and refugees;
- Seeing the point of view of foreigners, and also that of Italian residents; finding similarities between the migration stories told within the communities;
- Devoting "free" time to serving the community;

- Helping immigrants and refugees learn the Italian language and culture for everyday and more immediate uses.

3.2 Curricular Objectives and Key Competences

The following objectives were planned for a secondary school, although they could be used at and adapted to different school levels: history, economics and law: studying the history of migration flows towards Europe and Italy:

- Italian language: exploring linguistic and grammatical aspects of their own language
- Mathematics and science: reprocessing the data collected from Institutional websites
- ICT: learning coding languages
- Foreign languages: using a foreign language as a medium for translation and communication
- Creating digital artifacts (transferable skills).
- Citizenship skills

3.3 Expected Results

Service learning enhances 21st-century skills, and the project's results are expected to have an impact on the students' cognitive and emotional development.

- Learning skills through the maker approach;
- Critical thinking/problem-solving: to create their digital artifact, students will have to resolve an ongoing problem in their community;
- Creativity: demonstrating and expressing themselves creatively (thinking outside the box) by making a video, an app and suitable graphics to go into it;
- Communication: improving communication skills by meeting with, listening to and interviewing their school peers;
- Collaboration: this is an important life skill which can be cultivated by learning to collaborate with others and reaching an objective as a group;
- Literacy skills;
- Information literacy: students will collect and understand data, facts and statistics. Before any information can be used, students will be asked to check the data, facts and stories they have collected;

Technology literacy: thanks to their thorough understanding of today's technology, students will be capable of choosing the best (virtual or physical) tools to complete the task in hand.

4 Conclusion

As Canuto says, projects in service learning offer multiple opportunities to receive positive input, to practice a foreign language through authentic and meaningful interactions, to reflect and become aware of one's own identity and culture, and to appreciate the linguistic and cultural differences of others [11]. In short, it develops the intercultural abilities required in the eight key competences of the European Framework of Reference.

The main feature of service learning is working on social solidarity projects/actions that are not disconnected from school learning; rather, they are fully integrated into the curriculum, and thus help improve the well-being of the class.

Many service-learning–type projects have been tested in Italy since the setup of the SL (service learning) Working Group, approved by the Head of Education of the Ministry of Education, University and Research, which was formalized in November 2016, with the participation of three Regions (Calabria, Lombardy, Tuscany). The most recent ScAR (*Scuola Attiva Risorse*) project was based on the synergy between the Polytechnic University of Milan, schools in the Milan area, the Municipality of Milan and other public and private institutions [12].

The National Digital School Plan [13] also emphasizes the importance of advanced heritage education and suggests that all students should be offered an opportunity to experience the digital management of cultural heritage. Moreover, European policies [14] acknowledge the right of every citizen to access and participate freely in cultural life.

Given the experiences mentioned above, we hope service learning will also be tested in our communities to enable further research, experimentation and dissemination in this area.

References

1. Charles, L., Rankin, W., Speight, C.: Education, knowledge, and learning—an overview of research and theory about constructionism and making. PI-TOP **8** (2019)
2. Fiorin, I.: Oltre l'aula. La pedagogia del Service Learning, Mondadori (2016)
3. Donaldson, J.: The maker movement and the rebirth of constructionism. Hybrid Pedagogy. http://www.hybridpedagogy.com/journal/constructionism-reborn/
4. INDIRE: Avanguardie educative. http://innovazione.indire.it/avanguardieeducative/service-learning
5. Harris, J.D: Service-Learning and the Liberal Arts: How and Why it Works, pp. 21–39. Lexington Books (2011); Miettinen (2000)
6. Miettinen, R.: The concept of experiential learning and John Dewey's theory of reflective thought and action. Int. J. Lifelong Educ. 54–72 (2000)
7. Blumenfeld, P., Soloway, E., Marx, R., Krajcik, J., Guzdial, M., Palincsar, A.: Motivating project based learning: sustaining the doing, supporting the learning. Educ. Psychol. **26**, 369–398 (1991)
8. International Conference, the Future of Education 7th Edition, Florence, Italy Conference Proceeding by Pixel

9. Bullock, S.M., Sator, A.J.: Maker pedagogy and science teacher education. J. Can. Assoc. Curriculum Stud. **13**, 60–87 (2015)
10. Bruner, J: Life as narrative. Soc Res **71**(3) 691–710 (2004). https://ewasteschools.pbworks.com/f/Bruner_J_LifeAsNarrative.pdf
11. Canuto, L: Service learning: dai fondamenti teorici ai benefici per lo studente di lingua straniera. Univ. Br. Columbia **5**(2) (2016)
12. http://www.scar.polimi.it/wp-content/uploads/2018/11/Progetto-ScAR-sintesi.pdf
13. https://www.miur.gov.it/innovazione-digitale
14. The Faro Convention or the Council of Europe Framework Convention on the Value of Cultural Heritage for Society (2005)

14. [illegible]

Learning by Making. 3D Printing Guidelines for Teachers

Stefano Di Tore, Giuseppe De Simone, and Michele Domenico Todino

Abstract For many years now, and particularly since the 1930s, educational research has focused on the idea that all authentic education comes from experience. Nowadays, activism has found a natural affinity with the maker movement. Fab labs and creative ateliers have become more popular, especially for educational purposes, suggesting the coming of new types of "learning by doing." However, these new forms of "learning by doing" must take account of the technologies already present in a particular creative space used by makers. These technologies are mainly: 3D printers, CNC milling machines, 3D scanners, laser cutters, etc. This short paper begins with a premise of educational ergonomics, to introduce teachers, media educators and *animatori digitali* (digital coordinators) to the didactic implications of introducing different human–machine interfaces (HMI) into their practices. In particular we describe the main features of SLA and SLS 3D printing. The impacts we discuss of 3D printing are resolution, types of printing materials, average printing times, post-processing, and cost. We have selected these criteria because it has been documented that their impact is very heavy in certain school subjects. For example, an FDM 3D printer can be useful in terms of the ease of printing an object, but it may not reach the necessary level of detail for a meticulous reproduction of art objects or precision mechanisms that an SLA 3D printer can achieve.

Keywords 3D printer · Slicer · Didactics · Fab lab

1 Introduction

In recent years, educational research has seen that the technology used to design and support processes in development and engineering can also stimulate teaching–learning in general terms, but particularly in certain subjects. Moreover, the spread of digital manufacturing spaces and the increasing popularity of the maker movement [1] appear to have opened up new theoretical and practical possibilities for

S. Di Tore (✉) · G. De Simone · M. D. Todino
University of Salerno, Salerno, Italy
e-mail: sditore@unisa.it

D. Scaradozzi et al. (eds.), *Makers at School, Educational Robotics and Innovative Learning Environments*, Lecture Notes in Networks and Systems 240,
https://doi.org/10.1007/978-3-030-77040-2_24

pedagogical activism [2–5]. However, the educational potential of digital manufacturing spaces is contingent on a proper understanding of all the distinctive features of the technologies that are found in these spaces. This short paper aims to introduce teachers and, more specifically, media educators and *animatori digitali* (digital coordinators) to digital fabrication, through a focus on educational ergonomics [6]. We look at the didactic implications of employing the different forms of HMI [7] that can be found in 3D printing technology. This work analyzes two widely used 3D printing technologies, namely fused deposition modeling (FDM) and stereo lithography apparatus (SLA), and suggests what the implications of their educational use may be in terms of: resolution, possible printing materials, average printing times, post-processing, and cost. The criteria in this list were selected on the basis of documented experience [7]; they significantly affect applications in different subjects. 3D objects can be designed in CAD software of varying complexity, such as Autodesk Fusion 360 (a free version is available for students and educators), Rhinoceros 3D, SugarCAD, and Tinkercad. Fusion and Rhinoceros are designed for professional use, whereas SugarCAD and Tinkercad are more suited to educational use because they have easy-to-use graphic interfaces. After the 3D model is created, it must be imported (usually in.stl file format) into a slicing program, which will "slice" it into a series of thin layers or 2D levels, in preparation for 3D printing. Slicing programs also allow users to set basic parameters for printing, such as, the type of filling, the thickness of each level, the number and thickness of the external "walls," extruder temperature, etc. The most popular software programs for slicing are: Ultimaker Cura, Slic3r and Simplify3D (the first two are open source). After slicing the model, these programs create a G-code file, the de facto standard format, which can be used with any kind of 3D printer.

2 Fused Deposition Modeling (FDM) 3D Printers

Fused deposition modeling printers use thermoplastic material to print 3D objects, which they achieve by pushing melted plastic through an extruder and depositing it on a printing plate (see Fig. 1). There are different types of thermoplastic materials (generally filaments) on the global market, which differ in terms of their properties (i.e., resistance, elasticity, magnetic and aesthetic properties). FDM technology is probably the most widespread and inexpensive, on account of the printing materials it uses and the cost of the machine itself. This kind of printer is useful for creating 3D objects that need cantilever parts or support structures (scaffolds) during printing (see Fig. 1). However, the layer resolution is not very high, and printed objects can appear rough and unrefined (see Fig. 2). Support structures are substrates of material that are printed to enable the extruder to print curves and protrusions. This prevents molten material from falling onto the printer base. More specifically, support structures have predefined forms (determined by the software), or can be customized by the maker. Some FDM models have a double extruder and can print support structures from water-soluble materials. In this case, once the support structures have been

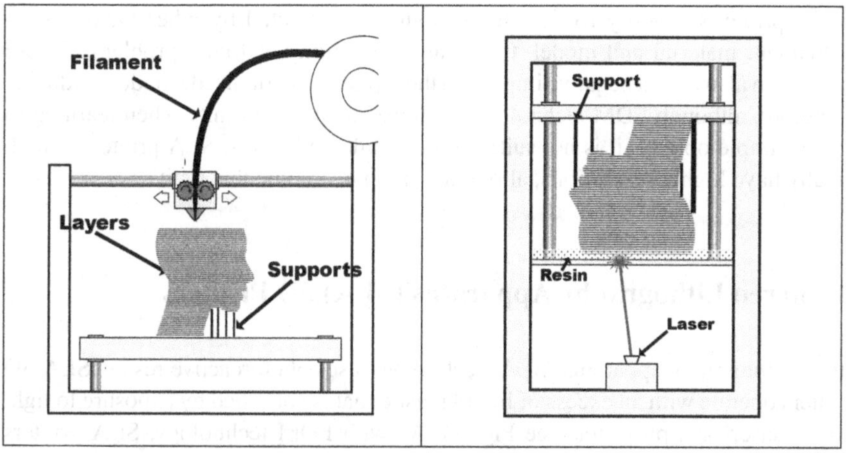

Fig. 1 FDM (right) and SLA (left) 3D printing technologies

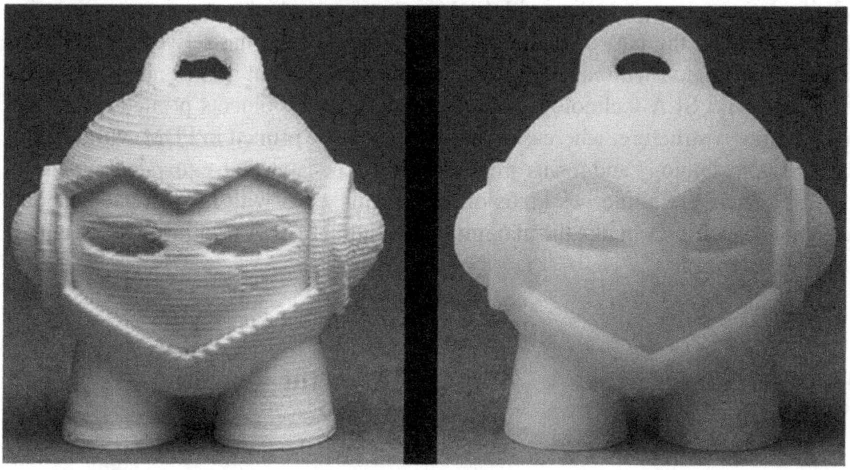

Fig. 2 The same object printed with FDM (left) and SLA (right). *Source* https://www.3dhubs.com/knowledge-base/key-design-considerations-3d-printing

printed, they can be dissolved in water, leaving no residues on the final model. It could be inferred from the above, that, given the many restrictions on printable shapes, the design process is not completely free with FDM technology. A double extruder completely eliminates the issue of residues and inaccuracies caused by support structures, but resolution and the properties of materials are still something to take into account (e.g., when printing a building model to scale, some elements, like pipes, cavities and other details, are too small for FDM 3D printer resolution, and have to be redesigned). On average, the maximum resolution of a non-industrial FDM printer in terms of layers is about 0.2 mm (not considering small movements on the printer's x and y axes). This resolution is affected by other factors, such as filament material and model, temperature, humidity, extruder problems. These factors can also have a negative impact on the replicability of the 3D model produced. Therefore, although FDM technology is an excellent opportunity when learning to design simple objects, it is not suitable for detailed elements. SLA printers, which usually have higher resolutions, allow makers to overcome this limit.

3 Stereo Lithography Apparatus (SLA) 3D Printers

Stereo lithography apparatus (SLA) technology uses photoreactive resins. SLA 3D printers operate with an excess of liquid plastic that is solidified by exposure to light from a laser or a projector (see Fig. 1). As with FDM technology, SLA printers build 2D layers to create 3D objects designed in CAD software, but the physical process is completely different. SLA printers have a laser that solidifies each layer, one after another, starting from the top of the 3D object. This means there are more issues relating to supports than with FDM technology, which require more expertise and attention on the part of the maker during the design phase. Even cantilevered parts can be printed without support structures in FDM, but almost all shapes need a scaffold with SLA technology. Indeed, almost all 3D objects printed with SLA need a support structure, whereas many objects can be printed in FDM without one. Also, SLA technology and resins are often more expensive than equivalent materials in FDM. There are numerous kinds of SLA resins, each with its own physical and aesthetic properties. Finally, the maximum resolution in SLA printing is (on average) 0.025 mm.

4 FDM Versus SLA: A Comparison for the Teaching Setting

Summarizing and extending our previous considerations, SLA technology offers makers more opportunities because it produces higher 3D resolution; however, more time has to be spent on the design of models. Another consideration is that SLA does not have some of the issues that FDM has: molten material deposited on the printer base; quantity of extruded material; filament pulling, etc. For these reasons SLA printer technology has a resolution of 0.025 mm instead of the 0.2 mm found in FDM technology. Consequently, SLA technology can print complex objects (such as dentures for dental technology courses). However, increasing the resolution of a 3D object means increasing post-production time.

Although FDM printing sometimes requires post-production processes to eliminate inaccuracies and vertical lines caused by the thickness of the layers or by non–water-soluble supports, the post-production and post-processing times for SLA printers tend to be longer on average. Models produced by SLA printers have to be "cured" in two stages. In the first stage, the 3D object is exposed to ultraviolet rays to finish the solidification process, and in the second, it is sanded to eliminate any marks where the supports were held. To summarize, makers should consider the final purpose of their 3D object when deciding which 3D printer technology to use. FDM printers with dual extruders provide greater design freedom and produce reasonable results when detail is not important (e.g., toys, medium-sized gears, boxes and containers, etc.). This makes FDM technology ideal for educational settings, where it can be used to introduce the basic concepts of CAD design, and for printing scientific or mechanical objects that do not exceed the maximum resolution parameters. SLA printers, on the other hand, do not provide maximum design freedom, because they rely on support structures that are sometimes difficult to remove. For this reason, SLA is not recommended for teaching CAD design, and SLA printers are recommended for 3D objects where detail is important (e.g., reproductions of monuments, jewelry for art institutes, and, as mentioned, dentures for dental technology courses, etc.).

5 Conclusion

We finish this short paper by mentioning some of the many repositories containing files of 3D printable objects that are available on the internet for free download. These include Thingverse, MyMiniFactory, Pinshape, Cults, GrabCad, CGTrader etc., and have entire sections dedicated to educational purposes. These repositories often contain information to help makers on: (1) which 3D printing technology is most suitable for the model; (2) what settings to use during printing. These online 3D model repositories are a good starting point for learning about and appreciating 3D printing, and for getting a better understanding of SLA and FDM technologies, their similarities, differences, opportunities and trends. Lastly, we would like to clarify that we have focused on FDM and SLA technologies in this paper because they are both suitable for mainly non-industrial uses. Other technologies, such as SLS, DLP, SLM, and MSLA, are too expensive for educational purposes.

References

1. INDIRE: Maker@scuola Stampanti 3D nella scuola dell'infanzia (2017). Link: http://www.indire.it/wp-content/uploads/2017/09/Libro-Maker-a-Scuola_2017.pdf. Last accessed 09 Dec 2019
2. Dewey, J.: Esperienza e Educazione. Cortina, Milan (2014)
3. Perla L.: L'agire didattico nelle teorie e nei modelli del Novecento. In: Rivoltella, P.C., Rossi, P.G. (eds.) L'agire didattico. La Scuola, Brescia (2012)
4. Falcinelli F.: Le tecnologie dell'educazione. In: Rivoltella, P.C., Rossi, P.G. (eds.) L'agire didattico. La Scuola, Brescia (2012)
5. Rivoltella, P.C.: Comunicazione e relazioni didattiche. In: Rivoltella, P.C., Rossi, P.G. (eds.) L'agire didattico. La Scuola, Brescia (2012)

6. Bonaiuti, G., Calvani, A., Menichetti, L., Vivanet, G.: Le Tecnologie Educative. Caroccio, Rome (2017)
7. Sibilio, M: La didattica semplessa. Liguori, Naples (2014)

Roboticsness—*Gymnasium Mentis*

Paola Lisimberti and Domenico Aprile

Abstract *Roboticsness* is an innovative teaching/learning approach based on the European Digital Agenda. The Lego robotics classroom is a living lab where students and teachers work together and share the enthusiasm of finding solutions thanks to new technologies. Project work and prototypes are produced, but the main objective is to train new prosumers to anticipate and prepare for the needs of a complex world.

Keywords Arduino · Coding · Creativity · Coworking · Problem-solving · Competences

1 The Project: LEIS Classroom

1.1 Goals

The Lego Education Innovation Studio (LEIS) robotics classroom was founded in 2014 at the Pepe Scientific High School in Ostuni (Puglia, Italy), where the Roboticsness [1] project (based on the Lego curriculum) has been conducted for five years. Here, a new philosophy of education is practiced: students are taught to think with machines. *Roboticsness* is an innovative teaching/learning approach based on the European Digital Agenda and civil law rules on robotics. The different manifestations of artificial intelligence have initiated a new industrial revolution and their diffusion has prompted the European Parliament to express itself on the matter through the Resolution of 16 February 2017 "Civil law rules on robotics". Rules to regulate the development of these technologies without conditioning the innovation process are necessary. The European document draws attention to the shortage of ICT professionals in education and in the workforce (no. 41) [2]. The growth in robotics requires Member States to develop more flexible education and training

P. Lisimberti (✉)
Liceo Pepe Calamo, Ostuni, Brindisi, Italy

D. Aprile
Liceo Scientifico "Fermi-Monticelli", Brindisi, Italy

D. Scaradozzi et al. (eds.), *Makers at School, Educational Robotics and Innovative Learning Environments*, Lecture Notes in Networks and Systems 240,
https://doi.org/10.1007/978-3-030-77040-2_25

systems, so as to ensure that the knowledge strategies for women match the needs of the robot economy (no. 42) [2]. Our goal is to move past the traditional concept of school as a system of linear transmission of knowledge from teacher to learner. Our school is based on learning by doing and on prototype construction.

Revolutions are made by people. Schools are made up of people: teachers and students.

How much distance is there between digital natives and teachers? Can schools take this route? Can we reach them, meet them?

1.2 Teaching Methods and Strategies

The main point of this experience is to learn in cooperative teams, where each member brings her individual skills to a project to build a new idea and a sense of community. Active learning, peer collaboration, personalized learning, learning by doing, and learning by mistake are the teaching practices of the robotics classroom. Teachers implement a process-driven approach, based on IBL (inquiry-based learning) and PBL (problem-based learning). This means starting with an analysis of (one or more) real problems (e.g., data log of acceleration in uniformly accelerated motion) and working backwards to an ideal model, in order to appreciate how it differs from reality. This approach leads students to analyze the whole problem, drawing on skills in mathematics, physics, engineering and technology.

1.3 Cooperative Learning and Cooperative Teaching

The teacher acts as a coach to support learning. No one is excluded during robotics activities: everyone works in groups, learning to interact with others. Building and programming a robot is the best way to promote digital culture, information technology, critical thinking, creativity [3], curiosity, initiative, persistence, and the value of error (Fig. 1).

2 Experiences

2.1 Curricular Robotics for First-Year Students (Aged 14–15, Science-Based High School)

The kinematics teaching unit (six hours) is led by two co-teachers in the robotics classroom and the physics laboratory. Students work in groups: each group builds a

Fig. 1 Human–machine interaction inspired by the Creation of Adam (Michelangelo Buonarroti)

robot (Lego Ev3) and programs it for an experiment in uniform linear motion. The activity consists of several steps:

- How to write a physics report
- What a Lego robot kit contains
- Let's make a robot
- How to program an Ev3
- Let's experiment: is uniform linear motion really uniform? Let's write a report.

2.2 STEM

The Lego environment provides an innovative learning and teaching approach to STEM (Science Technology Engineering Mathematic) subjects. The number of female students enrolled in the project has increased in five years. Girls find it hard to take the initiative; gender-based teams form and exclude girls; boys are reluctant to accept girls in a context where there are machines. Girls are autonomous in programming; their attitudes and qualities are different from those of male students; they are resilient; they know how to solve problems; they recognize strengths; they disassemble and reassemble robots; they program and reprogram machines; they reach objectives [4].

2.3 Participation in Exhibitions and Fairs

The project involves developing robotic applications and/or using robots for learning and for creating activities in which they are used. This has enabled students to take an active part in some of the most important exhibitions and fairs in Italy and Europe. Groups of students have attended:

- Maker Faire Rome, Europe's largest event on innovation (2015 and 2017)

- RomeCup, the event organized and managed by Fondazione Mondo Digitale, which offers an extraordinary immersive experience on the present and future (2014, 2015, 2017).

3 Results and Conclusions

During the 2014/2019 five-year period, the project showed significant results in terms of frequency, educational outcomes and overcoming the gender gap.

The indicators for identifying outcomes were:

- Improvement in school performance;
- Improvement in study method;
- Strengthening STEM skills in female students.

Throughout the project, the average attendance of enrolled students was around 98%.

The percentage of female students enrolled in the project grew exponentially (by up to 75%), fully in line with PNSD (National Plan for Digital Education) action #20 "*Girls in tech and science*" (reducing the gender gap) (Figs. 2 and 3).

Furthermore, the impact was positive on the subject performances of both male and female students. Performance in science subjects (particularly physics) improved for 70% of students compared to results in the first four months and the previous year.

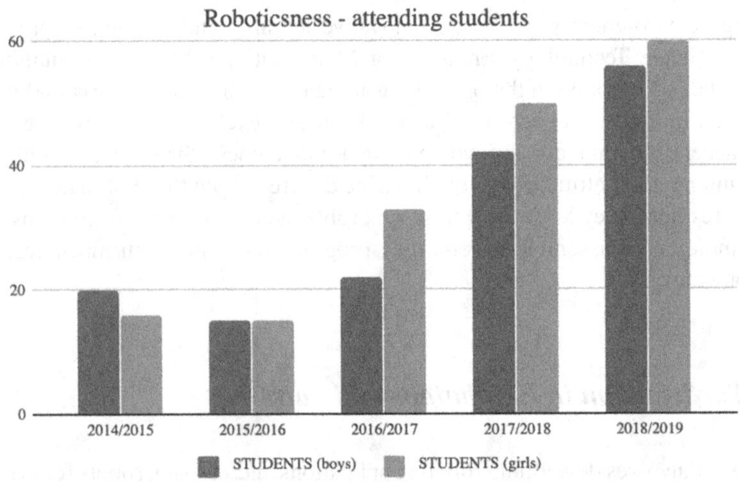

Fig. 2 Number of students during the 5-year project

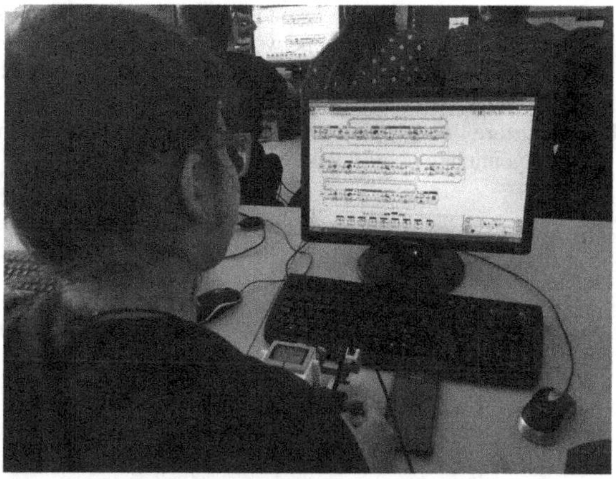

Fig. 3 Female students—programming step

The aim of the project is not to replace "obsolete" content and knowledge with something that is "fun and modern," but rather to make the transmission of "knowledge" more effective and to ensure that school can be closer to the everyday reality of digital natives, who are surrounded by captivating digital tools and robotics.

In keeping with the goals of the project and of several international ICT frameworks, the activities develop the following competences:

- Digital literacy (Recommendation 2006/962/EC, updated on 22 May 2018)
- Knowing how to deal with reality as compared with a theoretical reference model
- Knowing how to design strategies for solving problems through programming (DigComp 2.1, programming and problem-solving)
- Knowing how to make decisions consistent with goals and strategies
- Knowing how to handle errors
- Knowing how to process results and apply the necessary corrective measures
- Knowing how to develop simple prototype robot applications
- Knowing how to organize, analyze, summarize, infer, abstract and process
- Correlate, contextualize and generalize learned knowledge.

References

1. Roboticsness Homepage. http://www.liceopepe.wixsite.com/roboticsness
2. The European Parliament's "Civil law rules on robotics" document (European Parliament resolution of 16 February 2017 with recommendations to the Commission concerning civil law rules on robotics 2015/2103 INL). is available at the link URL: https://www.europarl.europa.eu/doceo/document/TA-8-2017-0051_IT.pdf. In particular "General principles concerning the development of robotics and artificial intelligence for civil use civil use", "Education and Work" n. 41 and n. 42
3. Gardner, H.: F ive Minds for the Future. Boston: Harvard Business School Press, tr. it. di E. Dornetti (2007), *Cinque chiavi per il futuro*. Milano: Giangiacomo Feltrinelli Editore, p.164:

Creative intelligence goes beyond existing knowledge and syntheses to pose new questions, offer new solutions, and shape works that push the boundaries of genres or shape new ones. Creativity is based on one or more recognised disciplines and requires judgements of quality or acceptability to be made by an informed environment (2006)

4. Campustore.: #womeninstem education edition, Paola Lisimberti, Girls do it better. (2017) https://youtu.be/SpkimcCWXGU

Curricular and Not Curricular Robotics in Formal, Non-formal and Informal Education

Educational Robotics and Social Relationships in the Classroom

Laura Screpanti, Lorenzo Cesaretti, Michele Storti, and David Scaradozzi

Abstract In a constructionist environment, robotics engagingly teaches traditional concepts, while applying them to compelling real-world problems. Educational robotics can help students develop soft skills, like teamwork, and improve the way they relate to each other. Researchers in different disciplines have devoted many efforts to exploring this dimension. One tool that may be useful for exploring the relational dimension of these activities is the sociogram. The case study reported in this paper presents findings from an experience which brought educational robotics, coding and tinkering to fourth graders in a primary school in Ancona (Italy). A questionnaire and a sociogram were administered to students, during curricular activities, before and after the project took place. The findings highlight some improvements in students' relations, but more investigation is needed into the process of describing students' relationships and their development in a project involving innovative methodologies and technology.

Keywords Educational robotics · Sociogram · Social network analysis · STEM · Constructionism · Coding · Tinkering

L. Screpanti (✉) · L. Cesaretti · D. Scaradozzi
Università Politecnica Delle Marche, Ancona, Italy
e-mail: l.screpanti@pm.univpm.it

L. Cesaretti
e-mail: l.cesaretti@pm.univpm.it

D. Scaradozzi
e-mail: d.scaradozzi@univpm.it

L. Cesaretti · M. Storti
TALENT Srl, Osimo, Italy

D. Scaradozzi
LSIS, CNRS, UMR 7296, Marseille, France

D. Scaradozzi et al. (eds.), *Makers at School, Educational Robotics and Innovative Learning Environments*, Lecture Notes in Networks and Systems 240,
https://doi.org/10.1007/978-3-030-77040-2_26

1 Introduction

Students' ability to learn is inextricably linked to the classroom environment. Developing and mastering new skills requires students to feel safe and supported. Educational robotics (ER) brings new tools and methodologies into the classroom for acquiring technological, social and multidisciplinary skills and competences [1–12]. During ER activities, students usually work in small teams, which encourages them to collaborate. This helps students who are isolated to improve their social relationships. Sociometric tests developed by Moreno [13–15] can help teachers examine the structure and interactions of a group, and get valuable data about social relationships in the classroom. Sociometric tests can help focus the teacher's awareness of students who may not feel connected and need extra attention, and this helps create a supportive learning environment.

Sociometry has been applied to educational research in many contexts. Exploring this approach within ER activities is rather uncommon, but one example can be found [10].

The aim of the present work is to explore the suitability of this approach for understanding classroom interactions.

2 Materials and Methods

2.1 *Participants and Procedure*

In early 2018, two classes at a primary school (ISCED 1, grade 4) in Italy were involved in a project in which educational robotics was brought into the classroom. An external educator carried out the activities and received support from the classes' regular teachers. Consent to participate was provided for 26 students in group A, and 22 students in group B. Students were involved in activities related to environmental issues, such as separate waste collection and forest conservation. First, they explored the environmental issues; then, they wrote stories about them and brainstormed to choose elements and ideas to create a tale invented by the whole classroom. After this preliminary phase, students assembled and programmed robots over four lessons (two hours per lesson) using Scratch and Lego WeDo 1. They were divided into groups of three or four and introduced to the subjects of robotics (mechanical structure of a robot, difference between robots and machines, sensors and motors) and programming (sequential instructions only). At the end of the short course, they were able to think about dramatizing the story on environmental issues through robots. The third phase of the project was dedicated to creating a stage background for the story through tinkering and using waste materials. The last phase of the project involved sharing the artifacts, the story and their work with parents and other friends. During a school festival the two groups showcased their experience by presenting their stories, short videos and pictures from the workshops.

2.2 Methodology

Students were asked to complete a questionnaire and a sociometric test at the start (BL) and end (PT) of the project. The questionnaire was administered on paper and the items were in Italian: three items asked about the relationship with the instructor and the methodologies used, two questions were about assembling and programming the robot, two were on their interest in this kind of activity in the future, and two were about having fun or cooperating with their companions.

The sociometric test consisted of four questions:

- Write the names and surnames of those classmates you would like as study companions for a particular school subject. You can write as many names as you like.
- Write the names and surnames of those classmates you would not want as study companions for a particular school subject. You can write as many names as you like.
- Write the names and surnames of those classmates you would like as playmates. You can write as many names as you like.
- Write the names and surnames of those classmates you would not want as playmates. You can write as many names as you like.

The sociometric test is a method used to get a description of interpersonal relations within a group and to stress the social status of each member. It mainly focuses on the affective-relational perspective (play) and on the perspective of a group working towards a common goal (study). The resulting sociogram represents members of the group as nodes and their relationships as edges. If there is no relationship between two nodes, no edge is shown on the graph. To examine the data from the sociometric test, a matrix (adjacency matrix) is built by replacing its elements with choices (value 1), rejections (value -1) or no choice or rejection (value 0).

Both the questionnaire and the sociometric test were developed by an expert educator and psychologist.

To check whether the structure of the social network changes over time, we analyzed BL and PT sociometric data and focused on the density, mean indegree and mean outdegree of the network. The density of the network is the ratio of the number of links to the number of possible links; it represents the connectedness of the network. Density is connected to distance, and the distance can represent the student's role in the network. Mean indegree and mean outdegree are measures of degree centrality that are appropriate for directed networks like the ones we obtained from the sociometric test. The sociometric test evaluated the students in two areas: play and study. For each area, students were able to jot down as many names as they wanted as "choices" (people they wanted to play or study with) or as "rejections" (people they did not want to play or study with). Paired t-tests of mean indegree and mean outdegree at BL and PT were performed in order to test the significance of the differences between the two paired samples. Change from BL to PT was assessed by means of the Wilcoxon signed rank test.

Table 1 Mean values of answers to questionnaire items

	A		B	
	BL	PT	BL	PT
Q1	2.14	2.44	2.55	2.5
Q2	2.04*	2.52*	2.18	2.33
Q3	2.90	2.92	2.95	2.95
Q4	3	2.8	3	2.81
Q5	2.71*	2.32*	2.59	2.59
Q6	2.33*	2.8*	2.18*	2.91*

*Paired data showing a statistically significant difference ($\alpha = 0.05$)

3 Results

Mean values and results from the Wilcoxon signed rank test for two paired groups of data for both group A and B are shown in Table 1. Specifically, the results from question 4 (fun with classmates) and 5 (good collaboration with my classmates) are shown in Fig. 1. The results from processing sociometric data are reported in Table 2.

A significant change in interest towards the robotics workshop was found (question 6) in both groups involved. No significant results were found for questions 4 and 5, but we can qualitatively observe that scores tend to be lower for PT. This negative trend seems to be corroborated by the sociometric analysis.

The results in Table 2 show that the project somehow increased each student's choices in the "play" area. No significant results were found in either group for the "study" area. The density of the network was not tested for significance, but it increased from BL to PT in both groups for both choices and rejections.

4 Conclusion and Future Work

Overall, the project recorded a high level of satisfaction in the groups involved (Table 1). Notably, the students' interest increased significantly (question 6). This result suggests that short workshop activities are effective in increasing interest in STEM subjects. Answers to questions 4 and 5 showed a decreasing or stable trend between BL and PT.

Conversely, the sociometric data in the area of play showed a significant increase in the mean number of choices. Moreover, although not significant, the same trend can also be observed in the areas of study and rejections (see Table 2). This is an exciting result that needs more investigation to uncover the underlying variables (for example, personal or group-related variables).

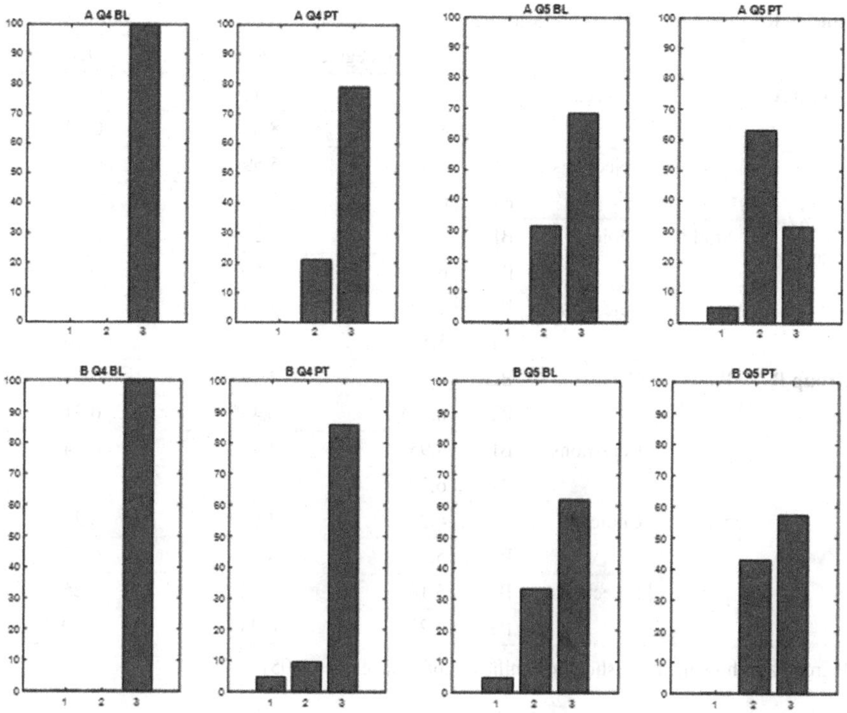

Fig. 1 Results from question 4 (Q4) for groups A and B before the activities started (BL) and after the activities ended (PT)

The overall picture of these two classes was that, in group A, there were more popular students (those who received more choices than others) and more rejected students (those receiving more rejections than others) than in group B. The density of the network was not tested for significance, but it showed an interesting trend: it increased from BL to PT, showing that more relations grew out of the collaborative environment of the ER activities. This result is not unforeseen, as some of the relevant literature advocates ER's capacity to stimulate social skills [1, 3–12].

Nonetheless, the present research, and other research like it [10], shows that more questions need to be answered on methodological research grounds and on the study of social relations within a study group. Some students, in fact, did not improve their status after the ER activities, as in [10]. This issue should be examined in depth, perhaps by including a number of background variables in the study, like gender, race or socio-economic status. Furthermore, variables like the duration of the activities or the way groups are formed for particular activities may affect the creation of social relations and thus deserve more attention in future studies.

Table 2 Results of the analysis of sociometric data

				Mean indegree	Mean outdegree	Density
Group A	Play	Choices	BL	7.45*	7.45	0.35
			PT	8.82*	8.82	0.42
		Rejections	BL	5.59	5.59	0.27
			PT	6	6	0.29
	Study	Choices	BL	6.27	6.27	0.30
			PT	6.50	6.50	0.31
		Rejections	BL	5.27	5.27	0.25
			PT	5.55	5.55	0.26
Group B	Play	Choices	BL	4.59*	4.59*	0.21
			PT	6.45*	6.45*	0.31
		Rejections	BL	4.95*	4.95*	0.24
			PT	6.5*	6.5*	0.31
	Study	Choices	BL	4.82	4.82	0.23
			PT	5.32	5.32	0.25
		Rejections	BL	5.18	5.18	0.25
			PT	6.23	6.23	0.30

*Paired data showing a statistically significant difference ($\alpha = 0.05$)

References

1. Benitti, F.B.V.: Exploring the educational potential of robotics in schools: a systematic review. Comput. Educ. **58**(3), 978–988 (2012)
2. Mubin, O., Stevens, C.J., Shahid, S., Al Mahmud, A., Dong, J.J.: A review of the applicability of robots in education. J. Technol. Educ. Learn **1**(209–0015), 13 (2013)
3. Cesaretti, L., Storti, M., Mazzieri, E., Screpanti, L., Paesani, A., Principi, P., Scaradozzi, D.: An innovative approach to school-work turnover programme with educational robotics. Mondo Digitale **16**(72), 2017–2025 (2017)
4. Kandlhofer, M., Steinbauer, G.: Evaluating the impact of educational robotics on pupils' technical-and social-skills and science related attitudes. Robot. Auton. Syst. **75**, 679–685 (2016)
5. Scaradozzi, D., Screpanti, L., Cesaretti, L.: Towards a definition of educational robotics: a classification of tools, experiences and assessments. In: Smart learning with educational robotics, pp. 63–92. Springer, Cham (2019)
6. Scaradozzi, D., Cesaretti, L., Screpanti, L., Costa, D., Zingaretti, S., Valzano, M.: Innovative tools for teaching marine robotics, IoT and control strategies since the primary school. In: Smart learning with educational robotics, pp. 199–227. Springer, Cham (2019)
7. Screpanti, L., Cesaretti, L., Marchetti, L., Baione, A., Natalucci, I. N., Scaradozzi, D.: An educational robotics activity to promote gender equality in STEM education. In: International Conference on Information, Communication Technologies in Education (ICICTE 2018) Proceedings, Chania, Greece
8. Scaradozzi, D., Screpanti, L., Cesaretti, L., Mazzieri, E., Storti, M., Brandoni, M., Longhi, A.: Rethink Loreto: we build our smart city!" A stem education experience for introducing smart city concept with the educational robotics. In: 9th Annual International Conference of Education, Research and Innovation (ICERI 2016), Seville, Spain (2016)

9. Scaradozzi, D., Screpanti, L., Cesaretti, L., Storti, M., Mazzieri, E.: Implementation and assessment methodologies of teachers' training courses for STEM activities. Technol. Knowl. Learn. **24**(2), 247–268 (2019)
10. Truglio, F., Marocco, D., Miglino, O., Ponticorvo, M., Rubinacci, F.: Educational robotics to support social relations at school. In: International Conference on Robotics and Education RiE 2017, pp. 168–174. Springer, Cham (2018, April)
11. Scaradozzi, D., Pachla, P., Screpanti, L., Costa, D., Berzano, M., Valzano, M.: Innovative robotic tools for teaching STREM at the early stage of education. In Proceedings of the 10th annual International Technology, Education and Development Conference, INTED (2016)
12. Scaradozzi, D., Sorbi, L., Pedale, A., Valzano, M., Vergine, C.: Teaching robotics at the primary school: an innovative approach. Procedia Soc. Behav. Sci. **174**, 3838–3846 (2015)
13. Moreno, J.L.: Sociogram and sociomatrix. Sociometry **9**, 348–349 (1946)
14. Moreno, J.L. (ed.): Sociometry Experimental Method and the Science of Society. Beacon House, New York (1951)
15. Moreno, J.L. (ed.): Sociometry and the Science of Man. Beacon House, New York (1956)

Analysis of Educational Robotics Activities Using a Machine Learning Approach

Lorenzo Cesaretti, Laura Screpanti, David Scaradozzi, and Eleni Mangina

Abstract This paper presents the preliminary results of using machine learning techniques to analyze educational robotics activities. An experiment was conducted with 197 secondary school students in Italy: the authors updated Lego Mindstorms EV3 programming blocks to record log files with coding sequences students had designed in teams. The activities were part of a preliminary robotics exercise. We used four machine learning techniques—logistic regression, support-vector machine (SVM), K-nearest neighbors and random forests—to predict the students' performance, comparing a supervised approach (using twelve indicators extracted from the log files as input for the algorithms) and a mixed approach (applying a k-means algorithm to calculate the machine learning features). The results showed that the mixed approach with SVM outperformed the other techniques, and that three predominant learning styles emerged from the data mining analysis.

Keywords Educational Robotics · Educational data mining · Problem-solving process identification

L. Cesaretti (✉) · D. Scaradozzi
Department of Information Engineering (DII), Università Politecnica Delle Marche, Via Brecce Bianche, 60131 Ancona, Italy
e-mail: l.cesaretti@pm.univpm.it

L. Screpanti · D. Scaradozzi
LSIS - Umr CNRS 6168, Laboratoire Des Sciences de L'Information Et Des Systèmes, Equipe I&M (ESIL), case 925 - 163, avenue de Luminy, 13288 Marseille cedex 9, France

L. Cesaretti
TALENT Srl, via Bachelet 23, 60027 Osimo, Ancona, Italy

E. Mangina
School of Computer Science, University College Dublin, Belfield Dublin 4, Ireland

D. Scaradozzi et al. (eds.), *Makers at School, Educational Robotics and Innovative Learning Environments*, Lecture Notes in Networks and Systems 240,
https://doi.org/10.1007/978-3-030-77040-2_27

1 Introduction

Educational Robotics (ER) is based on the constructionist learning theory proposed by Papert (1980). According to this theory, the construction of a personal and meaningful artifact promotes the construction of deep knowledge and provides a richer learning experience for students. When students design, build, program, debug and share robots, they are exploring the everyday phenomena in their lives in a new and playful way (Resnick et al. 1996). They are also improving their problem-solving and thinking skills (Scaradozzi et al. 2015, 2019b).

The spread of ER projects in schools all over the world (Miller and Nourbakhsh 2016) and the growing interest of teachers and policy-makers in this approach justify the research into new and more far-reaching ways to evaluate students' results. Researchers in the ER field have found a lack of quantitative analysis on how robotics can improve skills and increase the learning achievements of students (Benitti 2012; Alimisis 2013). For this reason, studies in recent years have focused on the evaluation of ER projects, employing qualitative (Liu 2010; Elkin et al. 2014), quantitative (Kandlhofer and Steinbauer 2016; Cesaretti et al. 2017; Scaradozzi et al. 2019a) and mixed methods (Kim et al. 2015; Chalmers 2018) to collect educational data. During a typical ER activity, students have to design a hardware or software solution to an open-ended problem and, generally, there are unlimited paths students can explore and implement. Using machine learning methods to collect and analyze ER data could help predict and classify students' behaviors, and discover latent structural patterns in large educational datasets (Berland et al. 2014).

Most studies in recent years (Berland et al. 2013; Blikstein et al. 2014; Chao 2016; Wang et al. 2017) have applied machine learning techniques to data gathered from students during programming activities without the physical presence of robots, and have obtained good results in identifying different patterns in specific coding tasks. This paper presents a research project that aims to analyze an introductory ER exercise using educational data mining techniques. The authors designed a tracking system to register log files containing the programming sequences created by (teams of) students in several secondary school classes. In this preliminary study, the collected data were used as inputs for machine learning algorithms, with two main objectives: 1—to identify different patterns in the students' problem-solving trajectories; 2—to accurately predict the final performance of the teams of students.

2 Methods

2.1 Procedure and Participants

Our experiment takes the existing research further by using educational robots. After upgrading the Lego Mindstorms EV3 programming blocks—using the Lego Mindstorms Developer kit (2019)—the authors were able to register several log files containing the programming sequences created by 197 Italian secondary school students (organized into 56 teams), while solving an introductory Robotics exercise. Using the data collected as input for machine learning algorithms, the authors tried to identify different patterns in the students' problem-solving trajectories and predict the final performance of their teams. The experiment was conducted from March 2018 to March 2019, and involved seven Italian lower/higher secondary schools (located in the Emilia Romagna and Marche regions). The authors used convenience sampling to involve students in the experiment. Data were collected from the robotics courses organized in the schools by the startup TALENT (a partner in the research project). Two of the schools decided to offer a curricular robotics class, five decided to offer a non-curricular robotics class. 18.78% of the participants had already used robotic kits at home or at school before the experiment.

2.2 *The Introductory Exercise*

One of the first steps in the "Introduction to Robotics" course, whose aim was to get students involved in the experiment, was an exercise with the Lego Mindstorms EV3 motors. They had to program the robot to cover a given distance (1 m), while trying to be as precise as possible. The educator suggested a specific block to turn on the robot's motors. Using the "Move Steering" block, students can select one of the four different options shown in Fig. 1 (On for Rotations, On for Degrees, On for Seconds, On). By modifying the numeric parameter highlighted in Fig. 1, the students' team can specify how many rotations the wheels have to complete. Following the same strategy, students can decide whether to define how many degrees or for how many seconds the motors should rotate. The first two numeric parameters are for steering the robot and the speed of the motor.

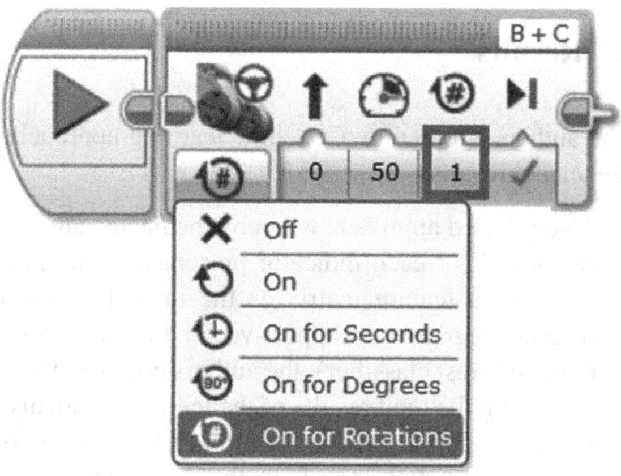

Fig. 1 Lego Mindstorms EV3 "Move Steering" block

To solve this exercise, the teams had to design their solution in a maximum amount of time (15 min for upper secondary school classes, 20 min for lower secondary school classes), using no measuring tools (set squares, rulers, etc.) to check the distance covered by the robot on the floor. The measurements only applied to some of the robot's parameters (e.g., the wheel radius). The instructor gave a final evaluation for this exercise based on how accurate the team was. If the error was $\leq$4 cm, the educator considered the **challenge completed**; if the error was $\geq$4 cm the educator considered the **challenge not completed.**

2.3 Data Preparation

The tracking system collected 1113 programming sequences created by the teams of students to solve the robotics exercise described in the previous section. Several technical steps had to be performed to transform these sequences into matrices. After this transformation, the following nine indicators were calculated for each programming sequence (using a function in Python that parses the log file): **Motors** (how many motor blocks are in the sequence); **Others** (how many blocks in the sequence belong to categories other than Motors); **Added** (how many blocks have been added, compared to the previous sequence); **Deleted** (how many blocks have been deleted, compared to the previous sequence); **Changed** (how many blocks have been changed, compared to the previous sequence); **Equal** (how many blocks have stayed the same, compared to the previous sequence); **Delta Motors** (the degree of change in Motor block parameters, compared to the previous sequence, and calculated only for blocks in the "Changed" category); **Delta Others** (the degree of change in Other block parameters, compared to the previous sequence); **Trials** (how many times students activated the programming sequence on the robot).

3 Results

The authors adopted two machine learning approaches, with the aim of predicting the teams' results:

- A supervised approach, whereby the mean value and the standard deviation were calculated for each indicator presented in the previous section, and were used to create a feature matrix as the input for four machine learning algorithms (logistic regression, support-vector machine (SVM), k-nearest neighbors, and random forest classifier); the authors compared how these algorithms performed, predicting the final results of the teams of students;
- A mixed approach, which combined the benefits of both the supervised and the unsupervised methods: a k-means algorithm was applied to calculate what clusters might form from the programming sequences. Then the log files were analyzed

to calculate the percentage of sequences belonging to each cluster: these new features were then used to create a feature matrix as input for the four supervised algorithms mentioned previously.

A repeated tenfold cross validation was performed (Kim 2009) to evaluate the effectiveness of the different strategies at predicting the educational results, and four parameters were calculated (Huang and Ling 2005): Accuracy, Mean Precision, Mean Recall, Mean F1—Score. The best performance was obtained with the SVM algorithms combined with the features calculated by the k-means algorithm (the mixed approach). This combination led to the following predictions of the students' performance at resolving the exercise (challenge completed or not completed): Accuracy = 0.89 (SD = 0.002), Mean Precision 0.84 (SD = 0.007), Mean Recall = 0.84 (SD = 0.011), Mean F1 Score 0.83 (SD = 0.011).

Three problem-solving strategies emerged from the dataset collected. The "**Mathematical strategy**" groups performed a low number of total tests (compared to the other strategies), all of them with a rotation parameter equal to a fixed value (see Fig. 2 in Appendix, which shows the trend of the rotation parameter set by a group of students taken as an example for this approach). They probably used a mathematical formula to choose the robot's parameters, and never changed the value in the Motor block. The "**Incremental Tinkering approach**" groups performed a medium number of tests (compared to the other strategies); they probably used a heuristic approach to choose the robot's parameters, refined them repeatedly (see Fig. 2 in Appendix) and analyzed the feedback from the robot (for a definition of tinkering see Resnick and Rosenbaum (2013)). The "**Irregular Tinkering approach**" groups performed a high number of tests (compared to the other approaches), and their strategy is characterized by a broken line representing the number of seconds set in the Move Steering block (see Fig. 3 in Appendix), and a broken line representing the Delta Motors value calculated between two contiguous tests (which not only included changes to the number of seconds, but also modifications to the speed and steering parameters, (see Fig. 4 in Appendix). This last group probably also used a heuristic approach to choose the robot's parameters, but a certain number of large modifications could indicate difficulty understanding the feedback from the robot. These three behaviors were the most common in the dataset analyzed.

4 Conclusions

This experiment presented a number of innovative features. To the best of the authors' knowledge, this is the first experiment involving a fair number of students (197) to gather and look at programming sequences designed by students during ER activities, and analyze them using machine learning techniques. Good results were obtained predicting the students' performance, especially in terms of Accuracy (0.89). The three styles presented in the Results section are close to that of the "planner scientist" (similar to the "Mathematical strategy") and the "bricoleur scientist" (similar to the "Incremental Tinkering approach" and to the "Irregular Tinkering approach") proposed by Turkle and Papert (1992). The promising results of this preliminary study have encouraged the authors to involve new groups of students in the experiment, and also to collect data for different types of challenges, in order to validate the approach.

Appendix

See Figs. 2, 3 and 4.

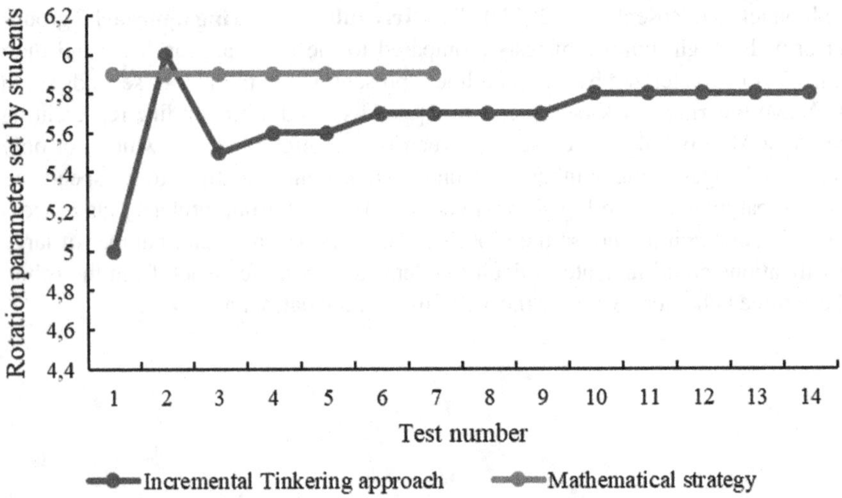

Fig. 2 Graph of rotations for two groups taken as examples of the "Incremental Tinkering" and "Mathematical" approaches

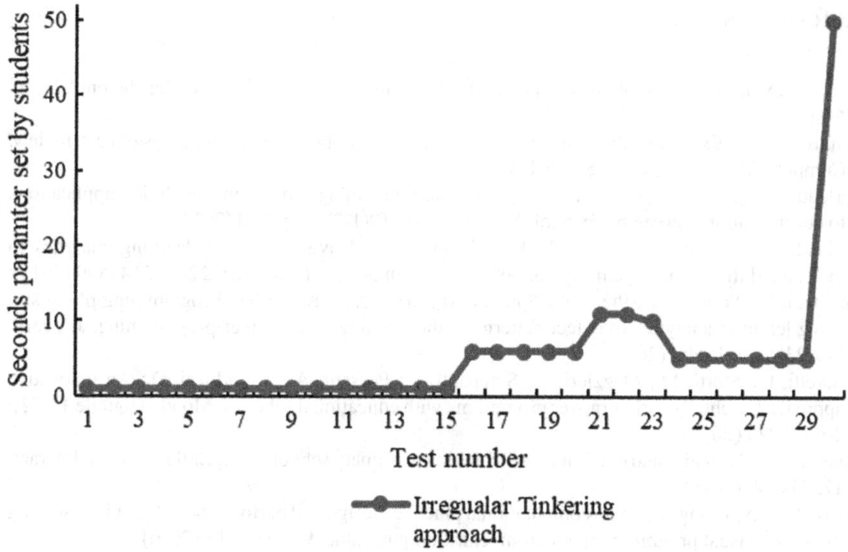

Fig. 3 Graph of seconds for the group taken as an example of the "Irregular Tinkering" approach

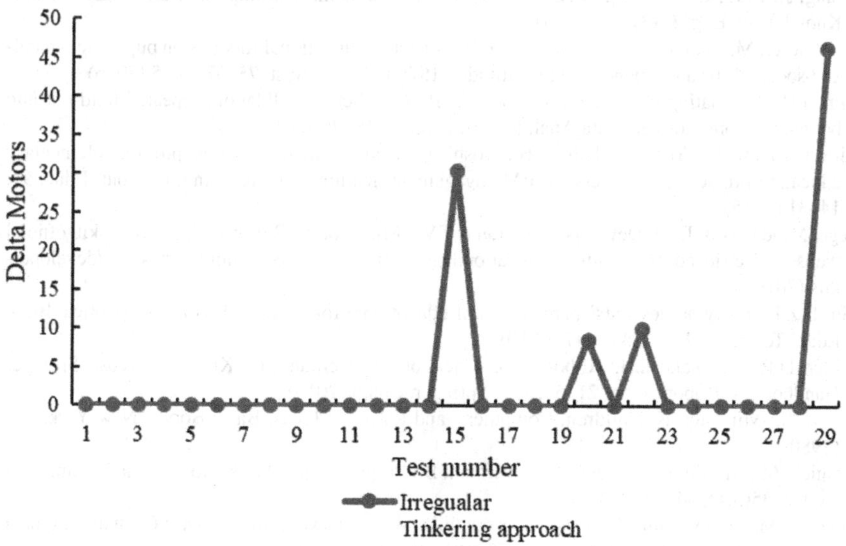

Fig. 4 Graph of Delta Motors for the group with an "Irregular Tinkering" approach

References

Alimisis, D.: Educational robotics: open questions and new challenges. Themes Sci. Technol. Educ. **6**(1), 63–71 (2013)

Benitti, F.B.V.: Exploring the educational potential of robotics in schools: a systematic review. Comput. Educ. **58**(3), 978–988 (2012)

Berland, M., Baker, R.S., Blikstein, P.: Educational data mining and learning analytics: applications to constructionist research. Technol. Knowl. Learn. **19**(1-2), 205–220 (2014)

Berland, M., Martin, T., Benton, T., Petrick Smith, C., Davis, D.: Using learning analytics to understand the learning pathways of novice programmers. J. Learn. Sci. **22**(4), 564–599 (2013)

Blikstein, P., Worsley, M., Piech, C., Sahami, M., Cooper, S., Koller, D.: Programming pluralism: using learning analytics to detect patterns in the learning of computer programming. J. Learn. Sci. **23**(4), 561–599 (2014)

Cesaretti, L., Storti, M., Mazzieri, E., Screpanti, L., Paesani, A., Scaradozzi, D.: An innovative approach to school-work turnover programme with educational robotics. Mondo Digitale **16**(72), 2017–2025 (2017)

Chalmers, C.: Robotics and computational thinking in primary school. Int. J. Child-Comput. Interact. **17**, 93–100 (2018)

Chao, P.Y.: Exploring students' computational practice, design and performance of problem-solving through a visual programming environment. Comput. Educ. **95**, 202–215 (2016)

Elkin, M., Sullivan, A., Bers, M.U.: Implementing a robotics curriculum in an early childhood Montessori classroom. J. Inf. Technol. Educ. Innovations Pract. **13**, 153–169 (2014)

Huang, J., Ling, C.X.: Using AUC and accuracy in evaluating learning algorithms. IEEE Trans. Knowl. Data Eng. **17**(3), 299–310 (2005)

Kandlhofer, M., Steinbauer, G.: Evaluating the impact of educational robotics on pupils' technical-and social-skills and science related attitudes. Robot. Auton. Syst. **75**, 679–685 (2016)

Kim, J.H.: Estimating classification error rate: Repeated cross-validation, repeated hold-out and bootstrap. Comput. Stat. Data Anal. **53**(11), 3735–3745 (2009)

Kim, C., Kim, D., Yuan, J., Hill, R.B., Doshi, P., Thai, C.N.: Robotics to promote elementary education pre-service teachers' STEM engagement, learning, and teaching. Comput. Educ. **91**, 14–31 (2015)

Lego Mindstorms EV3 Developer kit. Lego EV3 Mindstorms Software Developer kit official website. Retrieved from https://education.lego.com/en-au/support/mindstorms-ev3/developer-kits (2019)

Liu, E.Z.F.: Early adolescents' perceptions of educational robots and learning of robotics. Br. J. Educ. Technol. **41**(3), E44–E47 (2010)

Miller, D.P., Nourbakhsh, I.: Robotics for education. In: Siciliano, B., Khatib, O. (eds.) Springer Handbook of Robotics, pp. 2115–2134. Springer, Cham (2016)

Papert, S.: Mindstorms: Children, Computers, and Powerful Ideas. Basic Books, New York, NY (1980)

Resnick, M., Martin, F., Sargent, R., Silverman, B.: Programmable bricks: toys to think with. IBM Syst. J. 35(3.4), 443–452 (1996)

Resnick, M., Rosenbaum, E.: Designing for tinkerability. Design, make, play: Growing the next generation of STEM innovators, pp. 163–181 (2013)

Scaradozzi, D., Screpanti, L., Cesaretti, L., Storti, M., Mazzieri, E.: Implementation and assessment methodologies of teachers' training courses for STEM activities. Technol. Knowl. Learn. **24**(2), 247–268 (2019)

Scaradozzi, D., Screpanti, L., Cesaretti, L.: Towards a definition of educational robotics: a classification of tools, experiences and assessments. In: Smart Learning with Educational Robotics, pp. 63–92. Springer, Cham (2019b)

Scaradozzi, D., Sorbi, L., Pedale, A., Valzano, M., Vergine, C.: Teaching robotics at the primary school: an innovative approach. Procedia Soc. Behav. Sci. **174**, 3838–3846 (2015)

Turkle, S., Papert, S.: Epistemological pluralism and the revaluation of the concrete. J. Math. Behav. **11**(1), 3–33 (1992)
Wang, L., Sy, A., Liu, L., Piech, C.: Learning to represent student knowledge on programming exercises using deep learning. In: Proceedings of the 10th International Conference on Educational Data Mining (EDM), pp. 324–329. Wuhan, China (2017)

Learning Platforms in the Context of the Digitization of Education: A Strong Methodological Innovation. The Experience of Latvia

Arta Rūdolfa and Linda Daniela

Abstract The modernization of the education system, the digitalization of the educational environment and learning management systems (LMS), where one of the solutions is learning platforms, are the most urgent directions today's pedagogical work is taking to reap the benefits of the digital environment. Education quality can be improved in different ways: by changing the content of learning, forms of learning, learning methods and teaching aids; promoting the use of learning platforms in schools; introducing programming and robotics; using learning management systems and other systems. Technologies and digital solutions are transforming the educational landscape in technology-enhanced learning environments. On one hand, there are many possible solutions that provide technology-enhanced learning; while on the other, there is a need to transform educational processes, to transform competence in teaching, to analyze learning outcomes so that technology-enhanced environments can support knowledge construction. The authors of this paper analyze the results of research on learning platforms, in which several research methods were used: systematic literature analyses; development of learning platform evaluation tools; analyses of learning platforms; and surveys on teachers' attitudes to learning platforms. Altogether 705 teachers expressed their opinion on using learning platforms as a tool for enhancing knowledge construction, providing feedback and analyzing students' learning results. In this paper, the authors will discuss the results of analyses on nine learning platforms developed in Latvia, conducted using an evaluation tool with 22 criteria and 43 sub-criteria.

Keywords Learning platforms · Digital learning tool · Evaluation tool · Evaluation criteria · Technology enhanced learning · Knowledge construction

A. Rūdolfa (✉) · L. Daniela
Scientific Institute of Pedagogy, University of Latvia, Imantas 7th, 1, Riga, Latvia
e-mail: arta.rudolfa@lu.lv

L. Daniela
e-mail: linda.daniela@lu.lv

D. Scaradozzi et al. (eds.), *Makers at School, Educational Robotics and Innovative Learning Environments*, Lecture Notes in Networks and Systems 240,
https://doi.org/10.1007/978-3-030-77040-2_28

1 Terminology in the Field of Digital Learning

Digital solutions provide an opportunity for learning outside the specific boundaries of space. At the same time, pedagogical competences need to be developed so that digital solutions can be used in a targeted way to promote the acquisition of specific knowledge, skills and competences. One of the solutions for digital environments is the online learning opportunities that have entered the educational landscape thanks to the digital revolution. Digital solutions have advanced since 1971, with the mass production of the microprocessor by Intel [1], and continued with the definition of the World Wide Web in the 1980s and the first written web browser (WWW) in [2]—the result of work by Berners-Lee and Cailliau [2]—and its launch for public use in 1993. Since then, various online solutions have been developed which in turn led to the online learning solutions known to the world by different terms: *learning management systems (LMS), learning platforms (LP), massive open online courses (MOOC), online learning, digital education tools,* etc. In 1924, Sidney Pressey offered to test a machine that resembled a typewriter. It could be used to train memory processes and answer test questions and predict whether a specific answer would be chosen [3]. LMS are systems designed to capture student information digitally, create learning content for students and analyze student activity, and student learning outcomes [4]. They are excellent tools for educators, and their main difference from learning platforms (LP) is that the curriculum is created by the educators themselves. This approach has both positive outcomes, as educators are free to offer different learning solutions, as well as negative ones, since the learning offering depends on the individual educator's skill at designing the materials. Massive open online courses (MOOC) on the other hand, are courses in which a key emphasis is that the content participants have access to and learn is available online. A learning platform (LP) is a solution that is somewhere between an LMS and a MOOC. Unlike LMSs, where content is created by the educators themselves and available to students, the content of LPs is developed by the platform developers and the offer includes access to it [5, 6]. MOOCs, on the other hand, offer ready-made content but do not provide the management facility that is available for both LMSs and LPs. However, these boundaries become less distinct and unambiguous as the technological solutions continue to evolve. The authors of this study will use the term "learning platforms" as it is not yet synonymous with other terms used.

2 Teaching Conditions in Digital Learning Environments

The use of different digital learning platforms in the pedagogical process is also an increasingly urgent necessity. This is because students should not only acquire knowledge but should also become active participants in the learning process, and creators of new innovations, technologies, and technological solutions. One important factor that transforms the teaching environment is students' ability and willingness to self-manage their learning. This is essential in situations where students can themselves use a variety of online materials to organize their learning. The digitization of the learning process means students will increasingly take responsibility for their own learning, thus becoming responsible constructors of their knowledge, with the ability to function independently in the future, according to the demands of the employment market. The scientific databases on learning platforms reveal the following different formulations for these systems: *learning platforms; online learning; learning management systems; interactive learning,* etc. In many respects, this is an indication that similar online digital solutions are being discussed. However, we can conclude that digital forms of learning can be considered digital learning tools and that, in order to be defined as an online learning platform, they must meet the basic criteria of providing:

- A curriculum that is relevant to the education program;
- Effective use of ICT—online access (on the most popular smart devices);
- Effective use of ICT (ease of use and navigation; ability to upload and download documents; ability to create training content; data protection and network login;)
- Some interactivity;
- The possibility to ask for and receive feedback (answer—explanation; analyzing different dimensions of learner progress and data; two-way communication; opportunities for peer-to-peer learning);
- Connectivity with other widely used educational digitalization solutions (LMS);
- Self-directed learning (it should be accessible, understandable, usable without the presence of a teacher; opportunity for students/pupils to organize their own learning process);
- Different teaching methods (gamification principles) [5].

Learning platforms are one way for students to benefit from online digital solutions. Learning platforms are a digital learning tool that provides access to information across time and space. Students effectively have access to training platforms anytime, anywhere [7]. Only 9 (out of 39) of the learning platforms analyzed and assessed in this study meet the key criteria for learning platforms (immediate feedback, self-directed learning, a degree of interactivity, HTML—online access, etc.) and were evaluated and tested in depth. It is important to keep in mind the definition for learning platforms that was developed during this study, based on the main criteria for learning platforms in general. This will avoid any further confusion concerning what this study is about and what type of digital learning platform it refers to.

The learning platform is a digital, interactive online learning tool that incorporates learning content theory, exercises, tests, provides instant feedback and keeps track of a student's learning progress. Learning platforms should be easily accessible and easy to use for students, helping them to learn the content of a self-directed learning process and organizing teaching activities for teachers (Daniela & Rudolfa, 2019).

3 Methodology

A study entitled "Learning platforms as a learning tool in the context of digitalization of education in Latvia" was carried out to understand the opinions of Latvia secondary school teachers regarding the use of available learning platforms and to develop recommendations for improving their operation. The study summarized and analyzed the teacher survey (705 respondents, 9 questions) to find out:

- Teachers' attitudes to learning platforms;
- The most frequently used training platforms;
- The main benefits;
- The main reasons for not using them.

In addition, this study analyzed learning platforms against specific criteria, to assess their pedagogical and technical potential. The evaluation tool (43 criteria) was applied to 39 online digital sites (used by the teachers themselves and named in the questionnaires referred to above), however most of the online sites mentioned by the teachers are not considered learning platforms.

The research steps were the following:

- Surveying teachers on the use of learning platforms;
- Developing and approving the training platform evaluation tool;
- Evaluating training platforms developed or adapted in Latvia.

4 Learning Platform Evaluation Tool

As there are many suggestions on how to organize online learning, and teachers and school administrators need criteria for choosing an appropriate LP, we developed an original evaluation tool. For this kind of assessment tool, in addition to the **technical parameters** and possibilities, it is mainly important to look at the **pedagogical value** produced [8]. The databases of the scientific literature provide very little information on what forms of evaluation can be applied to learning platforms. There is greater emphasis on the technological evaluation of LPs, but little information can be found on assessing their pedagogical content (even taking into account their technological aspects) [4, 9–11].

This tool is useful for assessing the pedagogical potential of LPs, provided that the evaluation is carried out by people who already have a certain degree of skill in the assessment of pedagogical processes. Other tools should be used to evaluate whether a platform supports learning outcomes, and to measure it in terms of knowledge growth [8].

The main objective of developing the evaluation tool was to further the use of meaningful, convenient and effective LPs in the modern teaching process. Learning outcomes are more important than the way learners acquire their knowledge. Thus, it is essential not only to analyze the technical and visual parameters of the training platforms, but also their educational value and pedagogical potential.

The evaluation tool considers the visual perceptibility, accessibility and interactivity of LPs, whether or not students can receive feedback, whether the content is updated regularly, and whether learners are given self-regulated learning opportunities. Another aspect for assessment could be the manageability of LPs and the possibility of creating new content. Most of the criteria are scored 0, 1 or 2, but some are only evaluated in terms of whether the particular parameter is met (1 point) or not (0 points). The criteria for the evaluation tool came from analyzing the theoretical literature and various training platforms, and a combination of the authors' experiences. The tool was verified by using it to assess LPs used in Latvia [8].

The main factors to evaluate in LPs and their quality are:

- Compliance of content with the set curriculum
- Effective use of ICT—online availability (on the most popular smart devices)
- Usability and accessibility
- Degree of interactivity (interactive engagement is desirable)
- Provision of feedback (answer/explanation; analysis of student's progress)
- Compatibility with other digital tools
- Provision of two-way communication on the platform
- The use of the learning tool does not require the presence of adults, the tool is designed so that step by step "guides" the learner through the learning topics, it is easy to use (intuitive) [8].

Nine learning platforms used in Latvia and their score after being tested with evaluation tool (Fig. 1).

5 Research Results

The first quantitative method of data acquisition in empirical research is the survey. This method was chosen because the authors wished to get the opinions of as many respondents as possible regarding the use of learning platforms in Latvian secondary schools. The SPSS data mathematical statistical analysis program was used to analyze the data obtained from the questionnaire.

Fig. 1 Learning platforms used in Latvia and analyzed within the framework of this study

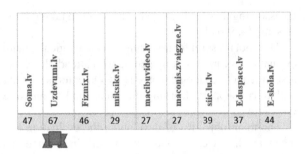

Soma.lv	Uzdevumi.lv	Fizmix.lv	miksike.lv	macibuvideo.lv	maconis.zvaigzne.lv	siic.lu.lv	Eduspace.lv	E-skola.lv
47	67	46	29	27	27	39	37	44

5.1 *Teachers Who Use Learning Platforms (N 573) Do So*

- To enable students to review what they have learned and consolidate their knowledge.
- To give students access to information in a convenient place and at a time that enables them to learn independently.
- To provide students with immediate feedback to promote their self-efficacy and thus encourage their willingness to engage in self-directed learning.
- Teachers who combine content on learning platforms with other forms of learning or who use content on learning platforms to assign individual tasks to students are also ready to develop new learning content; in other words, they take the current needs of their students into account and take advantage of the digital environment to instantly update curriculum content with the latest information.
- The most popular learning platform used by teachers is www.uzdevumi.lv (which can be translated as tasks.lv).

5.2 *Teachers Who* Do not *Use Learning Platforms in the Learning Process (N 79) Give These Reasons*

- Students already spend too much time on technology.
- Stereotypical statements that do not consider the benefits of using this learning tool, which is also associated with benefits for students, as evidenced by data correlation analysis.
- Claims of disliking the curriculum content; which, on the one hand, demonstrates a desire to retain control over student learning and distrust of self-directed learning, while, on the other, demonstrates stereotypical beliefs, since it is impossible for them to have a fact-based understanding of the content of learning platforms, if they do not use them.

(3) Despite the generally positive attitude to the place and role of learning platforms in the learning process, these experts who express their personal views on learning platforms also highlighted the issue that happens that learning platforms

offer low-complexity tasks that can sometimes mislead learners (e.g., in a situation when the student completes 5 tasks from a topic and he/she gets the idea that the topic has been mastered, because the platform does not offer a deeper learning/task level). This can lead to simplified memory-enhanced development (positive) but also to underdevelopment of higher thinking levels (metacognition), thus emphasizing the need for continuous critical evaluation of the content hosted on learning platforms. But it is important to understand that a much lower proportion of students requires the more difficult tasks, so creating such complex tasks would be disadvantageous for only a relatively small group of students. Uzdevumi.lv, for its part, is smartly adapted and performs the main functions successfully—*acquisition of new content, review, homework, tests, reinforcing knowledge, spirit of competition, feedback, diagnostic tests of previous years delivered interactively with various tasks,* and the pedagogical opportunities that have already been described.

5.3 *The Results from the Statistics on the Uzdevumi.Lv Learning Platform Show That*

- Students are prepared for a self-directed learning process, if they are given the opportunity; the data show that a large number of students connect to a learning platform after school, complete individually selected tasks and read the theoretical information there.
- Students using the PROF (paid) version spend twice as much time on learning tasks as students who have free access, which also confirms that they are able to self-learn if given the opportunity.
- Teachers using the PROF (paid version) develop more teaching materials and assign more tasks to students than users of the free version, suggesting that the former see the benefit of this type of learning and access to more interactivity on the learning platform.

6 Conclusions

1. Learning platforms are a way of promoting student-led learning and optimizing the work of teachers.
2. New and innovative teaching methods, forms of work, and transformed learning environments are necessary for preparing students for life in a constantly changing society, like today's.
3. Teaching platforms currently lack tasks that lead to sophisticated thinking operations; such tasks are expensive and are required by fewer students (possibly a niche in educational technology).

4. It is important to define basic terms before we can understand the digitalization solutions of the educational environment and how to use them in the pedagogical process.
5. Modern pedagogy should look for ways to bridge the gap between how students learn and how teachers teach. Today's students process information differently from our ancestors, and these differences are wider and deeper than teachers have anticipated.
6. Learning outcomes are more important than the way students achieve them. Thus, it is essential to analyze the educational value and pedagogical potential. Any learning platform is a tool in the hands of an indispensable component of education—the teacher.
7. Training platforms should be evaluated every year as technology advances very rapidly.

References

1. Chan, T.W., Roschelle, J., Hsi, S., Sharples, M., Brown, T., Patton, C., et al.: One-to-one technology-enhanced learning: an opportunity for global research collaboration. Res. Pract. Technol. Enhanc. Learn. **1**(1), 3–29 (2006)
2. Berners-Lee, T., Cailliau, R.: WorldWideWeb: proposal for a hypertext project (1990). Retrieved 12 July 2019
3. Watters, A.: The automatic teacher. Retrieved from: http://hackeducation.com/2015/02/04/the-automatic-teacher (2015)
4. Watson, W.R., Watson, S.L.: An argument for clarity: what are learning management systems, what are they not, and what should they become? TechTrends **51**(2), 28–34 (2007)
5. Latvian Information and Communication Technology Association (LIKTA). Recommendations for the Development and Evaluation of Digital Teaching Resources and Resources (2015)
6 El Emrani, S., El Merzouqi, A., Khaldi, M.: Massive online open courses platforms: analysis and comparative study of some pedagogical and technical characteristics. Int. J. Smart Educ. Urban Soc. (IJSEUS) **10**(1), 25–36 (2019)
7. OECD. Students, Computers and Learning. Making the Connection. Pieejams: https://read.oecd-ilibrary.org/education/students-computers-and-learning_9789264239555-en#page4 (2015)
8. Daniela, L., Rūdolfa, A.: Learning platforms—how to make the right choice. In: Daniela, L. (ed.) Didactics of Smart Pedagogy: Smart Pedagogy for Technology Enhanced Learning, Springer, pp. 191–212 (2019). ISBN 978-3-030-01550-3
9. Dabbagh, N.: Pedagogical models for e-learning: a theory-based design framework. Int. J. Technol. Teach. Learn. **1**(1), 25–44 (2005)
10. Dağ, F.: The Turkish version of web-based learning platform evaluation scale: reliability and validity study. Educ. Sci. Theor. Pract. **16**(5), 1531–1561 (2016)
11. Edmunds, B., Hartnett, M.: Using a learning management system to personalise learning for primary school students. J. Open Flex. Distance Learn. **18**(2), 11–29 (2014)

Educational Robotics: From Structured Game to Curricular Activity in Lower Secondary Schools

Alberto Parola, Elena Liliana Vitti, Margherita Maria Sacco, and Ilio Trafeli

Abstract Many attempts have been made to introduce robotics into curricular activities, although these have largely been occasional and discontinuous experiences. These experiments are often organized to teach specific know-how, enormously devaluating the technology's potential. Furthermore, they might be placed in the hands of experts who are not on the school's staff, and who conduct the entire project mostly without the involvement of teachers. Our research project is directed at lower secondary schools. Our aim is to try to move past the dictates of "teaching robotics" towards the less controversial vision of "teaching with robotics". Following the lead of Datteri and Zecca (Metodi e tecnologie per l'uso educativo e didattico dei robot. Mondo digitale 75, editorial (2018)) we propose robots as a mediation instrument for normal learning and for transversal competencies in the school setting. The proposed research is a three-year program within the host school's normal curricular subjects and timetable, in which our approach facilitates the learning process. The topics extend beyond coding, robotics, and STEAM in general, and aim to improve experiences in liberal arts subjects such as language, literature, and geohistory.

Keywords Coding · Robotics · Competences · Predictive ability · Problem-solving · Imaginative ability

1 Introduction

Curricular programs in compulsory education in Italy are often supplemented by structured extracurricular courses delivered by freelance tutors or organizations.

In recent years, such supplementary courses are mostly designed to develop skills in STEAM subjects (science, technology, engineering, art and mathematics). This is mainly due to the emerging interest in coding and in educational robotics (National

A. Parola (✉) · E. L. Vitti · M. M. Sacco · I. Trafeli
Centro Interdipartimentale Di Ricerca Per La Digital Education (Cinedumedia), Department of Philosophy and Educational Sciences, University of Turin, Via Verdi 8, 10124 Torino, Italy
e-mail: alberto.parola@unito.it

D. Scaradozzi et al. (eds.), *Makers at School, Educational Robotics and Innovative Learning Environments*, Lecture Notes in Networks and Systems 240,
https://doi.org/10.1007/978-3-030-77040-2_29

Recommendations for the School Curriculum, 2012[1]; National Recommendations and New Prospects, 2018[2]; National Plan for Digital Education[3]).

These private ventures are on the rise and are joined by PON[4] projects which are conceived to encourage transversal skills and to strengthen normal learning. Such projects usually include robotics in structured programs with cross-disciplinary aims (Scaradozzi et al. 2015). The school environment is very dynamic and engages teachers and students alike. Many refresher courses for teachers aim to respond to the needs arising from the ongoing pedagogical and sociological revolution.

The capacity of schools to find solutions for training needs is welcome, but it is important to understand whether these integrations are indicative of an acceptable level of pedagogical consciousness and effectiveness, as underlined by Datteri and Zecca (2018): "In an area of great, positive enthusiasm for the potential of new technologies for teaching, it is particularly important to conduct theoretical and empirical research that critically analyzes the validity of this potential" (own translation). It is thus essential to find educational contexts in which to carry out empirical research over a sustained period, to investigate the significance of the courses offered to teachers and students.

In the light of these needs, the Interdepartmental Center of Research for Digital Education (Cinedumedia), of the University of Turin's Department of Philosophy and Educational Sciences took a proposal to schools and educators to take part in a long-term evaluation of their activities and verify whether the courses can be offered during curricular school time.

The research context needed the maker/provider's willingness to get involved, and the educational institution's willingness to keep the research program going for a full training cycle.

After selecting two subjects—Istituto Sociale di Torino as the host school and Kidding Srl as the provider of the B4K supplementary course on science and robotics—the main task was to understand how to transfer B4K's robotics courses, which are structured as an extra-curricular activity, to a formal school setting with curricular goals.

[1] Decree no. 254 of the Ministry of Education, University and Research, 16 November 2012: Indicazioni nazionali per il curricolo della scuola dell'infanzia e del primo ciclo di istruzione.

[2] Note no. 3645/2018 of Ministry of Education, University and Research – Indicazioni nazionali e nuovi scenari.

[3] Piano Nazionale Scuola Digitale (PNSD).

[4] National Operational Programme 2014–2020 "For School - skills and learning environments" of the Ministry of Education, University and Research.

2 Methodology of B4K Extracurricular Courses

The Bricks4Kidz (B4K) method began in the US in 2008 and since then has spread to 43 countries. The main goal is to bring children closer to STEAM subjects using LEGO® Technic® and LEGO® Mindstorms®.

Tutors have B4K kits and slides. The sequence and the theoretical and practical aspects are already established and must be adhered to strictly, but tutors have the freedom to choose when and how to present the topics.

The B4K goals are different from a school's general curricular targets, and before this research, it was only ever offered as an extracurricular activity in all countries where it is used.

3 Adjustments to the Course

The research group set about establishing which of Istituto Sociale's curricular priorities could be the subject of the B4K procedure as an educational method. The most important adjustments were:

- Including the courses during school hours;
- Increasing the number of hours in the courses;
- Establishing comparison criteria against which to measure the students' performance and improvements achieved through the B4K's courses;
- Defining didactic and cross-curricular targets, based on the host school's three-year program of studies (PTOF);
- Encouraging a teaching style that builds knowledge actively and collectively;
- Suggesting a reorganization of the curriculum to reduce traditional classroom-taught lessons and promote the "Think-Make-Improve" approach.

To define educational targets consistent with the school's three-year study program, the research group considered the overall context of Italian schools under the existing national and European regulatory framework, beginning with the key learning competences recommended by the European Parliament (2006/962/EC).[5]

Specifically, the group identified the following main objectives: communication in the mother tongue; mathematical competence and basic competences in science and technology; digital competence; social and civic competences; sense of initiative and entrepreneurship.

The competence goals and the learning targets in science, technology and mathematics defined in the national recommendations for the school curriculum[6] were also included to ensure all curricular activities were accounted for.

---

[5] Recommendation of the European Parliament and of the Council of 18 December 2006 on key competences for lifelong learning (2006/962/EC).

[6] See notes 2, 3.

With these goals in mind, we proposed a robotics course in autonomous driving/self-driving to a third-year class of the lower secondary school. The topic of this course functioned on two levels: (1) it is a current topic and clearly useful for tomorrow's adults in view of the daily use of digital technology applied to mobility; (2) it allows each lesson to be structured as a problem-solving exercise applied to mobility issues that are continually changing, and for students to develop the ability to visualize space and predict the movement of an easy-to-program robot.

The traditional pattern of B4K lessons remained the same: each was divided into two parts, the first with a theoretical explanation and the second with personal experiments. However, the explanation part was designed to encourage participation, as well as active and collective knowledge construction.

4 Performance Review

During the fourth lesson, students took a test of the target knowledge, abilities and competences (Rychen and Salganik 2001) defined for the project. This evaluation allowed us to collect: (a) information about the students' knowledge, abilities and skills level after three lessons; (b) a summary of the results to compare against their normal performance; (c) any problematic areas that could be revised in the second part of the courses.

A code was applied to the results of these tests in order to define scores for each performance. These scores were converted into evaluations by the curricular teachers of mathematics, science and technology who were involved in the entire project. The grades assigned to special needs (SN) students were based on their individual education plans (IEP).

5 Results

The results show the B4K to be a viable method for reaching the knowledge targets, but the changes achieved do not sufficiently improve abilities and competences in these subjects.

Mid-course tests measure the students' competences and their ability to solve new problems, but this target was not reached. The research group assumes this failure depends on two factors: (1) students had never performed tests like this to evaluate their abilities and skills; (2) the tutor's teaching style was not designed to improve these targets.

In our opinion, in future research, the tutor's pedagogical style should be adjusted, as should the lesson structure. Less time should be devoted to theoretical explanations and more to active and collective knowledge construction. It is also important to change the duration of the course and the teachers' involvement: everyone involved should be given a role and the goal to learn the new STEAM teaching methods.

For next year's research, it will also be important to do more work on the metacognition targets. Students must be stimulated to become more aware of their knowledge and abilities. Indeed, this year, the tutors (chosen by the B4K manager, not by the university staff) did not work actively on improving and fostering this awareness.

6 Conclusions and New Prospects

After this first year, the research group of Cinedumedia and Istituto Sociale decided to renew the agreement for the next three years, to experiment a new version of the courses for a full instruction cycle.

Datteri and Zecca (2018) wrote: "Educational robotics is based on the idea that, in certain conditions, robotic construction and programming can 'train' multiple aspects of reasoning and creativity related to the observation, preview, discovery and explanation of phenomena, and the identification and execution of various kinds of problem-solving strategies" (own translation). The guidelines for this research were to create these conditions.

For the next three years' research, the Cinedumedia group is fully responsible for designing the new courses and has decided to focus on these targets:

(1) Improving thinking processes (not only computational thinking, but also systemic thinking and narrative thinking);
(2) Pursuing metacognition, the key skill of "learning to learn";
(3) Consolidating preview ability and imagination skills.

Our research project is directed at lower secondary schools. Our aim is to try to move past the dictates of "teaching robotics" towards the less controversial vision of "teaching with robotics". Following the lead of Datteri and Zecca (2018) we propose robots as a mediation instrument for normal learning and for transversal competencies in the school setting. The topics extend beyond coding, robotics, and STEAM in general, and aim to improve experiences in liberal arts subjects such as language, literature, and geohistory. As Ranieri states (2011), it is important, within educational technology research, to avoid the false hope that technology alone can produce improvements. Devices do not have an intrinsic power to change learning processes, they are merely vehicles, resources for reaching a goal. The focus of educational projects should not be technology itself, but rather designing a teaching method that can achieve set targets.

References

Brick for Kidz Homepage. https://www.bricks4kidz.com/. Last accessed 06 Sept 2019

Datteri, E., Zecca, L.: Metodi e tecnologie per l'uso educativo e didattico dei robot. Mondo digitale 75, editorial (2018)

Datteri, E., Zecca, L.: The game of science: an experiment in synthetic roboethology with primary school children. IEEE Robot. Autom. Mag. **23**(2), 24–29 (2016)
Ranieri, M.: Le insidie dell'ovvio. Tecnologie educative e critica della retorica tecnocentrica, Edizioni ETS, Pisa (2011)
Rychen, D.S., Salganik, L.H.: Defining and Selecting Key Competencies. Hogrefe & Huber Publishers, Ashland (2001)
Scaradozzi, D., Sorbi, L., Pedale, A., Valzano, M., Vergine, C.: Teaching robotics at the primary school: an innovative approach. Procedia Soc. Behav. Sci. **174**, 3838–3846 (2015)

Educational Robotics in Informal Contexts: An Experience at CoderDojo Pomezia

Lina Cannone

Abstract The project aims to develop technical skills in primary school students. Technical courses are usually not popular among schoolchildren. With the introduction of the National Plan for Digital Education (PNSD) by Italian Education Ministry since 2015, words like coding, robotics and computational thinking are used more frequently in primary school classrooms. CoderDojo is a worldwide movement working to introduce children to robotics and computer science. As CoderDojo Pomezia, we prepared a number of activities to encourage students' interest in computational thinking and robotics, and to improve their skills. In recent years, we have held several workshops for primary school students. The workshops have involved activities such as programming robotics kits, using robotics with six-year-olds, programming videogames, modelling 3D objects. The activities always have a hands-on approach. Over 100 primary schoolers have participated in the workshops. We have also held workshops to train teachers to introduce these technical skills in their classrooms. This paper presents the work of CoderDojo Pomezia to train children and demonstrate how students have improved their technical and social skills.

Keywords Coding · Educational Robotics · CoderDojo · Scratch · Primary school

1 Introduction

Today's citizens need specific occupational skills in professional, managerial and technical jobs to ensure economic growth and social well-being [1]. In 2018, the Council of the European Union issued "Recommendations on Key Competences for Lifelong Learning" (Fig. 1), aimed at preparing people for the skills needed for the world of work [2].

L. Cannone (✉)
Istituto Comprensivo Orazio Pomezia, Champion CoderDojo Pomezia, Pomezia, Italy
e-mail: lina.cannone@posta.istruzione.it

D. Scaradozzi et al. (eds.), *Makers at School, Educational Robotics and Innovative Learning Environments*, Lecture Notes in Networks and Systems 240,
https://doi.org/10.1007/978-3-030-77040-2_30

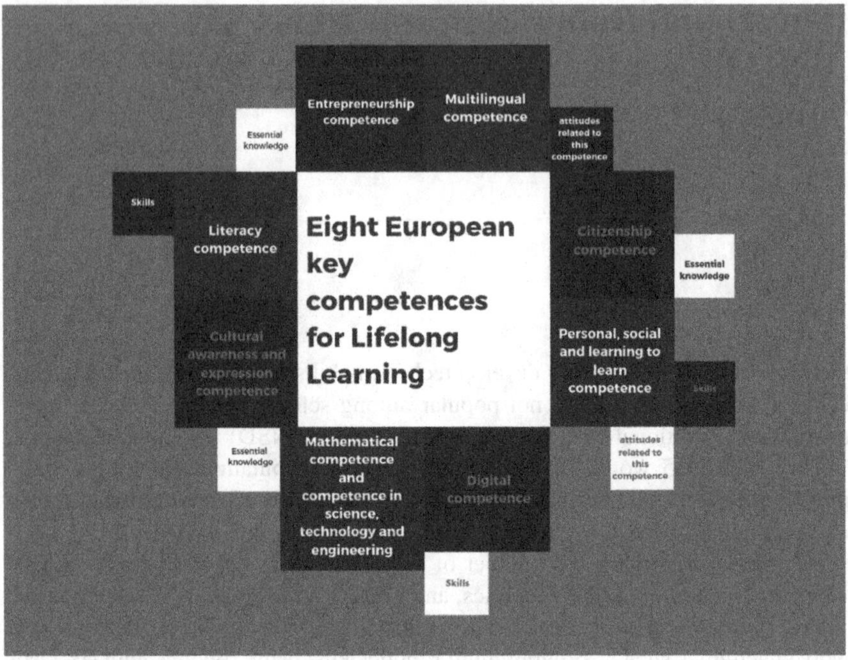

Fig. 1 Eight European key competences for lifelong learning

Several studies report examples of the effectiveness of new approaches to teaching and learning, like tinkering, coding and making, for improving creativity, computational thinking, problem-solving and cooperation [3]. Tinkering is an approach involving hands-on activities where students can create or modify objects using recycled materials alongside technological features such as batteries, LED lights, small motors. Tinkering is about learning from one's failures, and having unstructured time to explore and invent. The potential for innovation lies in the process of exploring and inventing [4]. Making has its roots in Papert's Constructionism, but the focus is on the creative process rather than merely on the finished product. This process has different aspects ranging from design thinking to physical creation. There is also a social aspect: as they work in teams, students have to manage conflicts. Another area of interest is coding. Like other subjects, learning to code involves various skills. Students have to resolve an issue, debug code and try to understand where it is failing, break down large problems into small ones. There are many organizations in Europe whose aim is to bring innovation into teaching and learning. One of these is the European Schoolnet Academy, a not-for-profit organization that promotes numerous projects, including Scientix and STEM Alliance, which promotes and supports a Europe-wide collaboration for STEM. Scientix projects include many resources for code learning and makerspace activities. In 2015, the Italian Ministry of Education, University and Research (MIUR) introduced the National Plan for Digital Education (PNSD) and the related Law no. 107. The Ministry's intention was to introduce the concepts of makerspace, school fab labs, robotics and coding to Italian schools. Since 2018, coding has been part of the national recommendations for the school curriculum in Italy. Several organizations around the world have established programs whose mission is to introduce computational thinking to children from an early age. Some examples are: Google CS First, code.org and its integration in the "Programma il uture" project developed by MIUR, MIT's Scratch, a powerful tool for creating artifacts, and CoderDojo.

CoderDojo is a global movement of free, volunteer-led, community-based programming clubs for young people [5]. Anyone aged 7–17 can visit a Dojo, where they can learn to code, build a website, create an app or a game, and explore technology in an informal, creative, and social environment. Within the CoderDojo Movement there is a focus on peer learning, youth mentoring and self-led learning that aims to help young people realize that they can build a positive future through coding and community. We have been volunteering for about four years with our local CoderDojo in Pomezia (Rome), where our work includes coding, robotics and tinkering. About 100 kids have attended our workshops. This paper will present our experience teaching robotics and coding.

2 Robotics Workshops

2.1 Technical Resources

Many studies have been conducted on the use of robots in educational environments, and the conclusions drawn are that robots help schoolchildren develop their logical, creative thinking, and problem-solving skills, and learn mathematics, science and programming [6].

mBot robotic kits are the key features of the activities of CoderDojo sessions. mBot is an Arduino-based robotics kit that enables students to assemble the robot with motors and sensors, without having to know anything about electronics. Sensors and motors can easily be plugged into mBot, and the device can be programmed in a Scratch-like environment. The Arduino programming language can also be used, as mBot is an Arduino-based platform. Assembly of mBot is extremely easy, as the platform's principal purpose is for learning programming. mBot can be controlled remotely. Figure 2 shows an mBot similar to those used during our workshops.

2.2 Students' Activities

Robotics workshops were attended by both male and female primary school students. Many workshops were held between April 2016 and June 2019. mBots were pre-assembled by mentors (CoderDojo volunteers). The robot is programmable and

Fig. 2 mBot robotics kit

students can control the motor (wheels, belts, robotic arm) and can read and process the data measured by the robot's sensors, such as distance, brightness, direction, etc. [7].

mBot can be programmed with mBlock, a free software similar to Scratch which allows the user to drag and drop blocks to create a code sequence. Figure 3 shows a sample program in which the red LEDs in mBot light up; then, if there is a black line on the left side, mBot turns left.

We proposed various activities during the workshops with different levels of difficulty. Children were divided into groups of four or five (Fig. 4); each group worked with one mBot. The students followed the instructions to connect the robot to the software and program their code in the following steps:

- mBot turns on and LEDs light up with a color sequence.
- mBot moves in regular geometric patterns (triangular, square, circle) and can repeat parts of the movement sequences.
- mBot's sensors detect obstacles and move it back away from them. If the robot turns 180° as it detects an obstacle, it can go back and forth endlessly between two obstacles facing each other. The robot can move into a space on the mat delimited by a dark line that is detected by the light sensor.
- mBot has a pen stuck to it and children can draw shapes by moving the robot.

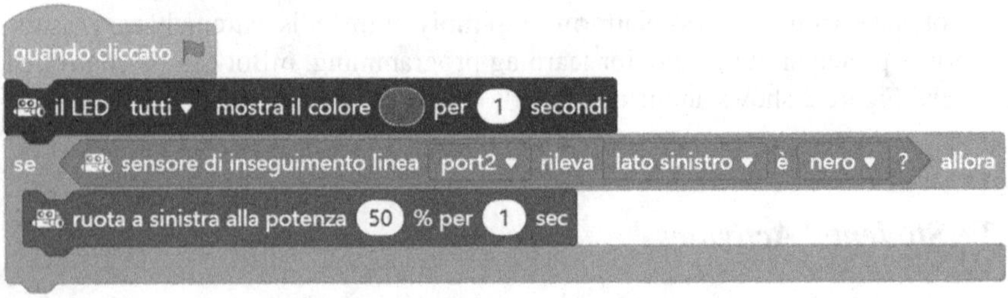

Fig. 3 A sample code created with mBlock

Fig. 4 mBot activity during a CoderDojo session

- Once the children had got to know the software, they created a code that would allow it to follow a complex path marked by a dark line and two flags acting as traffic lights, as shown in Fig. 5. The flags are controlled by an Arduino Nano with two ultrasonic sensors. Two mBots follow the dark line, but when the ultrasonic sensor detects movement, a servomotor lowers the first flag to stop the second mBot. Once it has passed, the flag is raised again.

Except for the last activity, which was very complex and needed the mentors' help, the participants enjoyed all the workshops. Since the children already knew Scratch, there were no great difficulties programming the mBots.

Fig. 5 mBot path with traffic flags

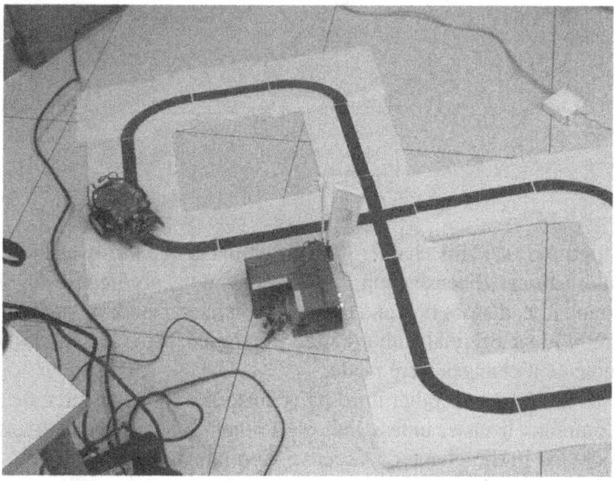

3 Conclusion

Students were enthusiastic about working with robots in particular and technology in general. The younger children drew and played with the robots. The older students appreciated the opportunity to create their own programs and to control the robot. They were all highly motivated to use mBots and to program them. They worked together, within their groups and with other participants. If one group improved something, other groups tried to emulate them, asked for information and suggested modifications. We observed an improvement in the students' technical skills, social and peer collaboration skills and, what is equally important, they had lots of fun.

References

1. Cesaretti, L., Storti, M., Mazzieri, E., Screpanti, L., Paesani, A., Principi, P., Scaradozzi, D.: An innovative approach to school-work turnover programme with educational robotics. Mondo Digitale (2017)
2. EC Europa EU website: https://ec.europa.eu/education/education-in-the-eu/council-recommendation-on-key-competences-for-lifelong-learning_en. Last accessed 10 Oct 2019
3. Scaradozzi, D., Screpanti, L., Cesaretti, L., Storti, M., Mazzieri, E.: Implementation and assessment methodologies of teachers' training courses for STEM activities. Tech. Know. Learn. **24**, 247–268 (2019). https://doi.org/10.1007/s10758-018-9356-1
4. Tinkerlab website: https://tinkerlab.com/what-is-tinkering. Last accessed 10 Oct 2019
5. Coderdojo website: https://coderdojo.com/movement/. Last accessed 10 Oct 2019
6. Scaradozzi, D., Sorbi, L., Pedale, A., Valzano, M., Vergine, C.:Teaching Robotics at the Primary School: An Innovative Approach, Procedia Soc. Behav. Sci. 174, 3838–3846 (2015)
7. Oujezdský, A., Nagyová, I.: ICTE J. (2016). https://doi.org/10.1515/ijicte-2016-0001

RoboCup@Home Education: A New Format for Educational Competitions

Luca Iocchi, Jeffrey Too Chuan Tan, and Sebastian Castro

Abstract In this paper we describe a novel methodology for holding educational robot competitions that has been developed within the RoboCup@Home Educational initiative. The methodology is based on two main elements: (1) a workshop + competition format where teaching material is interwoven with competitive tasks; (2) a team-centered approach, where rules and teaching material are tailored to participant teams to maximize the educational experience for students. The results and lessons learned after more than five years of organizing these competitions are also discussed in this paper.

Keywords Robot competitions · Educational robotics · Courseware for robot programming

1 Introduction

Robot competitions are a well-recognized way of fostering scientific, technological and educational activities in the fields of robotics and artificial intelligence, and are an engaging framework for researchers and students. We can broadly distinguish between scientific competitions—which mostly target university research groups—and educational competitions—which target undergraduates and high-school students. The most notable examples of scientific robot competitions are

L. Iocchi (✉)
DIAG, Sapienza University of Rome, Rome, Italy
e-mail: iocchi@diag.uniroma1.it

J. T. C. Tan
Nankai University, Tianjin, China
e-mail: jeffrey@nankai.edu.cn

S. Castro
Massachusetts Institute of Technology, Cambridge, USA
e-mail: scastro@csail.mit.edu

D. Scaradozzi et al. (eds.), *Makers at School, Educational Robotics and Innovative Learning Environments*, Lecture Notes in Networks and Systems 240,
https://doi.org/10.1007/978-3-030-77040-2_31

organized by the RoboCup Federation[1], and address four main domains: soccer, rescue, service robotics (@Home), and industrial robots. Noteworthy robot competitions include: RoboCupJunior, World Robot Summit (WRS) Junior, FIRA Youth, FIRST Robotics, VEX Robotics, etc.

The most significant difference between these two kinds of competition is in the knowledge and hardware requirements for participation in each: expensive, complex robot hardware and cutting-edge technological solutions are needed for research competitions, but are not required for student competitions. One important aspect to consider is the need to *bridge the gap* between these two kinds of competitions. This is important if more students and researchers are to get the opportunity to take part in them. However, there is currently very little effort being made to reduce barriers to entry, as both research and student competitions are organized roughly as follows:

1. Competition organizers define the rules of the competition (usually forming a technical committee).
2. Competition rules are distributed to the community.
3. Participating teams are formed and they are fully responsible for developing the solution for the problems mentioned in the rules. In some cases, the hardware is provided (standard platform) and teams only have to develop the software solutions.
4. Participating teams are selected to attend the competition (typically by qualifying in some way).
5. Competition event takes place.

The drawback of applying this same methodology to student competitions is that it requires them to solve problems that are usually too difficult for them. Student teams thus have to rely on teachers/mentors to provide technical guidance, and on online teaching materials which they usually have to find themselves. Moreover, most student competitions have participation restrictions (e.g., age limit), but once a student reaches this limit s/he is not necessarily ready to join a team for a research-focused competition. There is a gap in the "curriculum vitae" of robot competition participants, which may discourage students from taking part in robot competitions and from an education in AI and robotics.

To address this problem, we have devised a new format for a scientific educational competition that takes the following problem into account: *student teams may not have the resources (knowledge and hardware) to participate in a competition.*

In this paper, we describe RoboCup@Home Education[2] and present its innovative features: (1) a new method of organizing the competition that includes novel forms of support for teams, (2) open-source software and courseware to help teams solve competition tasks, and learn important concepts of artificial intelligence and robotics.

[1] www.robocup.org.

[2] www.robocupathomeedu.org.

2 RoboCup@Home Education Project

RoboCup@Home [1] is the largest worldwide scientific competition on service robots and human-robot interaction. It aims to foster research and development in service and assistive robotics, social robotics, human-robot interaction, etc. RoboCup@Home Education is an educational initiative that promotes education and development in service robots interacting with humans. It draws inspiration from RoboCup@Home, but has a special focus on education and supports new teams aiming to join the main RoboCup@Home competition in the future. RoboCup@Home Education activities started in 2015, with various competitions organized in Japan, and later reached many other locations worldwide (see details in the next section). RoboCup@Home Education includes three main streams of activities: (1) organization of workshops and challenges, (2) support for the development of open-source educational robot platforms, software and courseware, and (3) outreach programs (local workshops, international academic exchanges, etc.).

2.1 RoboCup@Home Education Methodology

The novel methodology for this project (still under experimentation) can be summarized in the following steps:

1. Definition of a general set of rules (based on the RoboCup@Home rulebook).
2. Announcement of the event and registering teams interested in participating.
3. Supporting teams during preparations (team formation, robot acquisition/build, suggestions for budget allocation and fund-raising, etc.).
4. Selection of participating teams.
5. Fine-tuning the competition rules and teaching materials. This step takes account of the experience of participating teams.
6. Monitoring the teams' development of solutions through remote assignments aimed at teaching the basic requirements of the competition tasks.
7. Workshop + Competition event take place.

As already mentioned, this method enables the teams to get more involved in the educational activities, by addressing most of the problems student teams typically have to deal with. It also helps them, not only with the technical developments, but also at the organizational level, to ensure team efforts are effective. More details about robot platforms, software, and courseware are given in the rest of this section.

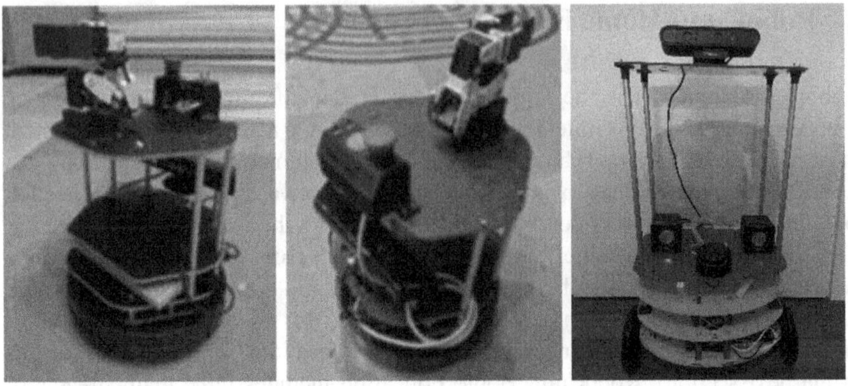

Fig. 1 Robots used in the competitions

2.2 *Robot Platforms*

Choosing suitable robot platforms for student competitions is crucial, as it has a big impact on the organization of the competition and on team formation. Two requirements need to be met for RoboCup@Home Education: (1) the cost of the robot, (2) the flexibility to implement advanced functionalities. To this end, we have chosen low-cost (open source) mobile robots based on ROS (Robot Operating System) middleware and have currently identified two platforms: TurtleBot2[3] and MARRtino[4] (see Fig. 1).

The modularity of ROS means that relatively little effort is needed to adapt the RoboCup@Home Education software and courseware to other ROS-based mobile platforms.

We provide three alternatives for running the robot software: (i) a laptop, (ii) a single-board computer (including Raspberry Pi), or (iii) both. A laptop running Linux OS and ROS is the best choice in terms of computational power and programming flexibility, but it increases the cost for the team and requires expertise to install and troubleshoot Linux and ROS. A Raspberry Pi is very low cost and comes with OS images, including all the software needed to participate in the competition, but has limited computational resources. Using a Raspberry Pi to execute the lower levels of software architecture and a laptop for the application level may be the preferred option. Although this configuration may appear more expensive, it is actually more convenient, since the laptop does not need to be configured to run ROS nodes and can be conveniently connected to the on-board Raspberry Pi over a wireless network. In this way, for example, a robot platform with a Raspberry Pi on-board running Linux and ROS modules can be controlled by an external laptop connected over a wireless network running, for example, MATLAB or a pre-configured Virtual

[3] www.turtlebot.com/turtlebot2.

[4] www.marrtino.org.

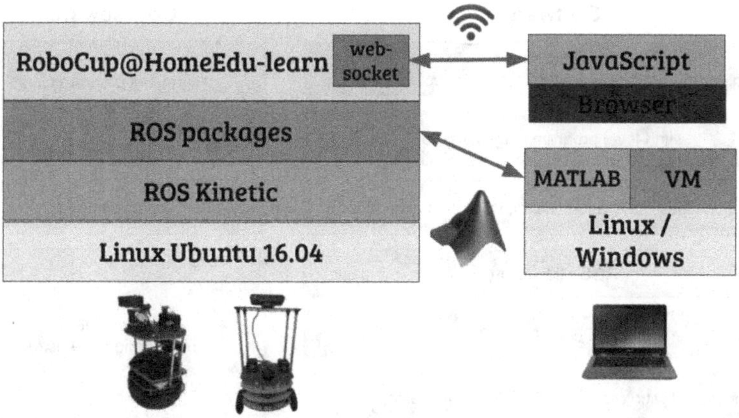

Fig. 2 Example of development architecture

Machine (VM) (see Fig. 2). Consequently, students can program the robot on any laptop (either Windows or Linux-based) with minimal configuration effort (e.g., installing MATLAB or a virtual machine), while all the robot software is pre-installed on the Raspberry Pi.

In addition to low-cost ROS-based robots, the RoboCup@Home Education Online Challenge 2020 has introduced the SoftBank Pepper humanoid robot[5] as an additional platform. In this case, we took advantage of using this standard platform, so that teams without a robot could develop software in a software development environment and using an emulator. After the developed code was sent, SoftBank staff transformed it into videos (running the code on their robots) and sent these videos back to teams so they could take part in the online challenge.

2.3 *Software*

The organization of the software of the robots used in the competition is shown on the left side of Fig. 3. The software stack is based on Linux OS and ROS middleware. The reason for this choice is the extensive support from the robotics development community in terms of developed modules, tools, libraries, and tutorials. On top of ROS middleware, we used several ROS packages to implement the robot's basic functionalities, such as navigation, vision, speech, and manipulation. Although ROS modules and tools are very flexible and powerful for the purposes of the competition, they are not easy to learn for high-school students and undergraduates. Therefore, we created another layer of software, a high-level programming interface, that is used to integrate the basic components at a higher level, hiding the complexity of ROS usage.

[5] https://www.softbankrobotics.com/emea/en/pepper.

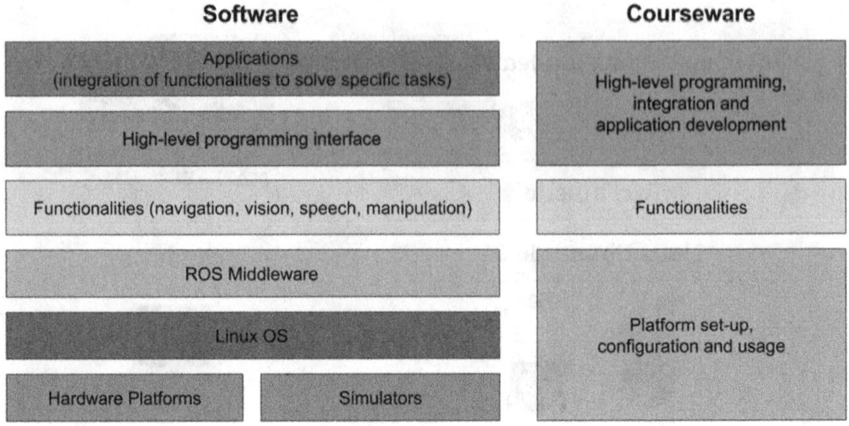

Fig. 3 Software and courseware organization

This higher-level layer of software currently supports two programming languages—Python and MATLAB—that are more suitable for students to learn. Additionally, Python and MATLAB are both considered among the top 10 programming languages by the engineering community in terms of career relevance [2].

To solve the competition tasks, students can develop their code on top of the previously mentioned stack of applications, without the need for low-level drivers for the robot hardware. Depending on the team's experience, we can organize workshops and competitions relating only to the application layer of this architecture (thus hiding all the details and complexity of ROS-based development), or we can provide material to use, and also extend the ROS components for more experienced teams. The creation of such a high-level programming interface is one of the novel contributions of RoboCup@Home Education, and it has been crucial for the development of the courseware, and the organization of the workshop and competitions.

2.4 *Courseware*

As already mentioned, the major differences between RoboCup@Home Education and other competitions is the development of courseware to teach the basic concepts needed to solve the competition tasks, and the fact that teams are involved in educational activities well before the competition dates. The courseware is organized in three sections (as shown on the right side of Fig. 3).

Platform set-up, configuration and usage contains instructions for installing the software, using the robot and the simulator, configuration and calibration, starting the platform services, and checking that everything is working. After this phase, teams are able to operate their robots and solve related problems.

Functionalities provides instructions for using and configuring individual functions (input/output/parameters). This part is platform-independent and uses high-level programming interfaces (when needed) to increase the learning curve for students. After this phase, teams have learned how to activate the basic functionalities that are already implemented in the software provided.

High-level programming, integration, and application development contains the basics of the high-level programming languages (Python, MATLAB) that are used by the teams to develop solutions for the competition tasks. Some examples are provided, although not the full solution to the competition tasks, which, of course, the teams themselves must provide during the competition.

3 Results, Lessons Learned and Future Work

We estimate that, since 2015, when the first challenge was organized in Japan, 13 competitions have been organized with the participation of around 100 teams and 500 students. Detailed information of some of the activities organized in the last few years are presented in [3].

At least seven events were planned for 2020, including the International Challenge during RoboCup 2020. However, all these events were canceled or postponed, owing to the COVID-19 pandemic. To keep teams involved, we organized the RoboCup@Home Education Online Challenge 2020,[6] which included online classrooms and project development in different categories. A total of 24 teams from several countries were selected for the finals, where teams presented their projects to a jury. The event was very successful and produced important teaching material that will be useful for future editions.

We have learned several lessons from all these events, which will serve to further improve the organization of future competitions.

It is important to mention that the teams operate in a realistic setting that represents the real world. Tight deadlines lead to students learning how to manage time and delegate tasks. The workshop is not constrained (many problems are unexpected and/or open-ended), and students need to solve problems, work in groups, discuss solutions, present their methodology in front of others, etc.

After each event, we collect feedback from students and teachers to improve future editions. Thus, the methodology, hardware, software and courseware described here are still evolving.

One main goal we are currently pursuing is the organization of workshops and challenges without the presence of expert organizers. For this we would use online material and call on the remote support of expert organizers. Other future objectives include self-paced teacher training, access to a support community (e.g., forum, social media page, etc.) for students and teachers, and scaling this model worldwide.

[6] https://www.robocupathomeedu.org/challenges/robocuphome-education-online-challenge-2020.

Finally, the experience of running online classes and online competitions was very successful in 2020, and we are planning to repeat and improve it in the coming years in parallel with physical meetings.

References

1. Iocchi, L., Holz, D., Ruiz-del-Solar, J., Sugiura, K., van der Zant, T.: RoboCup@Home: analysis and results of evolving competitions for domestic and service robots. Artif. Intell. **229**, 258–281 (2015)
2. Cass, S.: The top programming languages 2019. Retrieved September 16, 2019 from https://spectrum.ieee.org/computing/software/the-top-programming-languages-2019 (2019)
3. Tan, J.T.C., Iocchi, L., Eguchi, A., Okada, H.: Bridging robotics education between high school and university: RoboCup@Home education. IEEE AFRICON 2019 (2019)

Erwhi Hedgehog: A New Learning Platform for Mobile Robotics

Giovanni di Dio Bruno

Abstract Erwhi Hedgehog is one of the smallest mobile robots. It enables mapping and vision analysis, and also displays machine learning features. Behaving like a small, curious animal, eager to explore the surroundings, the robot can be used to test navigation, mapping and localization algorithms, thus allowing the prototyping of new hardware and software for robotics. This application is particularly handy for educational robotics, at both high school and university level. On the one hand, the project is fully open source and open hardware under MIT license and available on Github, so everyone can build his/her own Erwhi Hedgehog robot with the aid of a step-by-step guide. On the other hand, students with more advanced knowledge can use it as a prototyping platform for developing new software programs and features. Erwhi uses Intel RealSense, AAEON UP Squared and Myriad X VPU technologies, with software based on the Robotic Operating System (ROS), and implements SLAM algorithms, such as RTAB-Map. The machine learning aspect is based on the OpenVINO framework and a dedicated ROS wrapper was used. The software package includes all the programs needed to create a Gazebo simulation. In terms of hardware, motor control is based on an STM32 microcontroller and the Arduino software, and the robot works on the differential drive unicycle model. Finally, Erwhi is compatible with AWS RoboMaker tools.

Keywords Erwhi · ROS · Robotics · Machine learning · SLAM · Vision analysis

1 Introduction

Building a robot is an opportunity for students to learn several aspects of technology, from the design of functional hardware to the development of sophisticated software. Simultaneous mapping, vision analysis and machine learning are necessary features for several types of robots, and have countless applications. Students can currently build robots with these features using Lego kits or Robotis TurtleBot 3. While they

G. D. Bruno (✉)
Ass. Officine Robotiche, Rome, Italy
URL: http://gbr1.github.io

D. Scaradozzi et al. (eds.), *Makers at School, Educational Robotics and Innovative Learning Environments*, Lecture Notes in Networks and Systems 240,
https://doi.org/10.1007/978-3-030-77040-2_32

Fig. 1 Front and rear view of Erwhi Hedgehog robot

are both powerful tools, they also have their limitations. When it comes to building hardware, Lego kits provide flexibility, but they mostly rely on block programming, which can be limiting when developing more complex software. TurtleBots is flexible for both hardware and software, but requires expensive upgrading for special applications, such as object recognition.

Erwhi Hedgehog is one of the smallest mobile robots currently available. It allows simultaneous mapping, vision analysis and machine learning, and is completely open source and open hardware [1], so anyone can build or customize his/her own robot (Fig. 1). The robot can behave like a small, curious animal, exploring its surroundings, and can be used to test navigation, mapping, and localization algorithms. The aim of the entire project is to accelerate and simplify the development of robotics for researchers, educators, students and professionals, by providing a platform for prototyping new robotics hardware and software, and learning by doing.

2 Description of the Robot

2.1 Electronics

The main core of Erwhi Hedgehog's electronics (Fig. 2) is an UP Squared Atom x7-E3950 with 4 GB of DDR4 RAM and 64 GB of EMMC non-volatile memory. Vision is provided by an Intel® RealSense D435 depth camera. Machine learning runs on an AI Core X, Movidius™ Myriad™ X VPU on a PCIe board. Hard real-time controls are implemented on a custom carrier board called Sengi, and based on an STM32F103C8T6 micro controller. The main board is connected to Sengi via USB1.1. The micro controller is connected to a power motor driver and motor encoders. The carrier board includes an absolute Inertial Motion Unit BNO055 that is connected to the main core via a USB-serial adapter. Power is supplied by two

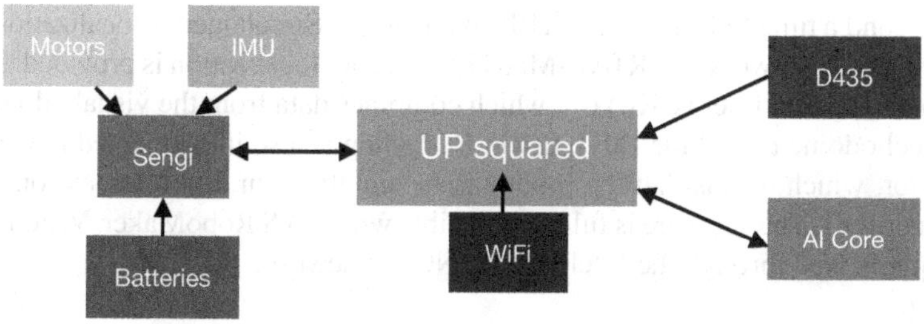

Fig. 2 Simplified diagram of the electronics of Erwhi Hedgehog

18,650 Li-Ion cells in series, with about 8.4 V at maximum charge. Power flows through a DC/DC step down converter which handles an output of 5.1 V and 8 A for a continuous load. The robot communicates over WiFi at 2.4/5 GHz, and Bluetooth.

2.2 Mechanical Parts

Mechanical parts are designed to be 3D printed using classical methodologies, such as the commonly used fused filament fabrication.

Erwhi Hedgehog is actuated by two DC gearmotors, with the robot moving on two casters, a ball caster at the front and an omniwheel at the rear.

2.3 Software

Erwhi Hedgehog runs on Ubuntu 16.04 LTS and ROS [2] Kinetic Kame (Fig. 3). The robot's mathematical model is implemented via the standard ROS control package [3], making the code very easy to adapt to any robot, simply by adjusting a few parameters. The joints are controlled by Sengi firmware, using the Arduino language and STM32duino libraries [4]. Its movements are based on the standard ROS navigation

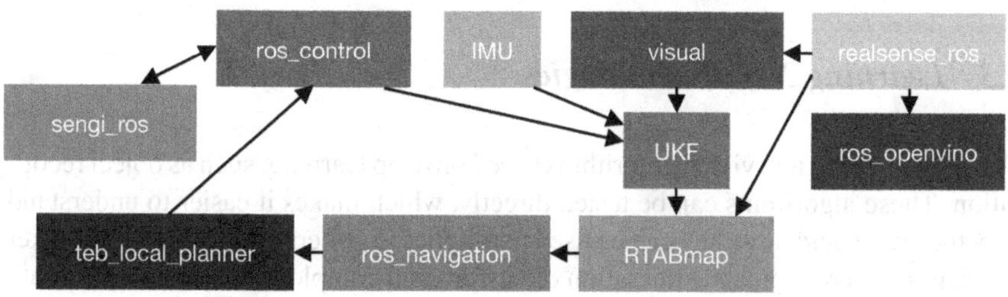

Fig. 3 Simplified diagram of the ROS architecture used in Erwhi Hedgehog

stack [5] and a timed-elastic-band (TEB) planner [6]. Simultaneous localization and mapping (SLAM) works on RTAB-Map [7] and robot localization is provided by the unscented Kalman filter (UKF) [8], which combines data from the visual odometry, the wheel odometry and the IMU. Erwhi Hedgehog is also implemented in Gazebo Simulator, which means that the robot can be simulated in different environments and conditions. The software is fully compatible with AWS RoboMaker. Vision tasks are implemented through the Intel OpenVINO framework.

3 Learning and Prototyping by Doing

3.1 Erwhi Hedgehog as a Teaching Tool

As a very recent project, Erwhi Hedgehog has not been used in schools yet. However, its small size has made it very popular at mass events, such as Maker Faires, where it facilitates the display of UGV (unmanned ground vehicle) technologies and functionalities to a wide audience, including families with young children. Erwhi Hedgehog has also been tested and used in ROS and advanced mobile robotics workshops as a benchmark platform for studying differences in various planners and SLAM algorithms. Hands-on testing of different planners and SLAM algorithms have proved to be an effective way of learning their distinctive features and weaknesses, such as their computational complexity. It also provides an opportunity for workshop attendees, mainly university students and freelancers, to use these platforms.

3.2 Learning by Modifying Hardware

Users can also add to or edit parts of Erwhi Hedgehog's hardware through the different sensors and actuators. For example, the robot can easily be connected to an Arduino board to make it usable with all Arduino-compatible sensors and actuators. This makes way for infinite application possibilities.

3.3 Learning New Technologies

The robot helps to test vision algorithms based on deep learning, such as object recognition. These algorithms can be tested directly, which makes it easier to understand how they work and how they can be used in robotics. Furthermore, AWS RoboMaker facilitates the creation and simulation of custom and complex software in the cloud, which eliminates the need for powerful local computers. In addition, it can also be used to deploy custom applications to Erwhi Hedgehog over the Internet of Things (IoT), making it is possible to work on a remote physical robot.

4 Conclusions

Erwhi Hedgehog is a good platform for gaining hands-on knowledge and experience of mobile robotics from the hardware to the software. It also represents the current status of mobile robotics, based on the latest industrial technologies. The platform should not be considered merely a standalone robot, but rather a reference design that students and educators can use to advance their projects, e.g., by adding computer vision to an existing project using only Erwhi Hedgehog software on different hardware.

The current focus of development in the project is updating and simplifying the robot.

References

1. Erwhi Hedgehog main Github repository. https://github.com/gbr1/erwhi-hedgehog (2019). Latest commit 930fb8c on 21 August 2019
2. Quigley, M., Conley, K., Gerkey, B., Faust, J., Foote, T., Leibs, J., Wheeler, R., Ng, A.: ROS: an open-source robot operating system. ICRA workshop on open source software, **3** (2009)
3. Chitta, S., Marder-Eppstein, E., Meeussen, W., Pradeep, V., Tsouroukdissian, A., Bohren, J., Coleman, D., Magyar, B., Raiola, G., Lüdtke, M., Perdomo, E.: ros_control: a generic and simple control framework for ROS. J. Open Source Softw. **2**, 456 (2017). https://doi.org/10.21105/joss.00456
4. STM32duino Arduino core Github repository. https://github.com/stm32duino/Arduino_Core_STM32 (2019). Latest commit 964576b 26 September 2019
5. Marder-Eppstein E., Berger, E., Foote, T., Gerkey, B., Konolige, K.: The office marathon: robust navigation in an indoor office environment. In: 2010 IEEE International Conference on Robotics and Automation, Anchorage, AK, pp. 300–307 (2010). https://doi.org/10.1109/ROBOT.2010.5509725
6. Teb_local_planner ROS package documentation. http://wiki.ros.org/teb_local_planner (2019). Latest access 6 October 2019
7. Labbé, M., Michaud, F.: RTAB-map as an open-source lidar and visual SLAM library for large-scale and long-term online operation. J. Field Robot. **36**(2), 416–446 (2019)
8. Robot localization ROS package documentation. http://docs.ros.org/melodic/api/robot_localization/html/index.html (2019). Latest access 6 October 2019

Educational Robotics and the Gender Perspective

Daniela Bagattini, Beatrice Miotti, and Fiorella Operto

Abstract In this paper we explore the role of stereotypes in educational choices. Data on secondary school enrollments show that girls are abandoning STEM subjects. There are many reasons for this, including social and family expectations, but also the perception that jobs and careers in technical and scientific sectors will make it hard to take care of a family. This is an important theme for the future. The number of jobs in ICT will increase, and the low quantities of women in these sectors will have a strong impact on the availability of skilled workers, as well as increasing the gender gap. What is the role of school in this context? What activities can get more girls interested in science? We focus, in particular, on how innovative approaches such as educational robotics can help girls engage with STEM subjects, as happened with the "Roberta" project, whose results will be illustrated in this work.

Keywords Coding · Educational robotics · STEM · Stereotypes · Gender perspective

1 What About the Gender Perspective?

One of the first objections we are used to hearing to this question is: "Isn't school already egalitarian? 82% of Italian teachers are female".

While this may be true, a high female representation does not automatically mean there is a special focus on gender issues.

D. Bagattini (✉) · B. Miotti
Istituto Nazionale di Documentazione, Innovazione e Ricerca Educativa (Indire), Firenze, Italy
e-mail: d.bagattini@indire.it

B. Miotti
e-mail: b.miotti@indire.it

F. Operto
Scuola Di Robotica, Genova, Italy
e-mail: operto@scuoladirobotica.it

D. Scaradozzi et al. (eds.), *Makers at School, Educational Robotics and Innovative Learning Environments*, Lecture Notes in Networks and Systems 240,
https://doi.org/10.1007/978-3-030-77040-2_33

Table 1 Percentage of female students by type of upper secondary school

Upper secondary school type	Girls
Music and dance; dance section	90.6
Human sciences	88.6
Languages	78.3
Human sciences—social economy option	71.1
Classics	70.1
Art	70.0
European/international	66.7
Technical institute—economics section	52.6
Science	48.9
Music and dance; dance section	47.6
Vocational	43.7
Vocational—3 years (IeFP)	34.9
Science—applied science option	32.8
Science—sports section	29.9
Technical institute—technology section	16.9

Source Ministry of Education, University, Research, https://dati.istruzione.it/opendata/

If we go back to degree courses by gender, we see that only 24% of teachers of technology subjects are female. This is the result of vertical segregation, which has continued through the years. Even in 2017/18, enrolments in upper secondary schools were heavily sectorialized by gender. (Table 1).

When we look at university careers, again we see gender differences in the various macro-areas, and even wider gaps on certain courses (Table 2).

On some degree courses within these macro-areas, the differences between the genders are even greater. For example, 75% of the teaching staff on the "Healthcare, nursing and midwifery" course (L/SNT1) are women.

Healthcare, however, is the area that has seen most change.

Science faculties are where the greatest differences are recorded (as are the humanities for males). The percentage of female students on the "Computer Science and Technology" degree, which includes courses in information science, and computer

Table 2 Percentage of women in university macro-areas

University macro-area	Female (%)
Humanities	78.8
Healthcare	69.5
Sociology	55.7
Science	37.9

Source Ministry of Education, University, Research, http://anagrafe.miur.it/index.php

technology, is 12%, whereas 21% of the students on the "Information Engineering" degree—which includes courses in electronic engineering, computer engineering, automation engineering, information engineering, etc.—are women.

Thus, there are still strong divisions, defined as "segregations" in educational choices, not because they are imposed, but, as analysis shows, because they are self-limitations on the range of possibilities. In these self-limitations, however, context and formal and informal guidance systems play a crucial role, particularly the expectations of parents and peer groups, but also those of the school system itself, and of career guidance. The gender perspective should also be taken into account in guidance actions.

A recent survey by Biemmi [1] investigated the motivations and paths of girls and boys enrolled in courses in which they are in the minority. Two very different universes emerge from the survey: for males, the decision to pursue a career in care is made for reasons that are endogenous to the school system and, often, it is a "second choice". Boys come to be there after starting other types of work or educational pathways, perhaps having come into contact with that type of work by chance. This is important for the success of training.

Girls on science courses, on the other hand, appear to be more determined. For some, choosing this type of path is almost a challenge. In general, these girls perform well at school and are less swayed by stereotypes, even if they wonder whether they will be able to reconcile their personal lives with their careers.

These girls have chosen to swim against the tide.

2 Why Talk About Gender Perspective and Educational Robotics?

The data in Tables 1 and 2 relate to enrollment in secondary schools and universities, which girls choose at an age when they and their families are already considering their future social aspirations based on the school they will attend. If we consider data on girls' involvement in STEM subjects in primary or middle school, we find this age range to be less open. This is the age at which these trends begin to manifest, as discussed by Banzato and Tosato, [2] "Several studies show that the differences between males and females begin to emerge in the transition of children from primary to secondary school" (p. 315).

The social expectations passed on by society and their families induce many girls to choose secondary schools on the basis of perceived opportunities for future employment. As support for working mothers is painfully inadequate in Italy, girls, consciously or otherwise, perceive that jobs and careers in technical and scientific sectors will make raising a family difficult, if not problematic [3].

Hence, the prophecy comes true: even girls who are good at mathematics or science, gradually turn away from engineering and physics towards a career in teaching, or the braver among them choose medicine, instead.

For girls and women, teaching is seen as a chance to combine work and a private life: part-time work with less career pressure.

In our case, the glass ceiling metaphor represents the invisible, transparent barriers that prevent girls from undertaking courses and careers in the technical and scientific sectors. So-called role models, that is, successful women in the fields of science and technology or entrepreneurship, are often so far removed from the average person and so unreachable that reference to them is sometimes discouraging. If you have to be a wonder woman to sign up for physics, how can normal girls break through the glass ceiling?

On the other hand, Italian families offer a disarming picture of what it means to be skilled in STEM subjects: the advice—which is unthought-of and unrelated to reality, which often comes from the families themselves—is that technical studies open up more opportunities for employment.

An overview of the situation in Italy will clarify many aspects.

The OECD Report "Education at Glance 2017/An eye on education" indicates that only 18% of Italians are graduates and in subjects not very closely related to economic developments. The OECD places Italy in second last place, only ahead of Mexico, whereas the average in the 35 most influential countries is 37% graduates [4].

3 A Multi-faceted Approach to a Complex Problem

As we can see, there are various reasons for the lack of women in Italy involved in technical and scientific fields and the solutions to these are not quick and simple.

School is often inaccurately thought to be the primary cause of girls' disaffection with STEM subjects. However, while it may not be the primary cause, school is certainly an important source of the problem.

For years, Italian schools have had projects dedicated to promoting STEM subjects among girls, whereas associations and centers for Girls in STEM, Girls for Coding have sprung up in society at large.

In 2008, Scuola di Robotica brought the Roberta project to Italy, aimed at getting girls to discover robots. The Roberta project was begun by the Fraunhofer Institute in Bonn, and is dedicated to developing a methodology for girls in STEM, using educational robotics [5].

The Roberta methodology included phases in which the genders were separate in the first few hours of the course. This was to enable the girls to learn to code the robot on their own, without, as often happens, delegating the technical aspects to their male counterparts. After this, male and female participants worked together again. Since the girls did not like football, competitions with robots or "robot wars", the mission of Roberta robots was to simulate the behavior and communication of animals. One of the best known missions was the "flight of the bees", where several robots simulated communication between bees informing their companions where there were many flowers.

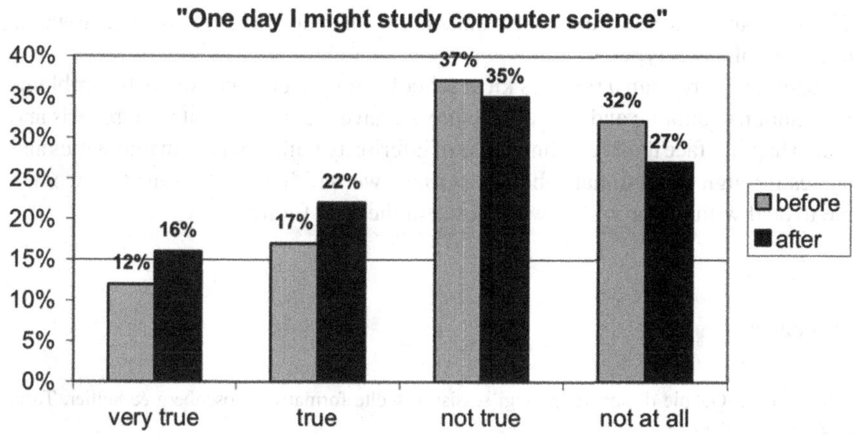

Fig. 1 Students' answers before and after a Roberta course (N = 499)

In 2007, the University of Bremen was tasked with evaluating the project's scalability and impact, with excellent results. In addition, many of the girls who participated in Roberta chose to study science. The table shows students' comments on the statement "I could study computer science", before and after participating in Roberta (Fig. 1).

The Roberta project and the many others that have been conducted in Italy to promote STEM among girls have shown that educational robotics is an excellent tool for this objective. Designing, building and programming a small robot involves many cognitive and social skills, and all students can in some way contribute to the project, which is a shared, group project.

4 Conclusion

Returning to our question, what do these data tell us? Why do we need them?

Stereotypes affecting educational choices restrict the freedom of both female and male students.

It is important to reflect on how educational choices are made and how they are affected by stereotypes. This enables us to broaden the spectrum of what is "thinkable" for male and female students alike, and removes the obstacles that limit their freedom of choice.

In addition to this, there are important consequences for the future of employment. The number of jobs in ICT will increase, and the low quantities of women in these sectors will have a strong impact on the availability of skilled workers, as well as increasing the gender gap [6]. Moreover, the new jobs of the future seem to go in the direction of greater interdisciplinarity, requiring more cross-disciplinary transversal skills compared to traditional sectorialization: one more reason to work towards getting rid of stereotypes.

Learning to program a robotics kit at school certainly does not solve the problems in the Italian economic and social structure we have mentioned, but if more girls and women happily face their fears and sense of inferiority with regard to mathematics and physics through educational robotics, perhaps we will have more sensitive women able to deal with the problems we will see in the near future.

References

1. Biemmi, I.: Gabbie di genere, retaggi sessisti e scelte formative. Rosenberg & Sellier, Turin (2016)
2. Banzato, M., Tosato, P.: Narrative learning in coding activities: gender differences in middle school. Formazione e insegnamento **1**, 339–354 (2017)
3. Ballona, E.S., Taviani, S.: Le Equilibriste. La maternità in Italia, save the children. https://s3.savethechildren.it/public/files/uploads/pubblicazioni/le-equilibriste-la-maternita-italia_1.pdf (2019). Last access October 2019
4. AAVV, OECD's annual education at a Glance. https://www.oecd-ilibrary.org/education/education-at-a-glance-2017_eag-2017-en (2019). Last access October 2019
5. Hartmann, S., Wiesner, H., Wiesner-Steiner, A.: Robotics and gender: the use of robotics for the empowerment of girls in the classroom. In: Zorn, I., et al. (eds.) Gender Designs IT. Construction and Deconstruction of Information Society Technology. Springer Link (2007)
6. Berra, Cavaletto: Scienza e tecnologia. Superare il gender gap. Ledizioni, Turin (2019)

European Recommendations on Robotics and Related Issues in Education in Different Countries

Michele Domenico Todino, Giuseppe De Simone, Simon Kidiamboko, and Stefano Di Tore

Abstract This short paper describes the preliminary phase in an innovative line of research comparing educational robotics in Italy and other countries, from the perspective of media education, and based on the European Parliament recommendations to the Commission on civil law rules on robotics. More specifically, all decision processes that affect digital citizenship should have the support of children and teenagers. For these reasons, this paper looks at the work of a group of Italian high school students in the fifth year of upper secondary school, who formulated a SWOT analysis to highlight their attitudes to robotics issues in relation to the European Union recommendations. This research started in 2018 and will be repeated this academic year with Italian and Congolese students—from the Institut Supérieur des Techniques Appliquées—with a qualitative analysis to establish student attitudes to robotics issues. Qualitative analysis was selected because the SWOT analysis is already divided into information categories, revealing a variety of concepts that are grouped together from the collected data. These results will be compared with any obtained in future years in Italy and other countries, to find further potential patterns.

Keywords EU recommendations · Robotics · Artificial intelligence · SWOT analysis on robotics · Media education

---

M. D. Todino (✉) · G. De Simone · S. Di Tore
University of Salerno, Fisciano, Italy
e-mail: mtodino@unisa.it

G. De Simone
e-mail: gdesimone@unisa.it

S. Di Tore
e-mail: sditore@unisa.it

S. Kidiamboko
Institut Supérieur Des Techniques Appliquées, Kinshasa, Congo
e-mail: simon.kidiamboko@ista.ac.cd

D. Scaradozzi et al. (eds.), *Makers at School, Educational Robotics and Innovative Learning Environments*, Lecture Notes in Networks and Systems 240,
https://doi.org/10.1007/978-3-030-77040-2_34

1 Introduction

This paper compares ICT education in robotics and artificial intelligence (AI) [1] in Italy and the Democratic Republic of Congo (DRC) purely in terms of media education [2]. It refers to other studies to describe the history [3], the service to citizens [4], the risks and myths [5] and the current and future levels of development in robotics and AI [6]. The starting point for this research is the recommendations to the European Commission on civil law rules on robotics [7] written in 2017 by the European Parliament's Committee on Legal Affairs. The focus of the recommendations is digital citizenship, which is needed for self-actualization, employability, social inclusion and responsibility. According to article 5 of the law, students should analyze, compare and critically evaluate data information, which today often comes from robots and machines that use artificial intelligence; students should identify appropriate forms of digital communication, robotics are a form of media; students should "be able to avoid risks using digital technologies for health and threats to one's physical and psychological well-being," and here it is important to know the risks linked to robots [8]. Thus, this law encouraged educational research to improve certain criteria for implementing these topics alongside students, teachers, families and industry experts. More specifically, all decision processes that affect digital citizenship should have the support of children and teenagers.

2 An Explorative Student SWOT Analysis on Robotics

For the above reasons, this paper has the following process: an educator in media studies in a fifth-year class (Table 1) at Istituto Statale Caravaggio upper secondary school in San Gennaro Vesuviano, Italy (www.iscaravaggio.gov.it) invited the students to prepare a SWOT analysis on the European recommendations on robotics. This would involve compiling a list of points highlighting their opinions and beliefs on robotics, based on their real-life experience. These ideas were then classified as strengths, weaknesses, opportunities and threats, and presented in table form along with direct quotes from the European document. Table 2 shows the results of the survey conducted in the 2018/19 academic year. The survey will be repeated in 2019/20 on new sets of students in two countries: (1) approximately 50 students in two fifth-year classes of the same upper secondary school in Italy; (2) approximately 75 students in their first year of university at Institut Supérieur des Techniques Appliquées in DRC; the research will take place in the fall semester to ensure the

Table 1 Details of fifth-year upper secondary class (2018/19)

Year of birth	Male	Girls
2000	13	2
2001	1	3

Table 2 Results of the SWOT analysis of Italian upper secondary school students (2018–19)

Dimension	Text quoted from the EU recommendations to summarize the students' opinions
Strengths	"automation of jobs has the potential to liberate people from manual monotone labour allowing them to shift direction towards more creative and meaningful tasks" (introduction, point J) [7]
Weaknesses	"whereas in the face of increasing divisions in society, with a shrinking middle class, it is important to bear in mind that developing robotics may lead to a high concentration of wealth and influence in the hands of a minority" (introduction, point K) [7]
Opportunities	Build "homecare and healthcare robots" (preamble to the code of ethical conduct for robotics engineers) [7]
Threats	"Considers that, as is the case with the insurance of motor vehicles, such an insurance system could be supplemented by a fund in order to ensure that reparation can be made for damage in cases where no insurance cover exists; calls on the insurance industry to develop new products and types of offers that are in line with the advances in robotics" (point 58) [7] and "establishing a compulsory insurance scheme where relevant and necessary for specific categories of robots whereby, similarly to what already happens with cars, producers, or owners of robots would be required to take out insurance cover for the damage potentially caused by their robots" (point 59-a) [7] besides "creating a specific legal status for robots in the long run, so that at least the most sophisticated autonomous robots could be established as having the status of electronic persons responsible for making good any damage they may cause, and possibly applying electronic personality to cases where robots make autonomous decisions or otherwise interact with third parties independently" (point 59-f) [7]

students are of comparable age. The Italian students were chosen for the research because there was open access to the school for their participation; the same is true for the students in Kinshasa. Researchers expect to find similarities between the Italian and Congolese SWOT analyses. Finally, this process aims to follow best practices in inclusive education [9], to find a proven approach to robotics, from a media education perspective, and share this approach with other countries.

As described above, a media educator first read the EU recommendations to the students. The students were then invited to state their opinions of them in terms of their strengths, weaknesses, opportunities and threats, for the purposes of the SWOT analysis. Classmates decided to confirm or reject each idea as a strength, weakness, opportunity or threat, or revise the opinion for inclusion in another section of the SWOT analysis. The reading of the recommendations and the students' work to produce the final version of the shared SWOT analysis with the quoted text took approximately eight hours.

In summary, Table 2 shows that students are wary of the threats of robotics, but they hope it will improve homecare, healthcare and liberate them from manual jobs, although this may not decrease divisions in society. These data will be compared with any obtained in future years in Italy and other countries, to find potential patterns.

3 Italy and DR Congo: A Predictive Comparison

Every year, the education system in DRC produces large numbers of graduates, who, unfortunately, usually struggle to get work. As a result, some of them try to start their own businesses to avoid unemployment. For example, one Congolese startup has developed "Taxmwinda," a platform for paying taxes. This platform uses a white-box approach [10] and includes four subsets: a database, an SMS handler, and a mobile payment layer. Another Congolese startup "AgroMwinda" launched a web platform to support small farmers and local organizations, by providing innovative solutions and attractive services that make them more visible to customers. Finally, a Congolese women's technology startup led by Thérèse Izay, an engineer from Institut Supérieur des Techniques Appliquées, designed a robot that controls traffic lights, and will replace some traffic police on the street. The government will partly fund the production of more robots. Interest in robotics in DRC has led to the idea of an international collaboration to repeat the research previously done in Italy, as described above, at the Institut Supérieur des Techniques Appliquées in the DRC's capital, Kinshasa. To start this collaboration, the Italian researchers asked their Congolese colleagues a number of questions (Table 3) that can be generalized for other international institutions that might decide to join this research.

4 Conclusion

In conclusion, in the 2019–20 academic year, the Italian and Congolese researchers will work in parallel, gathering new data from a total of 50 Italian students (who will be added to those involved in 2018–19) and 70 Congolese students. These data will undergo a qualitative analysis because the SWOT analysis is already divided into information categories, revealing a variety of concepts that are grouped together from the collected data. The data will also be analyzed on the basis of several research questions: (1) Do Italians and Congolese have common parameters for evaluating the strengths, weaknesses, opportunities and threats of robotics? (2) Are there any cultural factors that might align the previous considerations? Obviously, the limited number of students involved in this research means that the resulting data are not analyzed quantitatively. The aim of the study is to introduce the topic of media education in robotics, which needs to be explored in more depth. This paper simply describes the activities of the researchers and shows the preliminary results. It further aims to present a new perspective for the introduction of robotics to fifth-year upper secondary school classes, as part of media education, and shows an experience that can help advance theoretical knowledge in educational robotics, and its effect on contemporary society, which students should be learning at school.

Table 3 Answers provided by the researchers at the Institut Supérieur des Techniques Appliquées (ISTA). Please note that the Congolese research group is aware of other research published by their Italian colleagues

Question	Answer
(1) Is robotics regulated by laws in your country?	"No, till now there is no regulation about robotics in DR Congo even though robotics is being used in daily life in our country"
(2) Do high schools have robotics lessons in your country?	"No, robotics is studied, but this subject starts at university. Besides, white-box approaches in robotics could be useful in DR Congo, because it will help students to better understand complex systems based on artificial intelligence software"
(3) Would it be useful to read the European recommendations in high schools in your country?	"Yes, it is important to publish and share the European Parliament recommendations on robotics in Congolese universities, and discuss them, like it has been done in Italy so that students can be aware of them"
(4) How will you plan this research in your country?	We will be discussing the European Parliament recommendations on robotics with 75 students in the first year of the computer science and electrical engineering course, helped by researchers, during the academic year (2019/2020) in the Mechatronic Lab at the ISTA, Kinshasa. This project is important due to the fact that applications based on artificial intelligence (robotics and expert systems) are popular trends; thus, people who are not in touch need to start this contact to improve their technological skills and have more digital citizenship, also in DR Congo. Even though, there is no regulation, for the time being, of artificial intelligence in DR Congo, everyone tends to apply it when needed. At this point, ISTA has created an Artificial Intelligence work group in DR Congo (IARDC) as its contribution to push this field forward in the country

References

1. Russell, S.J.A., Norvig, S.J.: Artificial Intelligence: A Modern Approach. Prentice-Hall, Chicago (2003)
2. Rivoltella, P.C.: Media education. In: Rivoltella, P.C., Rossi, P.G. (eds.) Tecnologie per l'educazione. Pearson, Milan (2019)
3. Stanford University Report. One hundred year study on artificial intelligence. Report of the 2015 study panel (September 2016)
4. GID, The agency for digital Italy. White paper on artificial intelligence at the service of citizens. Version 1.0 March. https://ia.italia.it/assets/whitepaper.pdf (2019). Last accessed 12 September 2019

5. IBM. Demystifying artificial intelligence in risk and compliance. https://www.ibm.com/downloads/cas/DLJ28XP7 (2019). Last accessed 12 September 2019
6. IMCO Committee, European Parliament. State of the art and future of artificial intelligence. Department for Economic, Scientific and Quality of Life Policies Directorate-General for Internal Policies Author: Dr. Aleksandra Przegalinska PE 631.051—February (2019)
7. Delvaux, M., Mayer, G., Boni, M.: With recommendations to the commission on civil law rules on robotics (2015/2103(INL)). http://www.europarl.europa.eu/sides/getDoc.do?pubRef=-//EP//TEXT+REPORT+A8-2017-0005+0+DOC+XML+V0//EN (2019). Last accessed 12 September 2019
8. European Union. DigComp 2.0: the digital competence framework for citizens. https://publications.jrc.ec.europa.eu/repository/bitstream/JRC101254/jrc101254_digcomp%202.0%20the%20digital%20competence%20framework%20for%20citizens.%20update%20phase%201.pdf (2016)
9. Schwab, S.: Attitudes towards inclusive schooling. A Study on Students', Teachers' and Parents' Attitudes, p. 117. Waxmann, Münster (2018)
10. Todino, M.D.: Simplexity to orient media education practices. Aracne, Rome (2019)

Growing Deeper Learners. How to Assess Robotics, Coding, Making and Tinkering Activities for Significant Learning

Rita Tegon and Mirko Labbri

Abstract As more and more unstructured project-based activities fill the learning time of students, there is a growing need for assessment models for educational robotics, tinkering, making and coding activities. On one hand, there is a poor understanding or an underestimation of the need for evaluation, and its ability to improve systems and learning outcomes, while on the other, it is difficult to identify or devise suitable assessment frameworks. Examples from the international context are discussed more for their potential to raise awareness than as definitive answers.

Keywords STE(A)M · Assessment · Educational robotics · Tinkering · Making · Coding · Teacher training · Mindtools

1 Introduction

The spread of robotics, coding, making–tinkering activities in K-12 education in Italy is in part due to PNSD (National Plan for Digital Education, a policy launched to define a strategy for innovation within the school system and move it into the digital age) [1]. Its Action 15 established innovative scenarios for the development of applied digital skills. Since the 2014/15 academic year, the Ministry of Education has promoted its "Program the Future" (Action 17) [2] campaign to offer all students courses in making, robotics, and the Internet of Things.

These are also seen as key developments in STEM/STEAM (science, technology, engineering, art and mathematics) curricula focused on creating future employees in job markets with growing demands [3–5]. However, while the fact that they are enjoyable is certainly appreciable, it is not enough, since, according to the PNSD, efforts should be focused on the epistemological and cultural dimensions. These

R. Tegon (✉)
Liceo Ginnasio Statale "A. Canova", Via Mura San Teonisto, 16, 31100 Treviso, Italy

M. Labbri
Istituto Comprensivo Statale "S. Barozzi", Via Isidoro Mel, 8, 31020 San Fior (TV), Italy

D. Scaradozzi et al. (eds.), *Makers at School, Educational Robotics and Innovative Learning Environments*, Lecture Notes in Networks and Systems 240,
https://doi.org/10.1007/978-3-030-77040-2_35

are important methodological–didactic challenges for teachers that have no ready-made answers. On the other hand, the goal is completely clear: to promote learning, or rather, to grow deeper learners. To address this critical issue, teacher training programs should raise awareness of the fact that there is no such thing as a neutral medium [6]. All digital tools and all digital environments carry cognitive and social implications, and digital environments can provide the support for strategic and deep learning. Yet, while it is important to understand the crucial role of evaluation methods for improving the system's effectiveness [7], teaching practices and research (which is quite generous in other digital activities) both seem to fall short on suitable answers for evaluating robotics, coding, making and tinkering activities. We present several models here, to focus on this issue.

2 Robotics, Coding, Making, and Tinkering as Mindtools

Robotics, coding, making, and tinkering all involve the learning-by-doing approach and relate to the gluing action of computational thinking, which "represents a universally applicable attitude and skill set that everyone, not just computer scientists, would be eager to learn and use" [8]. They are all considered powerful tools that can even support learning in special education. However, in addition to displaying affinities, they also have their specificities.

Tinkering and making are both technology-based extensions of DIY (do-it-yourself) culture, which intersects with hacker culture, but they are more focused on physical rather than software objects and concepts. Moreover, tinkering sits at the more creative and improvisational end of the continuum.

Coding is linked to computer programming in English-speaking countries, whereas in some countries, like Italy and Spain, it has become a way of referring to visual and block programming, and a series of unplugged activities with educational objectives.

It is undeniable that there is a lack of agreement on the definition of robotics (educational robotics, educational robots, robots in education and robots for education), but consensus is strong about their potential. Angel-Fernandez and Vincze [9] suggest that educational robotics (ER) is a field of study that aims to improve the learning experience through activities, technologies, and artifacts in which robots play an active role. Therefore, the use of the *pedagogical activity* tag is suggested for activities in ER that have clear learning outcomes and evidence of learning.

A recent paper by Scaradozzi, Screpanti and Cesaretti [10] suggests that robots in education should be considered a broader field encompassing a wide range of applications, from assistive robotics to social robotics or socially assistive robotics [11]. Robotics in education (RiE), they emphasize, is not the same as ER, which is based on constructionism and the need for active exploration of the artifact. Robots playing an active role—as in social robots teaching or assisting teachers—cannot be defined as educational robots. Such activities can be beneficial for some aspects of learning, but they do not help children become "active prosumers of technology".

Hence, if, as Jonassen says, "mindtools are knowledge construction tools that learners learn with, not from" [12], ER, coding, making and tinkering can rightly be considered mindtools, insofar as they support knowledge construction [13].

3 The Underlying Pedagogies

Generally speaking, all these activities within the STEM/STEAM framework are a way to develop emotional, social and cognitive skills. Maker education is not about things, but rather connections, community and meaning. Also, maker-centered learning environments are built on educational theories like constructivism and constructionism, in order to develop knowledge and awareness through interactive, open-ended, student-driven, multi-disciplinary experiences. Although active learning is recognized as a higher-impact method in education, a recent study shows that most STEM teachers still choose traditional teaching methods (which are also preferred by students). This could be due in part to the increased cognitive effort required during active learning.

Researchers at Harvard University have recently discussed this issue in an article entitled "Measuring actual learning versus feeling of learning in response to being actively engaged in the classroom" [14].

4 Through the Maze of Assessment

School education is essentially about defining a curriculum, teaching and learning, and assessment. Of the three, assessment is the weak factor. Its use in formal assessment programs is for certification and accountability. But when it is used formatively (see feedback, in particular), it is one of the most powerful interventions found in the educational research literature that can improve learning and teaching and therefore the entire school system [15].

However, identifying models and tools to assess relatively new educational activities like robotics, coding, making, and tinkering is a an additional critical issue, because of a lack of models. Owing to their fundamental focus on collaborative, iterative, process-based learning, progress in the key 21st-century skills developed in maker education cannot be suitably tracked using traditional assessment methods. With their emphasis on cognitive, social and emotional aspects, traditional methods, on the other hand, are plentiful and often interconnected. Teachers, therefore, first have to identify the subject and transversal results they expect, define which dimensions need to be developed, and then observe them. After this, they have to choose an effective assessment model and, finally, evaluate. The challenge (but also the pedagogical benefit) can be understood by considering, for example, the five dimensions of The Tinkering Learning Dimensions Framework [16]: initiative and intentionality, problem-solving and critical thinking, conceptual understanding, creativity and self-expression, social-emotional engagement. Nevertheless, it is a challenge that needs to be met, in order to address the issue of how maker-centered learning can truly benefit the broader education system, for both students and teachers, and improve the quality of learning.

Below are presented validated evaluation frameworks from the international landscape.

4.1 Assessing Students' Work in Robotics (The Digital Technologies Hub)

Robotics-based tasks feature different types of knowledge and cognitive processes related to the digital, mathematical and socio-cultural contexts inherent to robotics-based learning. They are so tightly connected that they can be difficult to spot and observe. The Digital Technologies Hub developed by Education Services Australia for the Australian Government Department of Education [17] provides a large array of learning resources and services to support the implementation of quality Digital Technologies programs and curricula in schools. Among these can be found advice and resources to assist teachers in creating quality assessment tasks, that are flexible, but closely connected to the Australian Curriculum Achievement Standards. First, it is important to emphasize which elements of the achievement standard are to be targeted, and why and how.

4.2 Feedback, AfL, PASA

While the contexts and the key functions of evaluation (of the system, the school, for certification and selection purposes, and accountability) are clear, the common denominator of all of them is improvement of learning outcomes. As feedback strategies play a crucial role, in a frequently quoted article, Hattie and Timperley [18] present a conceptual analysis of feedback and analyze the evidence relating to its impact on learning and student performance. Furthermore, according to Popham, [19] feedback is more powerful when used by students themselves to adjust their learning strategies. By this he means peer feedback, and suggests it should be carried out from an early age. But feedback is what happens second, that is, after performance itself, through doing or making.

Starting from similar premises, in 1998, Black and Wiliam [15] had already developed the AfL (Assessment for Learning) strategies, which enabled students to become more independent in their learning, taking part in self-assessment and peer assessment. The AfL movement encourages educators to use assessment data primarily for formative purposes. This movement also features peer and self-assessment (PASA).

There are many PASA models, but it is interesting to observe that better learning outcomes are achieved if the evaluation criteria are negotiated with the students themselves.

Catlin [20] illustrates the role of PASA in the successful use of ERA (Educational Robotic Applications). In fact, PASA is an intrinsic aspect of educational robotics activities, because they normally involve students working in groups, sharing, discussing and evaluating each other's ideas on an almost continuous basis.

4.3 Assessing the Development of Computational Thinking: The ScratchEd Framework

Scratch is one of the best known environments employed to support computational thinking. Its first version was developed by the Lifelong Kindergarten Group at the MIT Media Lab, in 2003 [21]. A second version was released in 2013. In early 2019, the focus of energy shifted away from the ScratchEd online community to other ways of supporting educators working with Scratch [22]. Three key dimensions have been defined: computational concepts, computational practices, and computational perspectives. To assess them they have relied primarily on three approaches: artifact-based interviews, design scenarios, and learner documentation. The focus is on evolving familiarity and fluency through computational thinking practices.

Dr. Scratch, an analytical Beta tool that evaluates Scratch projects, should also be mentioned. Teachers can group their students to keep track of their progress rapidly and simply [23].

4.4 Assessing Collaborative Problem-Solving

Since a prominent feature of robotics, coding, making and tinkering activities is collaborative problem-solving, it may be interesting to look at it from the point of view of assessment. For this we will consider the framework developed by the Assessment Research Centre of Melbourne Graduate School of Education [24].

Collaborative problem-solving can be a simplifying element that makes the students collaborate on problems that lead them to learn higher-order skills in science, mathematics, history or even physical education. In other words, it is in itself a non-cognitive skill, or it is a skill that promotes other domain-specific and cognitive competences. The project identified many indicators for interpreting scales derived from five dimensions (participation, perspective-taking, social regulation, task regulation, and knowledge building), two dimensions (social, cognitive) or one general dimension of a collaboratively solved problem. The sound framework provides teachers with an opportunity to identify the student's Vygotsky zone of proximal development for instructional intervention. Even better, it allows the use of development progressions as an assessment strategy that can be explored.

5 Conclusions

Students and families seem to appreciate the spread of robotics, coding, making-tinkering activities in K-12 teaching practices, because of the element of fun and the commitment they promote, but also because they prepare them for employment opportunities.

The efforts of the school system should focus instead on the epistemological and cultural dimensions of these areas. Indeed, research shows they are rich in cognitive, social and emotional affordances that should be suitably encouraged. Their impact should be tracked to see whether the expected outcomes are reached.

As assessment is one of the most powerful interventions, according to the literature in educational research, and is crucial to improving learning, teaching and therefore the entire school system, appropriate assessment frameworks and tools should be introduced, to close achievement gaps and increase equity. However, identifying models and tools to assess relatively new educational activities like robotics, coding, making, and tinkering, all of which are collaborative, iterative and process-based forms of learning, is an additional critical issue, because of a lack of models. In addition to training in different pedagogical and educational paradigms, this critical issue should be addressed urgently through specific training in assessing activities carried out in the digital context and with digital tools [25].

References

1. PNSD. http://www.istruzione.it/scuola_digitale/index.shtml (2019). Last accessed 07 December 2019
2. Programma il Futuro Homepage, https://programmailfuturo.it/ (2019). Last accessed 07 December 2019
3. Caprile, M., Palmén, R., Sanz, P., Dente, G.: Encouraging STEM studies, labour market situation and comparison of practice target at young people in different member states. European Parliament, Policy Department A: Economic and Scientific Policy (2015)
4. Fayer, S., Lacey, A., Watson, A.: STEM occcupations: past, present, and future. US Bureau of Labor Statistics (2017)
5. Langdon, D., McKittrick, D.B., Beethika, K., Doms, M.: STEM: good jobs now and for the future. U.S. Department of Commerce, Economics and Statistics Administration (2011)
6. McLuhan, M., Fiore, Q.: The Medium is the Massage. Random House, New York (1967)
7. Hill, P., Barber, M.: Preparing for a Renaissance in Assessment. Pearson, London (2014)
8. Wing, J.: Computational thinking. Commun. ACM **49**(3) (2006)
9. Angel-Fernandez, J., Vincze, M.: Towards a definition of educational robotics. In: Austrian Robotics Workshop. Innsbruck University Press (2018)

10. Scaradozzi, D., Screpanti, L., Cesaretti, L.: Towards a definition of educational robotics: a classification of tools, experiences and assessments. In: Daniela, L. (ed.) Smart Learning with Educational Robotics, pp. 63–92. Springer, Cham (2019)
11. Siciliano, B., Khatib, O. (eds.): Springer Handbook of Robotics. Springer (2016)
12. Jonassen, D., Carr, C.: Mindtools: affording multiple knowledge representations for learning. In: Computers as Cognitive Tools, No More Walls, vol. 2, no. 6 (2000)
13. Mikropoulos, T., Bellou, I.: Educational robotics as mindtools. Themes Sci. Technol. Educ. **6**(1), 5–14 (2013)
14. Deslauriers, L., McCarty, L.S., Miller, K., Callaghan, K., Kestin, G.: Measuring actual learning versus feeling of learning in response to being actively engaged in the classroom. In: Proceedings of the National Academy of Sciences, vol. 116, no. 39, pp. 19251–19257. Kenneth W. Wachter, University of California, Berkeley, CA (Septemer 2019)
15. Black, P., Wiliam, D.: Inside the black box. King's College London (1998)
16. The Tinkering Learning Dimensions Framework. https://www.exploratorium.edu/ (2019). Last accessed 07 December 2019
17. DTH. https://www.digitaltechnologieshub.edu.au/ (2007). Last accessed 07 December 2019
18. Hattie, J., Timperley, H.: The power of feedback. Rev. Educ. Res. **77**(1), 81–112 (2007)
19. Popham, W. J.: Transformative assessment. Association for Supervision and Curriculum Development, Alexandria (2008)
20. Catlin, D.: Using peer assessment with educational robots. ICWL 2014 International Workshops. SPeL, PRASAE, IWMPL, OBIE, and KMEL, FET Tallinn, Estonia (2014)
21. MIT Teaching System Lab. Beyond rubrics: Moving towards embedded assessment in maker education. https://tsl.mit.edu/assessment-in-maker-ed/ (2019). Last accessed 07 December 2019
22. DrScratch. Homepage, http://www.drscratch.org/v2 (2007). Last accessed 07 December 2019
23. Scratch. http://scratched.gse.harvard.edu/ct/assessing.html (2019). Last accessed 07 December 2019
24. Griffin, P.: Assessing Collaborative Problem Solving. ARC, Melbourne (2014)
25. Eurydice Report. Digital education at school in Europe. https://eacea.ec.europa.eu/erasmus-plus/news/digital-education-at-school-in-europe_en (2019). Last accessed 07 December 2019

Buzzati Robots

Matteo Torre

Abstract This paper describes didactic and methodological choices on a course inspired by the story "The Seven Messengers" by Dino Buzzati. The aim is to investigate and introduce basic mathematics and physics skills. The project involves 35 students in the first year of the experimental "Liceo Matematico" school, promoted by the "G. Peano" Department of Mathematics of the University of Turin.

Keywords Robotics · Mathematics education · Mathematics and literature · Buzzati

1 Introduction

This paper describes a multidisciplinary didactic experience in mathematics and Italian literature that is based on the short story entitled "The Seven Messengers" by Dino Buzzati [1]. This idea has already been proposed by other authors [2], but the uniqueness of this version is the approach to the problem through educational robotics. The pedagogical principles of educational robotics can be found in the works of Seymour Papert [3, 4], who described the benefits of using construction and programming kits (in our case, Ozobots) for transforming students into creators of their own teaching tools and protagonists of their own learning. The strength of our teaching experience is, in my opinion, the introduction of robots to analyze and model a mathematical situation taken from a piece of literature.

The purposes are:

- To increase the time and quality of attention, compared to traditional lessons;
- To facilitate the learning of abstract concepts (space, time and speed) by representing them in a concrete context;
- To help students develop key scientific and linguistic skills by analyzing a literary text with a scientific theme.

M. Torre (✉)
Liceo Scientifico Statale "G. Peano", Tortona, Alessandria, Italy

D. Scaradozzi et al. (eds.), *Makers at School, Educational Robotics and Innovative Learning Environments*, Lecture Notes in Networks and Systems 240,
https://doi.org/10.1007/978-3-030-77040-2_36

This didactic experience took place during the 2018/2019 academic year and involved 35 students (15 male and 20 female) attending the first year of Liceo Scientifico "L. B. Alberti" in Valenza (Alessandria). The students are part of the national "Liceo Matematico" project proposed by the University of Turin, whose purpose is to enhance excellence among high school students, and increase their interest in technical and science subjects at university courses. The didactic experiment took place in five lessons over nine hours (morning and afternoon) and involved two high school teachers: my colleague, a computer teacher, and me.

2 The Ideas in the Project

The teaching project is based on the short story "The Seven Messengers" by Dino Buzzati. Below is a summary of the story, which will help the reader understand our intentions. The students were provided with the original complete version of Buzzati's text.

> A prince sets out to explore his father's kingdom and reach its outer frontiers. He takes with him seven messengers whom he sends back and forth between himself and the capital, to communicate with his family. To distinguish them more easily, he gives them these names: Alessandro, Bartolomeo, Caio, Domenico, Ettore, Federico, and Gregorio. As the days, months and years go by, the distance grows and communications become rarer. On the second day of the journey, the first messenger, Alessandro, returns to the capital. Meanwhile, the prince and his caravan continue traveling at a constant speed. In the following days the other messengers, Bartolomeo, Caio, Domenico, Ettore, Federico and Gregorio, leave consecutively following these rules: as one messenger catches up with the caravan (in the evening), he delivers the news to the prince, stops for the night and departs again in the opposite direction. The messengers travel at one and a half times the speed of the prince's caravan.

After reading the original text, we give the students two questionnaires. The first is on the textual analysis of the story and contains these questions:

1. What elements are relevant to understanding the text?
2. What elements are relevant to understanding it from a logical point of view?
3. Are there any elements in the story that explicitly refer to the fields of mathematics and physics? If so, do you consider these elements to be essential to your understanding of it?

The second questionnaire, on mathematics, has the following questions:

1. After how many days, from the start of the journey, will Alessandro be back?
2. More generally, if a messenger leaves on the *nth* day, what day will he return?
3. After how many days, from the start of the journey, will each of them be back?
4. Can the sequence of departures be expressed in a mathematical formula?
5. In the story the prince says he has traveled uninterruptedly for eight years, six months and fifteen days. How many leagues does he travel?
6. How many leagues do the seven messengers travel?

It took the students two and a half hours to complete the questionnaires (plus half an hour to read Buzzati's story) across two lessons. The remaining six hours were dedicated to "translating" the story through coding and robotics. Seven Ozobot robots simulated the seven messengers and one Ozobot was the king's son. Buzzati's short story was also transposed into a Scratch game. The students programmed the Ozobots to recreate the departures and speeds of the seven messengers. These data can be obtained directly from Buzzati's story and are the added value, and the innovation, of the teaching proposal. Indeed, this task requires students to have fully understood the story's mathematical elements and have transposed them into physical terms (i.e., distance covered and time spent). They should also have worked out the mathematical law that governs the messengers' departures and arrivals. These skills combine logical, mathematical, and also linguistic skills. The project goes even further by asking the students to apply these skills to writing the appropriate commands for the Ozobots in the programming language. The game was created in Scratch in the remaining hours of the project.

3 Educational Impact and Critical Analysis

Understanding the text of mathematics problems has always been one of the biggest obstacles to solving them. This educational project is an invitation to look beyond mathematics to language for ways to overcome this obstacle. The students analyze a narrative text to get mathematical information from it. The goal is to help students analyze a mathematical problem with a lengthier text than usual, and extrapolate problems from reality. Unexplained text information is exclusively mathematical and the second form should be used to focus on them. The students had no trouble determining when the first messenger, Alessandro, would return, because the original text states explicitly that the prince's caravan travels at 40 leagues a day, and the messengers travel at one and a half times the speed of the prince (60 leagues a day). However, imagining the messenger's path is less simple. Alessandro has to return to the castle (80 leagues), then go back to the point from which he started (another 80 leagues) and finally catch up with the prince, who has continued to walk at a speed of 40 leagues a day. The students suggest at least three possible approaches to this problem: *empirical* (simulating the path of the prince's caravan and the messenger on paper); *algebraic* (looking for the relationship between the messenger's progress and the caravan's progress); *physical-algebraic* (creating an equation of motion between the caravan's rate of travel and the messenger's).

Students understood that analyzing the text in logical and literary terms is key to their formulating the correct answers and the most appropriate mathematical approach. In fact, students used the physical-algebraic analysis of the rate of departures and returns and the journey times to answer up to the fifth question: they only have to multiply by five the number of the day on which the messengers left the prince to get the number of the day of their return from the castle (Fig. 1). The last two questions on the second questionnaire are easy to solve and open up possibilities for

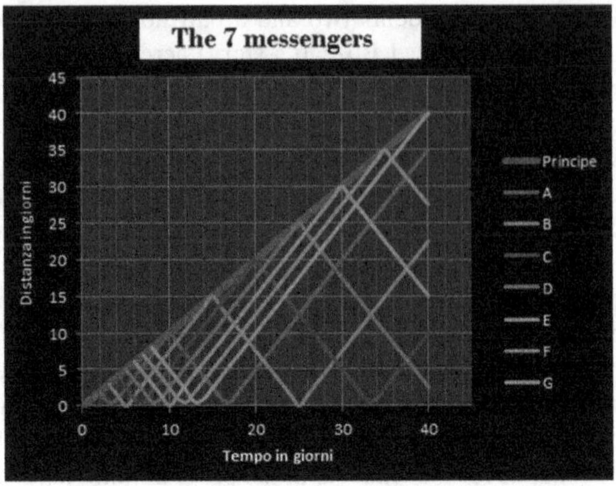

Fig. 1 Equations of motion for all messengers: diagram of the first 40 days

further interdisciplinary links. To find out how many leagues the prince has traveled, students have to decide how many days there are in 8 years, 6 months and 15 days. They pointed out the two leap years, and counted the six months with only 30 days in them, to reach a total of 3117 days. This reasonable hypothesis clashes with Buzzati's original text (*Domenico's return* on *the last day of the journey*) because, according to the calculations, Domenico should arrive on the 3125th day. It was fascinating to see the students' attempts to get past this discrepancy, by hypothesizing that the journey could have started at the end of a leap year, hence there would be three leap years in eight years of travel.

After this discussion, the students focused on robotics. The decision to transpose the story using Ozobots (Fig. 2) was made because the school had used these robots for several years for coding and block programming, and also to practice their mathematical skills. As they use the Ozobots, students develop a critical ability

Fig. 2 An Ozobot in action on a path created by a student

to appreciate their advantages and disadvantages. Ozobots also represent the active construction of knowledge through interaction with the world: students learn by manipulating objects [5] whose movements are decided in a simple programming language (Ozobots use a block programming language called Blockly that is very similar to what is used in its more famous counterpart Scratch[1]).

Ozobots are golf-ball–sized transparent spheres that can move along colorful paths, on paper or on digital surfaces. Students recreated the kingdom on paper and programmed seven Ozobots as the "messengers" and one as the "prince" moving at different speeds. They also decided that the programmed Ozobots should simulate the messengers' departures and arrivals during the first 40 days of the journey (Fig. 1).

Introducing robotics also increased inclusivity. Each student, including those with special needs, was assigned a specific role during the creation of the map of the kingdom. The robotics-related activities also helped all the students to learn how to handle errors virtuously. Errors were not seen as failures, but rather were exploited by all (teacher and students) as a learning tool that helps them try out new strategies and find the right solution. Errors move to the emotional dimension where, as is known from pedagogical research, they play a fundamental role in the knowledge process [6]. I believe this cross-disciplinary experiment involving mathematics and literature has had an impact on the emotional sphere of students, adding to their ability to solve non-trivial problems and also to think creatively. Assigning specific roles to all students based on their unique potential makes each of them responsible for their work and also helps students with special educational needs to reason more flexibly in unforeseen situations [7].

4 Final Remarks

This educational path differs from others where mathematics is a technique, pure syntax, separate from any historical or applicative context. Our approach to a mathematical problem involves transdisciplinary and transversal skills. Robotics is a resource that could be used more widely in teaching to solve mathematics (and physics), but should be applied with caution. The history of education in mathematics is littered with examples of what appear to be miracle tools and materials that are soon forgotten by teachers themselves [8]. If teachers reflect on the teaching and learning process, robots can serve as "objects with which to reason" [4, 9]. Indeed, as Damiani [10] argues, the body and the ability to manipulate are key not only as sensory mediators between the brain and the outside world, but also as the main device by which to learn and produce knowledge. According to Polya [11], solving problems is one of the most important activities of humans, which is why we teachers must offer our students numerous different contexts in which to mobilize solution strategies.

[1] https://ozoblockly.com/.

References

1. Buzzati, D.: I sette messaggeri. Oscar Mondadori, Milan (1968)
2. Arato, F.: Cosa racconta il numero. Belfagor **63**(1), 72–83 (2008)
3. Papert, S.: Mindstorms. Bambini computers e creatività. Emme edizioni, Milan (1984)
4. Papert, S.: Situating Constructionism. Ablex Publishing Corporation, Norwood, MA (1991)
5. Piaget, J.: The Construction of Reality in the Child. Basic Books, New York, NY (1954)
6. Cappello, S.: La dimensione emozionale nel processo di insegnamento-apprendimento. Formazione & Insegnamento **XI**(3), 233–238 (2013)
7. D'Amore, B., Pinilla, F.: La didattica e le difficoltà in matematica. Erickson, Trento (2008)
8. D'Amore, B.: Elementi di didattica della matematica. Pitagora, Bologna (1999)
9. Resnick, M., Martin, F., Berg, R., Borovoy, R., Colella, V., Kramer, K., Silverman, B.: Digital manipulatives: new toys to think with. In: Proceedings of the SIGCHI Conference on Human Factors in Computing Systems, pp. 281–287, Chicago (1998)
10. Damiani, P.: TCR e scuola: dallo strumento alla didattica. In: Grimaldi, R. (ed.) A scuola con i robot, pp. 95–131. Il Mulino, Bologna (2015)
11. Polya, G.: Come risolvere i problemi di matematica. Logica e euristica nel metodo matematico. Feltrinelli, Milan (Original work published 1945) (1967)

Escape from Tolentino During an Earthquake Saving as Many Lives and Cultural Objects as You Can

Paola Pazzaglia and David Scaradozzi

Abstract In the last five years, the Italian Ministry of Education has placed its focus on digital skills, recognizing them as fundamental and indispensable for the growth of future citizens in the information age. It has thus backed projects aimed at developing computational thinking and digital creativity at school. One of the highest-funded of such projects is "*Più vicini al nostro territorio—Valorizziamo i monumenti di Tolentino ... giocando con Scratch e App Inventor*" (Closer to our territory—appreciating the value of Tolentino's monuments ... while playing on Scratch and App Inventor). In this paper we describe this project, its vertical path, and the results of the first activities, which have already taken place in a primary school. These results show the progression of the skills and competences defined in the National Operational Programme document "For school 2014–2020" (Axis I Education, Objective 10.2, Improving students' key competences), and those set out in the National Plan for Digital Education.

Keywords Digital competences · Coding · Educational robotics · STEM · Primary school · Nintendo labo

P. Pazzaglia (✉)
Istituto Comprensivo "G. Lucatelli" Viale Benadduci, 23, 62029 Tolentino, Monte Cassino, Italy

D. Scaradozzi
Department of Information Engineering (DII), Università Politecnica Delle Marche, Via Brecce Bianche, 60131 Ancona, Italy

LSIS - Umr CNRS 6168 Laboratoire des Sciences de L'Information et des Systèmes, Equipe I&M (ESIL) case 925 - 163 avenue de Luminy, 13288 Marseille cedex 9, France

D. Scaradozzi
e-mail: d.scaradozzi@univpm.it

D. Scaradozzi et al. (eds.), *Makers at School, Educational Robotics and Innovative Learning Environments*, Lecture Notes in Networks and Systems 240,
https://doi.org/10.1007/978-3-030-77040-2_37

1 Introduction

Many researchers have been investigating the use of coding and robots to support education. Studies have shown that robots can help students develop problem-solving abilities and learn computer programming, mathematics, and science. The educational approach, based mainly on developing logic and creativity in new generations from the first stages of education, is very promising. Introducing coding and robotic systems is fundamental to the achievement of these aims, if begun at an early stage of education. At primary school, coding and robot programming are fun, making them excellent tools for introducing children to ICT and helping them develop their logical and linguistic abilities. In the last five years, the Italian Ministry of Education has placed its focus on digital skills, recognizing them as fundamental and indispensable for the growth of future citizens in the information age. It has thus backed projects aimed at developing computational thinking and digital creativity at school. While considering how to engage girls and boys in an activity with the potential to motivate them to consider a future career in STEM-related fields, the authors drew on personal experience (see from [1–9]) and on findings in the literature of great results obtained using educational robotics (ER) for the same aim. First, Benitti [10] reviewed ER experiences from around the world. Her results show that there is great potential in ER methodologies and activities in K12 classrooms, but the full power of these research outcomes still needs to be thoroughly exploited. Mubin et al. [11] highlight the great effort that research still needs to make to produce appropriate curricula that include robotics. Veselovská and Mayerová [12], on the other hand, present a qualitatively assessed ER activity in the context of an ER curriculum at lower secondary school. Sanders [13] considers robotics to be the real integrated approach to STEM education: rather than these four subjects addressing it separately as a mere series of notions, they could come together to re-elaborate these notions as part of active learning. Moreover, the benefits of ER education can also come in the form of cybersecurity [14], a STEM-related field that can empower future citizens of the digital world and protect today's boys and girls from the potential dangers of connected life. The authors have conducted coding and robotics teaching experiences in Italian schools since 2000–2001, when the first project was proposed. It was called "Building a robot" and a description of it can be found in [8]. This paper presents the project entitled "Più vicini al nostro territorio—Valorizziamo i monumenti di Tolentino … giocando con Scratch e App Inventor" (Closer to our territory—appreciating the value of Tolentino's monuments … while playing on Scratch and App Inventor). The main aim of this project is to encourage primary school students to experience the creative process, in spite of the earthquake, and to foster their literacy in their local heritage. The activities included the creation of a board game simulating Tolentino during an earthquake. Players have to escape from the town, saving as many lives and cultural objects as possible. In this paper we describe the project, the vertical path within it, and the results of the first activities, which have already taken place in a primary school. These results show the progression of the skills and competences defined in the National Operational Programme document "For school 2014–2020" (Axis I Education, Objective 10.2, Improving students' key competences), and those set out in the National Plan for Digital Education.

2 Material and Methods

The following subsections provide an account of the project activities.

2.1 *Underlying Pedagogical Approach*

The pedagogical theory used for the design and execution of the activities in this project is constructionism. Its premise is that knowledge building is the natural consequence of creating and experimenting. Students are encouraged to observe their own actions directly and to analyze the consequent effects. They are asked to share ideas in a highly motivating context. Here, technology and innovative learning environments give students room to learn in their own distinct style. Drawing on Howard Gardner's theory of multiple intelligences, students are encouraged to acknowledge their own skills and abilities. This can help them to think about their future, in terms of both their studies and their careers. Problem-based learning (PBL), a learner-centered approach, was employed. After a short explanation of the fundamental aspects of the project, students were given challenges to search collaboratively for effective solutions. Thus, they were also involved in project-based learning and peer tutoring. All the activities were designed on the TMI model (Think, Make, Improve), as suggested by Martinez and Stager [15]. First, students try to figure out what a solution to the problem might look like (Think); second, they try to implement the solution by building and programming the board game (Make); third, they look at their artefacts closely and try to debug or improve them, by using different points of view or by sharing their ideas with others (Improve). The methodology adopted is learning by doing, in other words, learning from experience. According to Dewey, learning should not be passive, but rather relevant and practical. This hands-on approach allows students to learn through practical tasks, after which their experience must be accompanied by thinking over and reflecting. To be effective, this methodology needs students to reflect, revise and integrate the experience. For this reason, each lesson included a session of sharing, when each group shared its work with the others and reflected on it. Moreover, peer tutoring was applied during coding activities (with Scratch), during which students worked collaboratively with others, improved relationships and increased motivation.

2.2 Tools and Materials

In the first module of the project, primary school students (aged 9–10) began to learn to code with Scratch 3, the free, easy-to-learn and visual programming language, and used Lego WeDo 2.0 for educational robotics activities. They went to school one afternoon a week to find out about the main principles of coding. They used their knowledge and applied their creativity to invent a board game that is both physical and virtual. The purpose of the game is to find out about the most important monuments in their town, Tolentino. Lessons were organized as hands-on learning laboratories in which students worked in teams of two or three sharing ideas and helping each other. The second part of the project started in October 2019 and is still in progress. Students from the lower secondary school are completing the board game using Scratch, APP Inventor and QR code.

2.3 Learning Objectives and Competences

The learning objectives and competences developed in the first part of the project were to: promote and develop computational thinking, coding and the use of digital technologies; encourage students' interest in STEM; improve soft skills, especially critical thinking, analyzing ability, problem-solving, planning skills, team working; reinforce logical reasoning; rediscover and enhance Tolentino's artistic and historical monuments.

2.4 Contents of the Activities

The activities included the creation of a board game simulating Tolentino during an earthquake. Players have to escape from the town, saving as many lives and cultural objects as possible. The module introduced the basic concepts of coding using Scratch. The teaching progression was:

- coding: introduction to Scratch: simple animations and experiments. Basic functions of Scratch and first sprite movements;
- coding: sequence of instructions and animations: drawing geometric shapes with Scratch. Starting with polygons (triangle, square …) students continued creating their own pictures;
- coding: cyclical repetition of instructions;
- coding and logic: synchronizing events: after the stories were created in analogical form (using paper, pen and storyboard), the children reproduced them with Scratch;

- robotics: basic principles of robotics; man–machine-robots: similarities and differences; building a robot using a Lego WeDo 2.0 kit and programming it with Scratch 3; programming challenges;
- robotics exercise: each group of children is asked to watch by a video captured during an earthquake and design and build a mechanism that reproduces the movement of the ground (undulation). Initially, some houses with weak foundations collapse, until the designed system is able to withstand powerful impacts generated by the WeDo engine;
- cultural heritage literacy: find information about Tolentino's most important monuments and their distinctive features;
- storytelling: be a tourist guide in Tolentino using Scratch;
- basic principles of gamification: how do you design a game? Brainstorming during which students discussed their points of view about a board game;
- creation of the board and markers from paper and recycled materials;
- final exhibition: students showed the board game to parents and teachers.

2.5 *Expected Results*

- Improving the learning process and school performance and motivating the students involved in the project.
- Developing new knowledge and competences in digital technologies, coding and computational thinking.
- Increased interest in STEM.
- Improved teamwork.
- Improved problem-solving and analytical skills.
- Learning the very basics of coding.
- Rediscovery and appreciation of Tolentino's most important monuments.
- Understanding how to behave during an earthquake.

3 Preliminary Results

The preliminary results are based on the activities described in Sect. 2, carried out over a given period of time. The proposed criteria turned out to be suitable for teaching new concepts to children, for attributing coherent purpose to building an application for tablets and a robot, and for teaching them the history of their town and the importance of the right behavior during an earthquake. Selected groups of professional operators were also invited to observe and to take part in the intermediate tests and experiments. Università Politecnica delle Marche will assess the activity by means of questionnaires [2, 5]. The first experimental project was carried out at Istituto Comprensivo "Lucatelli" in Tolentino, Italy. Primary School teacher Paola Pazzaglia, University professor David Scaradozzi, and Talent s.r.l. developer

Federico Camilletti developed the experimental project after their observation of primary school students. The project focused on increasing and pursuing logic and creativity as important educational skills in modern school. They were able to start the project.

The preliminary results are based on a comparison of the following items:

- higher marks (before and after the project) in some subjects, such as Italian, English, mathematics and science, showed the students had improved in those areas;
- the students' attitude, as they were always curious and grew more and more interested, especially when they understood the importance of working as a group to gain new skills while, at the same time, being able to deal with new problems;
- through systematic observation, it was possible to see an overall improvement in the EU competences, too (only those closely related to the project) (Figs. 1, 2 and 3).

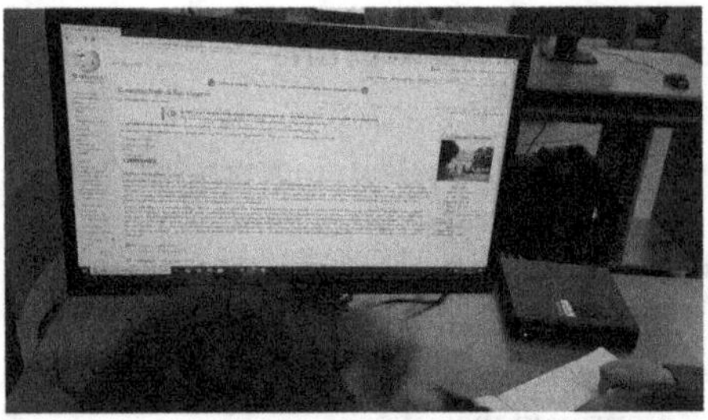

Fig. 1 Children looking for information about Tolentino's monuments

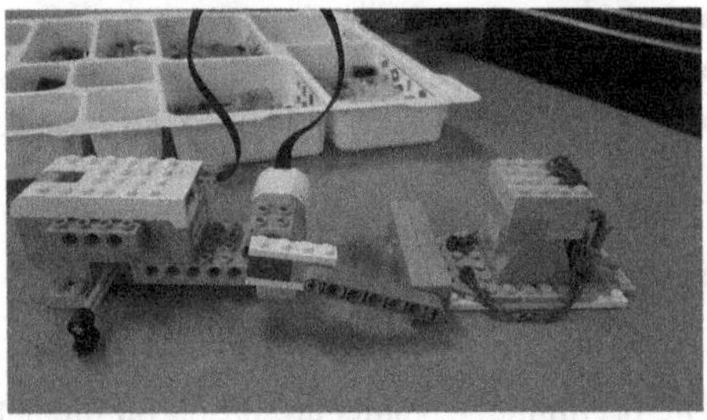

Fig. 2 One of the mechanisms children built using Lego WeDo 2.0 that reproduces the undulations of the earth

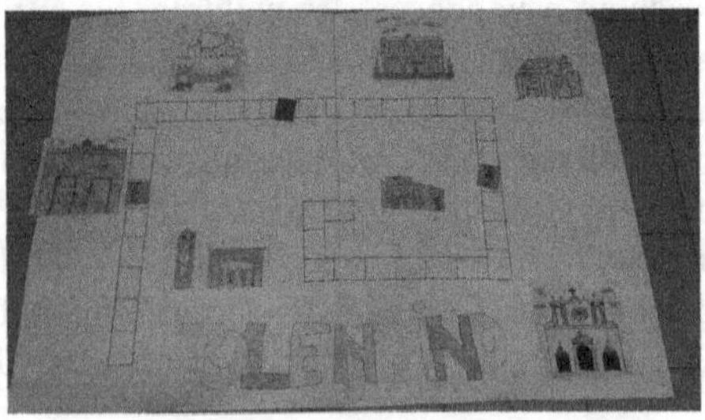

Fig. 3 The game's board

References

1. Scaradozzi, D., Cesaretti, L., Screpanti, L., Costa, D., Zingaretti, S., & Valzano, M.: Innovative tools for teaching marine robotics, IOT and control strategies since the primary school. In: Smart Learning with Educational Robotics, pp. 199–227. Springer, Cham (2019)
2. Scaradozzi, D., Screpanti, L., Cesaretti, L.: Towards a definition of educational robotics: a classification of tools, experiences and assessments. In: Smart Learning with Educational Robotics, pp. 63–92. Springer, Cham (2019)
3. Screpanti, L., Cesaretti, L., Marchetti, L., Baione, A., Natalucci, I.N., Scaradozzi, D.: An educational robotics activity to promote gender equality in STEM education. In: International Conference on Information, Communication Technologies in Education (ICICTE 2018) Proceedings. Chania, Greece (2018)
4. Scaradozzi, D., Screpanti, L., Cesaretti, L., Mazzieri, E., Storti, M., Brandoni, M., Longhi, A.: Rethink Loreto: we build our smart city!" A stem education experience for introducing smart city concept with the educational robotics. In: 9th Annual International Conference of Education, Research and Innovation (ICERI 2016). Seville, Spain (2016)
5. Scaradozzi, D., Screpanti, L., Cesaretti, L., Storti, M., Mazzieri, E.: Implementation and assessment methodologies of teachers' training courses for STEM activities. Technol. Knowl. Learn. **24**(2), 247–268 (2019)
6. Cesaretti, L., Storti, M., Mazzieri, E., Screpanti, L., Paesani, A., Principi, P., Scaradozzi, D.: An innovative approach to school-work turnover programme with educational robotics. Mondo Digitale **16**(72), 2017–2025 (2017)
7. Scaradozzi, D., Pachla, P., Screpanti, L., Costa, D., Berzano, M., Valzano, M.: Innovative robotic tools for teaching STREM at the early stage of education. In: Proceedings of the 10th Annual International Technology, Education and Development Conference, INTED (2016)
8. Scaradozzi, D., Sorbi, L., Pedale, A., Valzano, M., Vergine, C.: Teaching robotics at the primary school: an innovative approach. Procedia Soc. Behav. Sci. **174**, 3838–3846 (2015)
9. Morganti, G., Perdon, A.M., Conte, G., Scaradozzi, D.: Multi-agent system theory for modelling a home automation system. In: International Work-Conference on Artificial Neural Networks, pp. 585–593. Springer, Berlin, Heidelberg (2009)
10. Benitti, F.B.V.: Exploring the educational potential of robotics in schools: a systematic review. Comput. Educ. **58**(3), 978–988 (2012)
11. Mubin, O., Stevens, C.J., Shahid, S., Al Mahmud, A., Dong, J.J.: A review of the applicability of robots in education. J. Technol. Educ. Learn. **1**(209–0015), 13 (2013)
12. Veselovská, M., Mayerová, K.: Programming constructs in curriculum for educational robotics at lower secondary school. In: International Conference EduRobotics 2016, pp. 242–245. Springer, Cham (2016)
13. Sanders, M.E.: Stem, stem education, stemmania (2008)
14. Kasemsap, K.: Robotics: theory and applications. In: Moore, M. (ed.), Cybersecurity Breaches and Issues Surrounding Online Threat Protection, pp. 311–345. IGI Global (2017)
15. Martinez, S.L., Stager, G.: Invent to learn: making, tinkering, and engineering in the classroom. Constructing modern knowledge press, Torrance, CA (2013)

Ten years of Educational Robotics in a Primary School

Mariantonietta Valzano, Cinzia Vergine, Lorenzo Cesaretti, Laura Screpanti, and David Scaradozzi

Abstract Many researchers and teachers agree that the inclusion of science, technology, engineering, and mathematics in early education provides strong motivation and greatly improves the speed of learning. Most primary school curricula include a number of concepts that cover science and mathematics, but less effort is placed in teaching problem-solving, computer science, technology and robotics. The use of robotic systems and the introduction of robotics as a curriculum subject educates children in the basics of technology, and gives them additional human and organizational values. This paper presents a new program introduced in an Italian primary school, thanks to a collaboration with National Instruments and Università Politecnica delle Marche. Specifically, the project's curricular aim was to improve logic, creativity, and the ability to focus, all of which are lacking in today's generation of students. The subject of robotics will be part of the primary school's curriculum for all five years. The program has delivered training to the teachers, and a complete program in which children have demonstrated great learning abilities, not only in technology, but also in collaboration and teamwork.

Keywords Educational robotics · Primary school · Curricular robotics

M. Valzano (✉) · C. Vergine
ICS Largo Cocconi, Largo G. Cocconi, 10 – 00171, Roma, Italy

L. Cesaretti · L. Screpanti · D. Scaradozzi
Dipartimento Di Ingegneria Dell'Informazione (DII), Università Politecnica Delle Marche, Via Brecce Bianche, 60131 Ancona, Italy
e-mail: d.scaradozzi@univpm.it

D. Scaradozzi
LSIS - Umr CNRS 6168, Laboratoire Des Sciences de L'Information et des Systèmes, Equipe I&M (ESIL), case 925 - 163, avenue de Luminy, 13288 Marseille cedex 9, France

D. Scaradozzi et al. (eds.), *Makers at School, Educational Robotics and Innovative Learning Environments*, Lecture Notes in Networks and Systems 240,
https://doi.org/10.1007/978-3-030-77040-2_38

1 Introduction

Many researchers have been investigating the use of robots to support education [1–4]. Studies have shown that robots can help students develop problem-solving abilities and learn computer programming, mathematics, and science [5–8] and [9]. The educational approach is based mainly on fostering logic and creativity in today's generation of students, who are at the beginning of their educational journey. Learning to program robots is an opportunity for primary school students to develop their linguistic and logical skills [10, 11 to focus on pedagogical rather than technological issues. This paper presents an innovative program that was developed to teach the basics of robotics as a curricular subject in primary schools. The same methods are used in other subjects (Italian, mathematics, science, etc.) to provide multidisciplinary validation and motivation. Education in Italy is compulsory from the ages of 6–16, and is divided into five stages: kindergarten (*scuola dell'infanzia*), primary school (*scuola primaria*), lower secondary school (*scuola secondaria di primo grado* or *scuola media*), upper secondary school (*scuola secondaria di secondo grado* or *scuola superiore*) and university. The main subjects are Italian, English, mathematics, natural sciences, history, geography, social studies, physical education and visual and musical arts. The five-year program for primary schools is divided into two main blocks. During the first two years, students are introduced to robotics, including learning what a robot is, what kinds of rules it follows, how to build a puppet, then a machine using everyday materials, and finally using the Lego WeDo system [12]. In the last three years, children have been building and programming robots with the Lego NXT system. LEGO Education WeDo is an easy-to-use robotics platform that introduces young students to hands-on learning through LEGO bricks. It is the easiest graphical programming software that National Instruments has to offer. It is a fun and simple way to expose young students to basic engineering concepts at an early age. LEGO Education WeDo provides a hands-on learning experience that actively engages children's creative thinking, teamwork, and problem-solving skills. It is a hands-on platform that primary school students can use to build simple robotics applications which are controlled in a simplified version of LabVIEW installed on a personal computer. By combining the intuitive and interactive interface of LEGO Education WeDo software with the physical experience of building models out of LEGO bricks, students can bridge the physical and virtual worlds for the ultimate hands-on, minds-on learning experience (LabVIEW Graphical System Design) [13]. The system has been applied in primary schools in other countries, and the potential benefits for children's education have been studied. Mayerovà's study [14], for example, analyzes the first encounter of third-grade primary school students with LEGO WeDo. Romero [15] describes a pilot study of robotics in primary schools, together with the reasons for choosing LEGO WeDo for the children's activities, namely, the programming language's shallow learning curve (visual programming rather than code writing), and the educational content provided. In recent years, National Instruments, Università Politecnica delle Marche and primary schools have collaborated on improving the use of new technologies from the first grade of school. One of the schools involved is Istituto Comprensivo Largo Cocconi.

Following previous pedagogical, technological and educational experiences in numerous institutions, this project arose from a collaboration between researchers at Università Politecnica delle Marche (UNIVPM) and teachers at Istituto Comprensivo Largo Cocconi. This paper will present the different aspects of the project and the preliminary results. The presentation is organized as follows: Sect. 2 describes the objectives and the expected results of the project; Sect. 3 explains the tools used and the timeline of the project; Sect. 4 illustrates the preliminary results; the conclusions and future developments are illustrated in Sect. 5.

2 Objectives and Expected Results

The first and main aim of this project concerns the introduction of robotics into primary schools as a standard curricular subject, rather than as an extracurricular activity outside of school hours. The specific goals of the project include increasing children's abilities, teaching them to program a machine, and getting them to consider robotics a normal method of working, rather than an exceptional activity. Robotics gives students another opportunity to develop their logical skills and their creativity [16], both key to reasoning and critical thought. The first stage of the experiment, which took place between 2010 and 2015, covered an entire five-year period of primary school. The priority set was to introduce ROBOTICS as a curriculum subject, improving the usual competences reported in the regular national plans for education with a new teaching theme. Now the classes of the ROBOTIC Institute run throughout the school year, and teachers hold regular meetings to compare notes on their skills and results. Each class can schedule and organize teaching units that are specific to their students, by matching them to the topics of ordinary lessons. The new study program is presented below for the five years that make up primary school. The main objectives of the project are as follows:

1. Grade I
    - Getting to know the concepts of robotics with the introduction of Asimov's literature and the three laws of robotics;
    - Learning about individual mechanical elements through simplified ordering and planning programs: learning the differences between the shapes, materials, colors and functionalities of the elements available on the market;
    - Planning a model in the LEGO system using a simplified program;
    - Understanding the concepts of verification and validation of the model in the work environment.
    - Introduction of Lego NXT-EV3 and different devices.
2. Grades II–III
    - Acquiring the ability to attribute coherent purpose to a constructed robot;

- Introduction to the concept of the ROBOT as a machine that has to complete a specific task;
- Studying sensors and actuators by comparing them with systems in the human body;
- Introduction to the programming environment for the LEGO WeDo system.
- Building a simple robot that can interact with its surroundings.

3. Grades IV–V

- Acquiring the ability to attribute coherent purpose to a complex constructed robot;
- Acquiring the ability to build a robot with specific and relatively complex objectives.
- Planning a robot for a specific research purpose, which can live in a defined environment.
- Introduction to software for robot analysis.
- Creating a technical manual for the final operator explaining how to design and build a robot.

In the last two school years, the project introduced students to distributed control in the robotic environment using the SAM LABS educational kit [17]. The new activity begins in fourth grade classes by teaching students the differences between central control and distributed control in the environment. They analyze real situations like smart traffic lights, heating and air-conditioning systems, etc. [18]. In the fifth grade, students can use the same material to design their own distributed control inventions. These robotics units are designed with an interdisciplinary approach. They are linked to other subjects—physical education, history, geography and language—which helps to introduce an integrated and combined path to knowledge construction. The method is applied to the same scientific procedures: observing, thinking, verifying. The children's work teams are organized by roles and skills, and change during the year, improving inclusive behaviors and cooperative learning.

3 Instruments and Time

These activities take place during the hours set aside in the weekly timetable "for optional subjects", as required by the Ministry of Education. The program includes educational trips related to the topics the students are learning. These trips begin in the second and third grades of primary school, and may be to science museums or research institutes, where students get involved in more specific workshops about mechatronics and robotics. The hardware and software for this project include five kits of LEGO WeDo for the first and second grades and five Kits of LEGO MINDSTORMS NXT in the other grades.

Table 1 INVALSI [19] tests of 2015 experimental fifth class

Average percent score (net of cheating, 2019)	Students' score (net of cheating, 2019) in the same scale of the national report	Percent score difference comparing classes/schools with similar familiar background	Average familiar background	Percent observed score	Percent cheating
64.1	212.4	6.0	Low	66.5	4.0

4 Results

The results are based on the activities described in Sect. 2. The proposed criteria taught new concepts to children, enabled them to attribute coherent purpose to a constructed robot, and taught them robotics, including an introduction to Asimov's three laws of robotics. Selected groups of professional operators will also be invited to observe and to take part in the intermediate tests and experiments. Observations from the first experiment of robotics taught as a curricular subject led by primary school teacher Mariantonietta Valzano and Professor David Scaradozzi at the Istituto Comprensivo Largo Cocconi in Rome are that logic, attention and focus improved significantly in these primary school students, as did their team work skills. After evaluating these results and comparing them with different students in classes of the same level, it was seen that the children who completed the entire robotics program attained higher levels of competence, and acquired the skills that were planned for them faster and more profoundly. This in turn changed behaviors within the class and improved social inclusion.

The first result was clear when comparing INVALSI [19] 2015 tests with those of the fifth-grade participants in the experiment (see Table 1).

Average Score Lazio: 56.3

Average Score Central Italy: 57.4

Average Score Italy: 56.6.

According to these results, students who completed an educational robotics course attained a higher degree of competence compared with students in classes from the same school and the countrywide average.

5 Conclusion

The teachers in the project have recorded significant improvements in their students, which demonstrates the important value of using robotic systems in all aspects of teaching. Students have been very curious and passionate during these activities, exploring tasks like robot design and programming, and developing important life skills like teamwork and problem-solving.

This curriculum project has focused on increasing and pursuing creativity and logical thinking as important educational skills in modern school. The good results obtained since the start of the project in 2010 have convinced the school's teachers and head teacher to involve other classes in the experiment. Now nine classes have introduced robotics as a regular subject.

Finally, SAM LABS and IoT have also been introduced, allowing students to learn what these innovative technologies are bringing to our lives.

References

1. Scaradozzi, D., Cesaretti, L., Screpanti, L., Costa, D., Zingaretti, S., Valzano, M.: Innovative tools for teaching marine robotics, IoT and control strategies since the primary school. In: Smart Learning with Educational Robotics, pp. 199–227. Springer, Cham (2019)
2. Scaradozzi, D., Screpanti, L., Cesaretti, L.: Towards a definition of educational robotics: a classification of tools, experiences and assessments. In: Smart Learning with Educational Robotics, pp. 63–92. Springer, Cham (2019)
3. Screpanti, L., Cesaretti, L., Marchetti, L., Baione, A., Natalucci, I.N., Scaradozzi, D.: An educational robotics activity to promote gender equality in STEM education. In: International Conference on Information, Communication Technologies in Education (ICICTE 2018) Proceedings. Chania, Greece (2018)
4. Scaradozzi, D., Screpanti, L., Cesaretti, L., Mazzieri, E., Storti, M., Brandoni, M., Longhi, A.: Rethink Loreto: we build our smart city!" A stem education experience for introducing smart city concept with the educational robotics. In: 9th Annual International Conference of Education, Research and Innovation (ICERI 2016). Seville, Spain (2016)
5. Atmatzidou, S., Demetriadis, S.: Advancing students' computational thinking skills through educational robotics: a study on age and gender relevant differences. Robot. Auton. Syst. **75**, 661–670 (2016)
6. Cesaretti, L., Storti, M., Mazzieri, E., Screpanti, L., Paesani, A., Scaradozzi, D.: An innovative approach to school-work turnover programme with educational robotics. Mondo Digitale **16**(72), 2017–2025 (2017)
7. Kandlhofer, M., Steinbauer, G.: Evaluating the impact of educational robotics on pupils' technical-and social-skills and science related attitudes. Robot. Auton. Syst. **75**, 679–685 (2016)
8. Sullivan, F.R.: Robotics and science literacy: thinking skills, science process skills and systems understanding. J. Res. Sci. Teach. Official J. National Assoc. Res. Sci. Teach. **45**(3), 373–394 (2008)
9. Scaradozzi, D., Screpanti, L., Cesaretti, L., Storti, M., Mazzieri, E.: Implementation and assessment methodologies of teachers' training courses for STEM activities. Technol. Knowl. Learn. **24**(2), 247–268 (2019)
10. Scaradozzi, D., Pachla, P., Screpanti, L., Costa, D., Berzano, M., Valzano, M.: Innovative robotic tools for teaching STREM at the early stage of education. In: Proceedings of the 10th annual International Technology, Education and Development Conference, INTED (2016)
11. Scaradozzi, D., Sorbi, L., Pedale, A., Valzano, M., Vergine, C.: Teaching robotics at the primary school: an innovative approach. Procedia Soc. Behav. Sci. **174**, 3838–3846 (2015)
12. Lego Wedo Education (2019), https://education.lego.com/en-gb/product/wedo
13. LabVIEW Graphical System Design—From Kindergarten to Rocket Science, http://www.ni.com/newsletter/50596/en/
14. Mayerová, K.: Pilot activities: LEGO WeDo at primary school. In: Proceedings of 3rd International Workshop Teaching Robotics, Teaching with Robotics, pp. 32–39 (2012)

15. Romero, E., Lopez, A., Hernandez, O.: A pilot study of robotics in elementary education. In: 10th Latin American and Caribbean Conference for Engineering and Technology. Panama City, Panama (2012)
16. Eguchi, A.: Educational robotics for promoting 21st century skills. J. Autom. Mobile Robot. Intell. Syst. **8**(1), 5–11 (2014)
17. SAM LABS (2019) https://samlabs.com/
18. Morganti, G., Perdon, A. M., Conte, G., Scaradozzi, D.: Multi-agent system theory for modelling a home automation system. In: International Work-Conference on Artificial Neural Networks, pp. 585–593. Springer, Berlin, Heidelberg (2009)
19. Net of Cheating (2019), http://www.invalsi.it/invalsi/ri/sis/documenti/022013/falzetti.pdf

Educational Robotics at Primary School with Nintendo Labo

Mauro Gagliardi, Veronica Bartolucci, and David Scaradozzi

Abstract In the last five years, the Italian Ministry of Education has focused on digital skills, recognizing them as fundamental and indispensable for the growth of the future citizens of the information age. Numerous requests have come from the European Commission, the Italian Ministry of Education and the employment world regarding the introduction of new technologies in schools, whether or not this is part of curricular activities. National guidelines for kindergarten and primary school curricula promote the introduction of new tools and new multimedia languages as fundamental for all disciplines. The idea of the National Operational Programme (PON) and the National Plan for Digital Education (PNSD) is to boost digital knowledge and participation in STEM subjects. The project presented in this article was launched in this context and was a collaboration with the Nintendo company to evaluate the "Nintendo Labo" product at educational level. This trial was conducted in a third-grade class at the "Allegretto di Nuzio" primary school in Fabriano (AN). The kit, an evolution of the Nintendo Switch console, was initially created for recreational purposes. The advantages and limitations of the product came to light during the few months of the experiment. The "Nintendo Labo: assembly—play—discover" educational project allowed students to merge theoretical and practical aspects of their knowledge, and understand complex systems through design and simulation.

Keywords Digital competences · Coding · Educational robotics · STEM · Primary school · Nintendo labo

M. Gagliardi (✉)
Istituto Comprensivo "Imondi Romagnoli", Largo Fratelli Rosselli, 13, 60044 Fabriano, AN, Italy

V. Bartolucci · D. Scaradozzi
Dipartimento Di Ingegneria Dell'Informazione (DII), Università Politecnica Delle Marche, Via Brecce Bianche, 60131 Ancona, Italy
e-mail: d.scaradozzi@univpm.it

D. Scaradozzi et al. (eds.), *Makers at School, Educational Robotics and Innovative Learning Environments*, Lecture Notes in Networks and Systems 240,
https://doi.org/10.1007/978-3-030-77040-2_39

1 Introduction

Many researchers have investigated the use of coding and robots to support education, showing that robots can help students develop problem-solving skills and learn computer programming, mathematics, and science. For this reason, the authors conducted an experiment in primary schools using innovative kits that can be customized with cardboard. While considering how to engage girls and boys in an activity with the potential to motivate them to consider a future career in STEM-related fields, the authors drew on personal experience (see [1–6]) and on the great results found in the literature achieved using coding and educational robotics (ER) for the same aim. First, Benitti et al. [7] reviewed ER experiences around the world, showing that there is great potential in ER methodologies and activities in K12 classrooms, but the full power of these research outcomes still needs to be thoroughly exploited. In fact, [8] Mubin et al. highlight the great effort that research still needs to produce appropriate curricula that include robotics. Veselovská et al. [9], on the other hand, present a qualitatively assessed ER activity in the context of an ER curriculum at lower secondary school level. Sanders et al. [10] consider robotics to be the integrated approach to STEM education: rather than these four subjects addressing it separately as merely a set of notions, they could come together to re-elaborate these notions as part of active learning.

2 Material and Methods

The pedagogical theory used for the design and execution of the activities in this project is constructionism. Its premise is that knowledge building is the natural consequence of creating and experimenting. Students are encouraged to use the materials at their disposal, directly observe their actions and analyze the effects. For these reasons, Nintendo created a kit with cardboard cut-outs and other materials, such as laces, elastic bands and infrared-reflective stickers. All the materials have to be assembled using the Nintendo Switch console display and Joy-Con controllers to create a "Toy-Con" that can interact with the included software. Nintendo designed Labo as a customizable technological game. In Italy, marketing for the product began in April 2018. The application to begin the project—to test whether the product could be used in Italian primary schools—was submitted in July for the 2018/2019 academic year. Although the product was originally designed for recreational purposes, its functionalities have since been appreciated and deemed useful for teaching the principles of engineering, physics, and basic programming. In each of the aforementioned "Toy-Cons", the paper parts and the controllers called "Joy-Cons" enabled each artifact to become programmable and interactive, thanks to augmented reality. During the project, students invented new ways of using the cardboard, programming and placing the controllers in housings other than those designed by the manufacturer. At the end of each design phase, the console was connected to a projector and all the groups could invite other classmates to the "game". The classroom experienced extraordinary levels of cooperation, which in turn reduced distraction and bullying, and increased participation. Nintendo's images were projected onto an erasable wall. This had the advantage of involving students in a phase of reengineering, drawing, commenting, and reporting. The company's assembly instructions for the Toy-Con were loaded into the software. The projected images were surprisingly engaging for students. The various steps were described in three-dimensional animations and there were special functions to stop, rewind or fast forward the film. In Nintendo Labo, all actions on the software could be linked to the hardware, so it was easy to let the students proceed by trial and error. This characteristic helps students who have motion or reaction difficulties. The "Garage" feature enabled students to program according to the principle of "if… then…", which is typical of cooperative-learning, using the: "Causal Node" and "Effect Node" functions. Students could create the Causal Node by simply placing elastic bands on the programmed junctions on the console (i.e., like guitar strings). Once the "if you touch" command was assigned and activated by plucking the elastic, an Effect Node was activated (i.e., the console emitted a sound when the "if you touch" command was assigned to the "play note" command in the corresponding Effect Node). The junctions between the nodes, which appear as icons, can be displayed by tracing a line on the console touch screen with a finger. This helps children become accustomed to recognizing nodes graphically as relationships in a mind map that becomes more complex with the addition of the "and" connection.

2.1 Learning Objectives and Competences, Expected Results

The learning objectives and competences considered in this first part were:

- Promoting and developing new knowledge and competences in computational thinking, coding and the use of digital technologies.
- Improving soft skills, especially critical thinking, analytical, problem-solving, and planning skills, and teamwork.
- Reinforcing logical reasoning.
- Increasing interest in STEM, robotics, and coding.
- Improving the learning process and school performance to motivate the students involved in the project.
- Improving teamwork.
- Learning the very basics of coding.
- Experimenting with the "Toy-garage" software implementation strategy.
- Creating a simple robotic artifact that can interact with the environment.
- Acquiring the ability to build cardboard robots for specific and relatively complex purposes.
- Presenting artifacts that can represent a "causal node" or an "effect node".

- Creating a PowerPoint technical manual for the final operator explaining how to design and implement the objects created.

Throughout the project, the use of poor quality, structured and unstructured material helped increase awareness of different materials, and the ability to predict their characteristics and effects. The main objective was to increase logic in the consequentiality of the actions undertaken, so that students would become accustomed to verifying scientifically on the basis of how materials behave. Another important goal was to get the students used to visual programming based on conceptual maps, to promote metacognition, which makes students aware of their choices. No less important are the cross-cutting objectives: collaborating with classmates, making positive contributions to the group, learning to accept others through cooperative-learning and problem-solving.

2.2 *Contents of the Activities*

The first and main objective of this project is introducing robotics and coding to primary school curricula, and proposing them as extracurricular activities. Robotics give students a different opportunity to develop their logical skills and creativity, which are essential features of reasoning and critical thinking.

2.2.1 The Radio-Controlled (RC) Car Activity

The first activity in the experiment was constructing a radio-controlled (RC) car. Each student had his own cardboard kit, which they folded to assemble, learning the importance of being accurate and precise at each step. They had to manipulate the die-cut cardboard sheets while avoiding folding it in the wrong place, which could compromise the result. Only when the Joy-Con was inserted did students have proof that their artifact was correctly assembled. Two Joy-Cons (controllers) are needed for this Toy-Con, and can also be customized with cardboard and programmed remotely from the portable console. Acting on the frequency ("Hertz") of the two devices, placed on the sides of the car, students see the latter move in different directions. Vibrations will affect the cardboard parts that are in contact with the support surface. Once the structure was completed and the functionality of the movements was verified, students personalized their cars by adding colored cards, acetate sheets and yarn. The innovative nature of cardboard-based robotics undoubtedly attracted students to the customization phase, motivating them through trial and error, without the frustration of the feeling of failure. Next, in the speed competition, students discover how different materials and thicknesses behave, depending on the vibrations they are subjected to. Those who used thicker or heavier materials definitely found they lost speed or accuracy in their movements. However, the desire to find errors and ask questions about the unforeseen events stimulated them to try different things and explain the solutions they found to the class. It was easy to link the objects created during the project to topics studied in the third grade. After the RC car, the students were asked to build a project linked to their subjects, and create a description in Italian with an explanatory drawing or diagram showing how to use it. Many children redesigned the RC toy car as a dinosaur and linked it to the prehistorical period they had recently studied. The background flora was created from the same materials using the snap-fit mechanism, and a storytelling activity was set up about prehistoric times. Groups of students took turns in a storytelling and coding activity with the console, and applied it to the dinosaur paperboard. When the dinosaur got to a certain place, the child controlling its movements had to answer questions to be able to continue. From time to time, the path and the setting changed, depending on how the workgroups were organized. Stimuli were given to encourage the students to cooperate and help each other. Along the way, students added reflective labels to trees or bushes built to create "autopiloted" functions. They simulated food on the trees (cubes that could be pushed to show new variables and activate mechanisms, or to trace pathways in places where there was no light). The background and the path were enhanced with questions and mathematics exercises related to the school program. As the activities progressed, students created new robots autonomously, to which they added Joy-Cons. Some of them transformed a truck into a mammoth or a unicorn into an excavator. At the end of the RC car activities, the infrared camera was tested. A new background was designed, with a path added to a box. Students observed from different points of view, that is, no longer from top-down but also from the front. This led them to experience positions in space and points of view based on what they had studied in geography. The RC car activity was also easy to link to the topic of planting strategies they had learned in history. Students developed an RC seeder, enabling them to add a seed compartment with bottlenecks to drop the seed according to the vibrations generated remotely. The seeder was tested in a new scenario with hills, mountains and a maritime setting (subjects studied in third-grade geography).

2.2.2 The Fishing Rod Activity

Constructing the fishing rod was useful practice for some of the theories students learn in science and mathematics. Students built the robotic rod and "fished" interactively. As they held the rod they could observe the images of the marine environment and their hook projected against the wall. The fishing activity exploits the differential command of the controllers. Two controllers were applied to the handle and the reel of the rod to capture movements during the virtual fishing exercise. The fishing simulation allowed students to explore the behavior of different species of fish at different depths. Their motor coordination was exercised as they moved the rod and the resistance sensors responded. Furthermore, for each fish they caught, the scientific name was highlighted, its weight was shown in multiples and sub-multiples of kilograms, as was the total weight of the catch. The activity involved practicing sums, equivalences and transforming units of weight. Students learned about gross, tare, and net weights. All this took place in a playful atmosphere, with the class divided into teams. Those who could not answer quickly were helped by the other members of the group. Using the tools in the kit, the students drew their fish on cardboard and twinned them into the virtual sea at a precise depth where they are supposed to live. The exercise was also useful for practicing weight estimation.

2.2.3 The Motorcycle Handlebar Activity

Designing and building a motorcycle handlebar with controllers inside the knobs gave students a full immersion into a virtual world. Taking advantage of the immersive potential of Nintendo, students created a teaching aid for students with dysgraphia or taking their first steps in handwriting. A group of students prepared a physical track for the motorbike in the shape of a letter. Meanwhile, other groups created the digital counterpart of the track. The virtual letters (tracks) had a motorbike on them that was controlled via the physical handlebar. As they drove the virtual motorbike along the tracks, the students drew virtual letters, a more engaging experience for them than doing traditional writing exercises using their fingers or a pencil.

2.2.4 The Piano Activity

Designing and assembling a cardboard piano brought the Toy-Con concept to music lessons, where the new technological instrument could be used alongside traditional instruments. The piano plays because a controller with an infrared sensor detects the movement of the cardboard keys, which have reflective labels, and the console processes the right note. Each step in the procedure was projected onto a screen for the entire class to view. The students programmed tones, sounds, distortions, and recorded audio. The last activity saw the class create a virtual solfeggio by holding a controller programmed to modulate the tone. Each movement of the hand corresponded to a change in sound.

3 Developments and Preliminary Results

The experiment took place in the 2018/2019 academic year and validated quantitatively the introduction of the Nintendo Labo system to the primary school curriculum as an educational tool. After getting acquainted with the commercial tool and its proposed activities, it was decided to set aside the suggested transmission approach in favor of an interactive approach. Teachers and students customized their artifacts with a variety of objects, such as blocks of wood, sheets of acetate, paper and cardboard of different thicknesses, straws, yarn, colors, aluminum foil and objects easy to find at school. In this way, learners could put their creative skills to work personalizing these artifacts, through what M. Resnick calls the "spiral of creative learning". Students held various objects in their hands, identified their characteristics and imagined what to create. As they played with the kit and the available material, they shared ideas with other group members, and thought over what they could do. Although the materials differed, the essence of the learning process was the same. Children perfect their ability to think creatively and develop ideas by writing down projects and illustrating them to others; in turn they learn to consider alternatives and to take cues from others. Programming through the "Nodes" enables them to acquire new knowledge and computer skills while playing. When a child tells others about the new technological features they have discovered, or a new way of connecting the cardboard pieces, she gives others the opportunity to share in the creative moment, not only incrementally but also cooperatively. The teacher's function is not only to transmit knowledge and facilitate a learning dynamic respectful of the rules, but also to co-experiment. Students then took part in a STEAM-related experiment in ER, with structured and non-structured materials, which combined tinkering and making. Two-monthly reports were sent to the company on the work done by the students. The experimental work took place in the multi-purpose laboratory of the "Allegretto di Nuzio" primary school, created in collaboration with the Loris Malaguzzi International Center, and Reggio Children. The laboratory was designed as a creative atelier by a team of pedagogues, architects and experts in education. The colors of the walls, the mobile and flexible furniture, and the technological instruments, such as cutting plotters, 3D printers and magnetic walls, were beneficial for the aims of the work. One of the purposes of the technology was to foster the conscious and intelligent use of the resources by the children. The activities show that this methodology encourages them to master the fundamental concepts of technology and how they relate to each other: needs, problems, resources, processes, products, impacts, control. Italian Law no. 107/2015 and Legislative Decree no. 62/2017 call for activities related to computational thinking. Thanks to the programming of the console, which is interfaced with the sensors in the paperboard, the students' mental processes were challenged to solve problems. From time to time, solutions arose from specific methods and tools and through strategy planning. This was particularly the case for those situations that called for the construction of a procedure, a series of operations to solve a problem, or the establishment of a network of connections.

References

1. Scaradozzi, D., Screpanti, L., Cesaretti, L., Mazzieri, E., Storti, M., Brandoni, M., Longhi, A.: Rethink Loreto: we build our smart city!" A stem education experience for introducing smart city concept with the educational robotics. In: 9th annual International Conference of Education, Research and Innovation (ICERI 2016), Seville, Spain (2016)
2. Scaradozzi, D., Screpanti, L., Cesaretti, L., Storti, M., Mazzieri, E.: Implementation and assessment methodologies of teachers' training courses for STEM activities. Technol. Knowl. Learn. **24**(2), 247–268 (2019)
3. Cesaretti, L., Storti, M., Mazzieri, E., Screpanti, L., Paesani, A., Principi, P., Scaradozzi, D.: An innovative approach to school-work turnover programme with educational robotics. Mondo

Digitale **16**(72), 2017–2025 (2017)
4. Scaradozzi, D., Pachla, P., Screpanti, L., Costa, D., Berzano, M., Valzano, M.: Innovative robotic tools for teaching STREM at the early stage of education. In: Proceedings of the 10th annual International Technology, Education and Development Conference, INTED (2016)
5. Scaradozzi, D., Sorbi, L., Pedale, A., Valzano, M., Vergine, C.: Teaching robotics at the primary school: an innovative approach. Procedia Soc. Behav. Sci. **174**, 3838–3846 (2015)
6. Morganti, G., Perdon, A. M., Conte, G., Scaradozzi, D.: Multi-agent system theory for modelling a home automation system. In International Work-Conference on Artificial Neural Networks, pp. 585–593. Springer, Berlin, Heidelberg (June 2009)
7. Benitti, F.B.V.: Exploring the educational potential of robotics in schools: a systematic review. Comput. Educ. **58**(3), 978–988 (2012)
8. Mubin, O., Stevens, C.J., Shahid, S., Al Mahmud, A., Dong, J.J.: A review of the applicability of robots in education. J. Technol. Educ. Learn. **1**(209–0015), 13 (2013)
9. Veselovská, M., Mayerová, K.: Programming constructs in curriculum for educational robotics at lower secondary school. In: International Conference EduRobotics 2016, pp. 242–245. Springer, Cham (November 2016)
10. Sanders, M.E.: Stem, stem education, stemmania (2008)

Educational Technologies and Assistive Robotics

Study and Development of Robust Control Systems for Educational Drones

Maria Letizia Corradini, Gianluca Ippoliti, Giuseppe Orlando, and Simone Terramani

Abstract This paper considers the problem of attitude and altitude control of quadrotors using the sliding mode control theory. The mathematical model of the quadrotor is derived using the Euler-Newton formalism. The sliding-mode is applied to the Parrot Mambo minidrone, which is a strong example of bringing educational robotics to formal (MATLAB, Python, JavaScript), non-formal (Tynker, Blockly, Swift Playground) and informal education. The control considered shows good performance and enhanced robustness.

Keywords Robust Control · Drones · Sliding-Mode Control · Education · Robotics

1 Introduction

The Parrot Mambo minidrone is an example of a new way of bringing the teaching of STEM (science, technology, engineering, mathematics) subjects into the classroom [1]. It offers a variety of tools and approaches that make it suitable for students from primary school to PhD level. In fact, it supports the best coding platforms, providing the end user with plenty of choices for tackling heterogeneous STEM problems. In particular, for primary schools it offers block coding with platforms like Tynker and

M. L. Corradini (✉)
School of Science and Technology, University of Camerino, Via Madonna delle Carceri, 62032 Camerino, MC, Italy
e-mail: letizia.corradini@unicam.it

G. Ippoliti · G. Orlando · S. Terramani
Dipartimento di Ingegneria dell'Informazione, Università Politecnica Delle Marche, Via Brecce Bianche, 60131 Ancona, Italy
e-mail: gianluca.ippoliti@univpm.it

G. Orlando
e-mail: giuseppe.orlando@univpm.it

S. Terramani
e-mail: S1085358@studenti.univpm.it

D. Scaradozzi et al. (eds.), *Makers at School, Educational Robotics and Innovative Learning Environments*, Lecture Notes in Networks and Systems 240,
https://doi.org/10.1007/978-3-030-77040-2_40

Blocky; for secondary schools and universities it offers text coding with JavaScript and Python and also compatibility with MATLAB & Simulink [2]. Simulink is the environment used to implement a robust nonlinear control algorithm (sliding mode control), in order to demonstrate the flexibility and the versatility of this novel educational ecosystem: from block coding to state-of-the-art prototyping, implementation and testing of experimental control approaches. This paper considers a mathematical model of the Parrot Mambo, which was used to modify the classical sliding mode control algorithm in order to get enhanced robustness against disturbances.

2 Quadrotor Mathematical Model

As shown in Fig. 1, the quadrotor contains four motors placed in a cross-configuration (front and rear propeller rotate counterclockwise, while the right and the left ones rotate clockwise) [3]. Each rotor is linked to a propeller. The quadrotor is an underactuated system: in fact, it has *6 DOF* (degree of freedom), three Cartesian coordinates (*X, Y, Z*) and three attitude angles, ϕ, θ and ψ (roll, pitch and yaw, respectively) but only four rotors/motors. However, it is possible to choose four controls to set the altitude (vertical distance from the ground) and the attitude (angular position) of the drone, that is, *Throttle* U_z: vertical force obtained by increasing or decreasing the speed of the four propellers by the same amount; *Roll* $U\phi$: torque with respect to the x axis of the body frame, obtained by increasing or decreasing the speed of either the left or the right propellers; *Pitch* U_θ: torque with respect to the y axis of the body frame, obtained by increasing or decreasing the speed of either the front or the rear propellers; *Yaw* $U\psi$: torque with respect to the z axis of the body frame, obtained by increasing (or decreasing) the speed of both the front and the rear propellers, while decreasing (or increasing) both the left and the right propellers. Two coordinate frames are used to describe the motion of the drone: the earth-centered inertial frame and the body-fixed frame. The earth frame $\left(O_e\ x_e\ y_e\ z_e\right)$ is used to define the

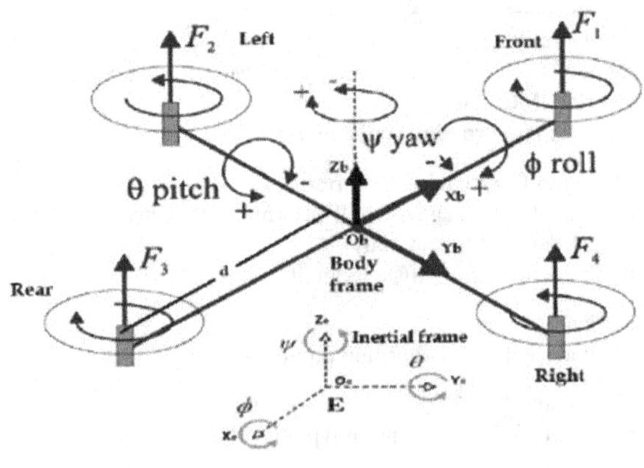

Fig. 1 Quadrotor structure representation

position vector of the quadrotor $\xi = \left[X\ Y\ Z\right]^T$ expressed in meters and the vector of Euler angles $\sigma = \left[\phi\ \theta\ \psi\right]^T$ expressed in radians. The body frame $\left(O_b\ x_b\ y_b\ z_b\right)$ defines the vector of linear velocities $\mu = [u\ v\ w]^T$ expressed in meters per seconds, the vector of angular velocities $\omega = \left[P\ Q\ R\right]^T$ expressed in radians per seconds, the forces F expressed in newton and the torques τ expressed in newton meters. The dynamic model was derived using the Newton–Euler formalism. As seen in [4], the dynamics of the *6 DOF* rigid body under the effect of external forces and torques are:

$$\begin{bmatrix} mI_{3,3} & 0 \\ 0 & J \end{bmatrix}\begin{bmatrix} \dot{\mu} \\ \dot{\omega} \end{bmatrix} + \begin{bmatrix} \omega \times m\omega \\ \omega \times J\omega \end{bmatrix} = \begin{bmatrix} F \\ \tau \end{bmatrix} \tag{1}$$

In order to develop Eq. (1), the different forces and torques that affect the quadrotor need to be determined. The main forces that affect the system are:

- *Input forces and torques*: directly produced by the four rotors. As a result of rotation, each motor exhibits an aerodynamic moment M_i and a lift force F_i given by [5]:

$$F_i = C_T\Omega_i^2,\ M_i = C_q\Omega_i^2 \tag{2}$$

where C_T and C_q are the thrust and the torque coefficient respectively, Ω_i is the velocity of the ith rotor. As reported in [6], the resultant input torque vector is:

$$\tau_{in} = \begin{bmatrix} U_\phi \\ U_\theta \\ U_\psi \end{bmatrix} = \begin{bmatrix} C_T d\left(\Omega_4^2 - \Omega_2^2\right) \\ C_T d\left(\Omega_3^2 - \Omega_1^2\right) \\ C_q\left(-\Omega_1^2 + \Omega_2^2 - \Omega_3^2 + \Omega_4^2\right) \end{bmatrix}. \tag{3}$$

The vector of input forces with respect to the body frame is:

$$F_{in}^B = \begin{bmatrix} 0 \\ 0 \\ U_Z \end{bmatrix} = \begin{bmatrix} 0 \\ 0 \\ C_T\left(\Omega_1^2 + \Omega_2^2 + \Omega_3^2 + \Omega_4^2\right) \end{bmatrix}. \tag{4}$$

- *Gyroscopic effect*: this is produced by rotation of the propeller and it is given as:

$$\tau_g = \left[-\sum_{i=1}^{4} J_{tp}\left[\omega \times \begin{pmatrix} 0 \\ 0 \\ 1 \end{pmatrix}\right] - (-1)^i\Omega_i \right] = \begin{bmatrix} -Q\Omega J_{tp} \\ P\Omega J_{tp} \\ 0 \end{bmatrix} \tag{5}$$

where J_{tp} is the total rotational moment of inertia around the propeller axis and Ω is the sum ($\Omega = -\Omega_1 + \Omega_2 - \Omega_3 + \Omega_4$) of the four propeller speeds expressed in radians per seconds.

- *Gravitational force*:

$$F_G = \begin{bmatrix} 0 \\ 0 \\ -mg \end{bmatrix} \tag{6}$$

where g is the gravitational acceleration and m the mass of the drone.

Replacing Eqs. (2)–(6) into Eq. (1), and expressing the rotational subsystem with respect to the earth frame, the dynamics equations become [6]:

$$\begin{cases} \ddot{X} = (S_\phi S_\psi + C_\psi S_\theta C_\phi)\frac{U_z}{m} \\ \ddot{Y} = -(S_\phi C_\psi + S_\psi S_\theta C_\phi)\frac{U_z}{m} \\ \ddot{Z} = -g + C_\theta C_\phi \frac{U_z}{m} \\ \ddot{\phi} = \frac{1}{J_{xx}}(U_\phi + (J_{yy} - J_{zz})\dot{\theta}\dot{\psi} + \dot{\theta}\Omega J_{tp}) \\ \ddot{\theta} = \frac{1}{J_{yy}}(U_\theta + (J_{zz} - J_{xx})\dot{\phi}\dot{\psi} - \dot{\phi}\Omega J_{tp}) \\ \ddot{\psi} = \frac{1}{J_{zz}}(U_\psi + (J_{xx} - J_{yy})\dot{\phi}\dot{\theta}) \end{cases} \tag{7}$$

3 Sliding Mode Controller

The sliding mode control technique consists of a high-speed switching control law that forces the state trajectories to follow a manifold called switching surface, and keeps the trajectories on this surface [7]. Sliding mode control design is a two-step procedure: sliding surface synthesis to provide the desired performance, and control design to ensure that all motions slide along the desired surfaces. The sliding mode control (SMC) is composed of equivalent control and corrective control [8]. The former ensures that the trajectories stay on the surface. The latter ensures that the trajectories reach the surface and compensates for any variations around the surface, whether due to external disturbances or to unmodeled dynamics. The sliding control for a given state is:

$$U_i = U_i^{eq} + U_i^C \tag{8}$$

where U_i^{eq} is the equivalent control and U_i^C is the corrective control. As in [9], let the sliding surface be of the form:

$$S_i = \dot{e}_i + \alpha_i e_i \tag{9}$$

where S_i is the sliding surface for a given state, e_i is the error between the actual state and the desired value, and α_i is a positive proportional gain. As for most first order systems, the chosen Lyapunov function is:

$$V_i = \frac{1}{2}S_i^2. \tag{10}$$

To ensure system stability, the time derivative of the Lyapunov function $\left(\dot{V}_i = S_i\,\dot{S}_i\right)$ must strictly be negative. To ensure that, we can set [6]:

$$\dot{S}_i = -\beta_i S_i - \lambda_i sign(S_i). \tag{11}$$

by the corrective control. As is well known, sliding mode control is robust against matched bounded disturbances. However, as proposed in [6], a more realistic assumption on unmodeled dynamics and external disturbances can be represented as:

$$d_i = \rho_i |S_i|^{p_i} |U_i|^{v_i} + d_{0i} \tag{12}$$

where 0 < vi < 1, 0 ≤ ρi ≤ 1, pi is a positive real number and d0i is a bounded disturbance. With corrective control the surface dynamics take the form [6]:

$$\dot{S}_i = (-\beta_i |S_i|^{m_i} - \lambda_i) sign(S_i) + d_i \tag{13}$$

where m_i is a positive integer. To respect the condition of a strictly negative derivative of Eq. (10) we can take [6]:

$$m_i = \frac{p_i + 1 - v_i}{1 - v_i}. \tag{14}$$

and the control algorithm is [6]:

$$\begin{cases} U_z = \dfrac{m}{\cos\theta\cos\phi}\left(\ddot{Z}^d + g + \alpha_z\,\dot{e}_z + \beta_z |S_z|^{\frac{p_z+1-v_z}{1-v_z}} sign(S_z) + \lambda_z sign(S_z)\right) \\ U_\phi = J_{xx}\left(\ddot{\phi}^d + \beta_\phi |S_\phi|^{\frac{p_\phi+1-v_\phi}{1-v_\phi}} sign(S_\phi) + \lambda_\phi sign(S_\phi) + \alpha_\phi\,\dot{e}_\phi\right) \\ \quad - \left(J_{yy} - J_{zz}\right)\dot{\theta}\dot{\psi} - \dot{\theta}\Omega J_{tp} \\ U_\theta = J_{yy}\left(\ddot{\theta}^d + \beta_\theta |S_\theta|^{\frac{p_\theta+1-v_\theta}{1-v_\theta}} sign(S_\theta) + \lambda_\theta sign(S_\theta) + \alpha_\theta\,\dot{e}_\theta\right) \\ \quad - (J_{zz} - J_{xx})\dot{\phi}\dot{\psi} + \dot{\phi}\Omega J_{tp} \\ U_\psi = J_{zz}\left(\ddot{\psi}^d + \beta_\psi |S_\psi|^{\frac{p_\psi+1-v_\psi}{1-v_\psi}} sign(S_\psi) + \lambda_\psi sign(S_\psi) + \alpha_\psi\,\dot{e}_\psi\right) \\ \quad - \left(J_{xx} - J_{yy}\right)\dot{\phi}\dot{\theta} \end{cases} \tag{15}$$

4 Experimental Results

The performance of the control algorithm deployed on the Parrot Mambo drone are shown below. The reference trajectory is shown in red (dashed line) and the actual trajectory is shown in blue (continuous line). In Fig. 2, the weight of the drone has been modified (its weight has been increased by 16% compared with the nominal mass) and we can see that the control algorithm, with surface dynamics of Eq. (11), cannot compensate for this weight increase, since there is an offset of about 10 cm in altitude (see Fig. 2a). On the other hand, the control algorithm with surface dynamics of Eq. (13), after a transient, is able to reach the desired altitude (see Fig. 2b). In Fig. 3, the yaw angle is shown, and the drone is in an environment with disturbances that are created artificially (wind fan). In Fig. 3a, spikes and an oscillatory behavior are seen for the control algorithm with surface dynamics of Eq. (11). In Fig. 3b, the control algorithm, with surface dynamics of Eq. (13), eliminates the spikes, and the behavior of the drone does not differ more than 0.1 radians with respect to the reference signal.

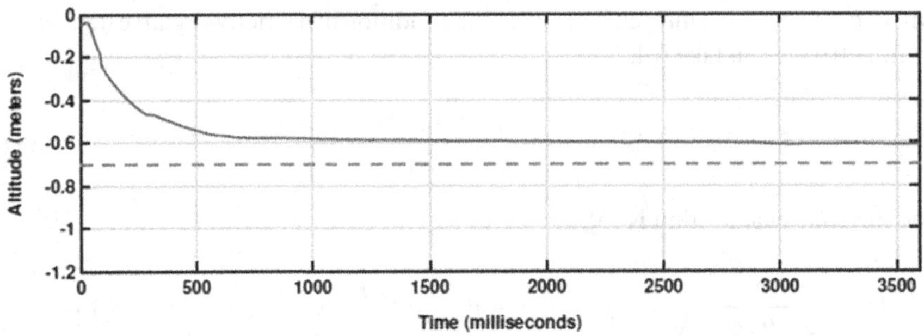

(a) Control algorithm with surface dynamics of Eq. (11)

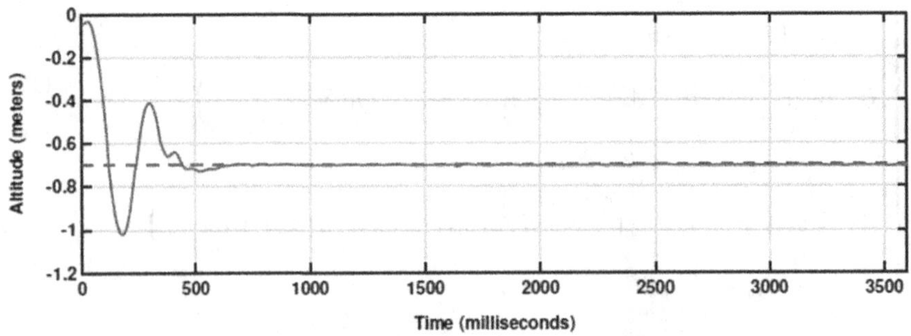

(b) Control algorithm with surface dynamics of Eq. (13)

Fig. 2 Altitude control

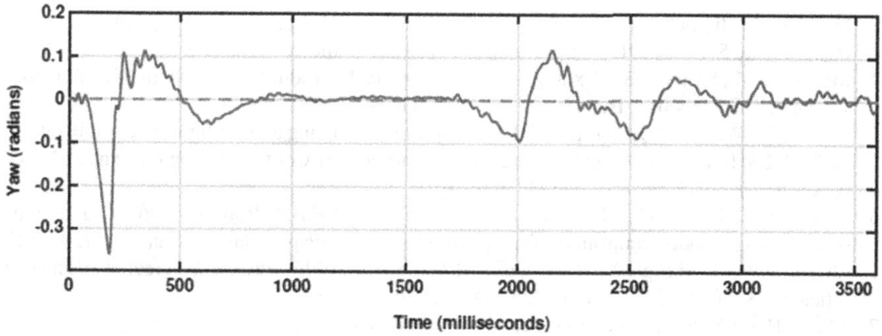

(a) Control algorithm with surface dynamics of Eq. (11)

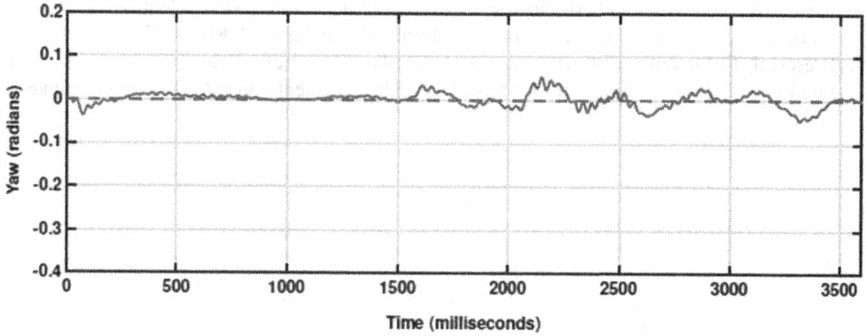

(b) Control algorithm with surface dynamics of Eq. (13)

Fig. 3 Yaw control

5 Conclusions

This paper has proposed a sliding mode control to control the altitude and the attitude of the quadrotor. The derived quadrotor model and the controller have been tested on a real drone, and the results demonstrate the robustness and the effectiveness of the control algorithm. What has been shown so far also demonstrates the flexibility and versatility of this new educational ecosystem: from block programming (via Simulink) to prototyping, to implement and test cutting-edge experimental and research control approaches.

References

1. Parrot Educational.: (2019). https://edu.parrot.com/
2. Parrot Minidrones Support from Simulink.: (2019). https://it.mathworks.com/hardware-support/parrot-minidrones.html

3. Fum, W.Z.: Implementation of Simulink Controller Design on Iris+ Quadrotor. Naval Postgraduate School (2015). https://calhoun.nps.edu/handle/10945/47258
4. Murray, R.M., Sastry, S.S., Zexiang, L.: A Mathematical Introduction to Robotic Manipulation, 1st edn. CRC Press Inc., Boca Raton, FL, USA (1994)
5. Azzam, A., Wang, X: Quad rotor arial robot dynamic modeling and configuration stabilization. In: 2010 2nd International Asia Conference on Informatics in Control, Automation and Robotics (CAR 2010), vol. 1, pp. 438–444 (March 2010)
6. Abci, B., Zheng, G., Efimov, D., El Najjar, M.E.B.: Robust altitude and attitude sliding mode controllers for quadrotors. IFAC-PapersOnLine **50**(1), 2720–2725 (2017). https://doi.org/10.1016/j.ifacol.2017.08.576, http://www.sciencedirect.com/science/article/pii/S2405896317309473, 20th IFAC World Congress
7. Khalil, H.K.: Nonlinear Systems. Pearson, 3 edn (2001)
8. Singh, G.K., Hole, K.E.: Guaranteed performance in reaching mode of sliding mode controlled systems. Sadhana **29**(1), 129–141 (Feb 2004). https://doi.org/10.1007/BF02707005
9. Khalid, K.M.: Design and Application of Second Order Sliding Mode Control Algorithms. Ph.D. thesis. University of Leicester (2003). http://hdl.handle.net/2381/30209
10. Bresciani, T.: Modelling, Identification and Control of a Quadrotor Helicopter. Master's thesis. Lund University Libraries (2008). https://lup.lub.lu.se/student-papers/search/publication/8847641

Arduino: From Physics to Robotics

Irene Marzoli, Nico Rizza, Alessandro Saltarelli, and Euro Sampaolesi

Abstract This paper discusses how a microcontroller, like Arduino, can improve laboratory practice in Italian upper secondary school and change students' attitudes towards STEM subjects. Since 2015, we started a close and fruitful collaboration with several high school teachers in the Marche region to introduce microcontroller programming to the physics lab. Notably, the project also involved teachers of other subjects, such as computer science, and with different backgrounds, for example electronic engineering, thus showing the inherently interdisciplinary character and versatility of Arduino. Students were engaged in hands-on activities, working in small groups of four to five people, supervised by learning assistants and teachers. Arduino was used to interface with sensors, to control the experimental setup, and for data acquisition. Finally, we could also make contact with robotics, by building a simple prototype of a rover.

Keywords Arduino · Physics education · Laboratory practice · Microcontroller

1 Introduction

If you ask Italian teachers about experimental activity in their classes, the vast majority of them will complain about the lack of laboratory space, the old-fashioned equipment, and the almost non-existent technical support. Of course, there are a few notable exceptions, but too often science, in particular physics, are taught as abstract disciplines. Rote learning and the apparent lack of connection to everyday life are some of the reasons why Italian students are neither proficient in STEM[1] nor highly

[1] The acronym STEM stands for science, technology, engineering, and mathematics.

I. Marzoli (✉) · N. Rizza · A. Saltarelli
School of Science and Technology, University of Camerino, Camerino, Italy
e-mail: irene.marzoli@unicam.it

E. Sampaolesi
Liceo "Giacomo Leopardi", Recanati, Italy

D. Scaradozzi et al. (eds.), *Makers at School, Educational Robotics and Innovative Learning Environments*, Lecture Notes in Networks and Systems 240,
https://doi.org/10.1007/978-3-030-77040-2_41

motivated to pursue a career in science and research [1]. This is especially true for women [2].

How can laboratory practices in secondary school be improved, given the limited budgets and facilities available? Can we present science in a more appealing way to our students? In the era of smartphones, computers, and information technology, does it make sense to use a stopwatch to measure the oscillation period of a pendulum? Domotics is changing our homes and lifestyles. We drive intelligent cars equipped with all kinds of sensors, but this digital revolution is only very slowly entering our classrooms. Indeed, there are only a few examples of attempts to include programming and sensor development in laboratory practice. This is true not only for high schools, but also at university level. For instance, one initial proposal for a curricular framework for introducing microcontroller programming to the physics lab at Winona State University is reported in [3].

In 2015, we began to devise a series of experiments, based on the Arduino platform, suitable for high school teachers and students. This pilot project involved 4 high schools, 10 teachers, and 150 students (58 female) all in their final year. Students were directly engaged in the experimental activity, which took place after school on a voluntary basis. In a student-centered educational perspective, they were able to build and run the experiments on their own, with the scaffolding provided by teachers and learning assistants. Students worked in small groups of four to five people, in order to foster peer-interaction and teamwork. To fully exploit the potential of a microcontroller, one should know its hardware, its software, and how to connect it to sensors and external circuits and devices. Hence, many skills and competences are requested and trained when operating such a platform: basic knowledge of electronics and circuitry, programming and coding, the ability to use a breadboard and make connections with jumper wires, sensors, and power supplies. Of course, not all students, especially at high school level, have already acquired background knowledge of this kind. So, it may be necessary to provide a brief introduction to scientific programming and to bring forward part of the physics curriculum (voltages, currents, and the basics of circuits). On the other hand, this is a good opportunity to show the interrelationship between different scientific disciplines. Finally, data analysis and a public discussion of the results are important parts of the project, in order to promote critical thinking and meaningful learning. Although there was no formal assessment, students were highly motivated and devoted several hours to carefully preparing their group report and final presentation.

2 What is Arduino?

Arduino is an open-source project started in 2005 at the *Interaction Design Institute Ivrea* by an international team composed of Massimo Banzi, David Cuartielles, Tom Igoe, Gianluca Martino, and David Mellis. Their aim was to realize an inexpensive, easy-to-use electronic platform, that was able to interact with the environment, receive inputs and provide outputs to actuators and other devices. Arduino's hardware and software are both open source: construction files and instructions are freely downloadable [4]. In principle, an expert could even build the Arduino board from scratch, with a breadboard and the basic electronic components. However, a solid background in electronics or engineering is not necessary to start tinkering with Arduino. Indeed, the best way to enter the Arduino worldwide community is to buy the starter kit—for less than €100—and then learn from tutorials, examples, and projects already developed and shared by other makers. Arduino is instructed to perform tasks of varying sophistication by a code, called *sketch*, written in a programming language, the Integrated Development Environment (IDE) software, similar to C and C++ . Most importantly, Arduino's software IDE is not only open source but also cross-platform, as it runs on Windows, Macintosh OSX, and Linux operating systems.

There is an ever-growing number and variety of Arduino boards and shields, developed to meet the most diverse needs. Nowadays, Arduino boards can be connected to the internet, controlled remotely, and can even send messages via Twitter. Despite the different capabilities and performances, all these boards share the same underlying structure and working principles. Throughout this paper we will concentrate on the most popular board: Arduino Uno, based on the ATmega328P, an 8-bit microcontroller with a clock frequency of 16 MHz, 32 kB of flash memory for program storage, and 2 kB SRAM for program execution (see Fig. 1). The board can be connected to a computer via a USB cable and communicate using a virtual serial port.

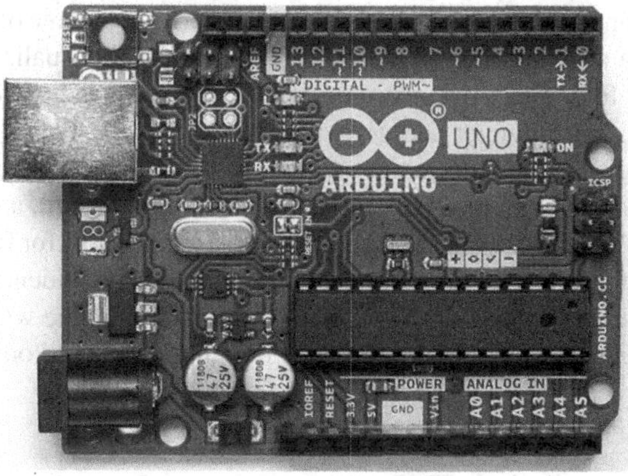

Fig. 1 Front view of Arduino Uno board. Analog pins are numbered from A0 to A5 (bottom right), whereas digital pins go from 0 to 13 (top row). The symbol " ~ " denotes pins that can deliver an analog voltage by means of the pulse width modulation (PWM) technique

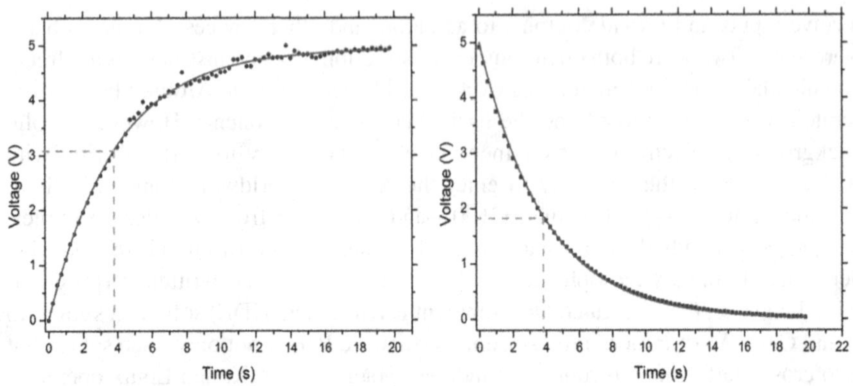

Fig. 2 Charge (left) and discharge (right) process of an *RC* circuit: $R = 3850\ \Omega$ and $C = 1$ mF. The theoretical time constant is $\tau = RC = 3.85$ s. The experimental values are, respectively, $\tau = (3.87 \pm 0.08)$ s and $\tau = (3.83 \pm 0.08)$ s. Experimental data (blue dots) are fitted with an exponential curve (red solid line)

3 Arduino in the Physics Lab

One of the simplest experiments that can be performed using Arduino is measuring the *RC* circuit time constant [5]. A typical experimental setup would include a power supplier connected to an *RC* circuit, consisting of a resistor in series with a capacitor. During the charge or discharge process, the voltage across the capacitor is usually monitored with an oscilloscope. Arduino is able to perform the role of a square-wave generator, a data acquisition system, and a real-time signal visualization tool. The advantage is twofold: expensive instruments, like the stabilized power supplier and the oscilloscope, are no longer necessary and virtually every student can build their own setup, using a breadboard, jumper wires, and the basic electronic components, write down the *sketch* with the instructions for the Arduino board, and visualize the data on the serial monitor. Data are then copied to a spreadsheet for further analysis, plotting, and fitting. The results found by the high school students are shown in Fig. 2. The calculated values of the *RC* time constant are in line with the expected theoretical value, thus proving the validity of our approach based on Arduino.

4 A First Approach to Robotics

When interfaced to an ultrasonic sensor (like HC-SR04), Arduino can be used to investigate kinematics [6]. Examples of possible experiments are observation of a free-falling body or an oscillating spring. The ultrasonic module consists of two piezoelectric devices acting, respectively, as transmitter and receiver. It has four pins: two of them (GND and VDC) are used to power it and must be connected to the corresponding GND and 5 V pins on the Arduino board. The other two pins are

Fig. 3 Basic components of a rover (left) and the final result after assembly (right)

called *trigger* and *echo*. When a square wave of 10-μs width is sent to the *trigger*, the transmitter emits a train of ultrasound waves. At the same time the *echo* port is raised to HIGH. As soon as the reflected wave is detected by the receiver, the *echo* pin returns to LOW. The elapsed time Δt between the two events is measured by Arduino, using the built-in *pulseIn* function. The distance d from the obstacle is then easily calculated using the formula $d = c\ \Delta t/2$, where c is the speed of sound.

The same sensor and an Arduino motor shield are the essential elements for controlling a small rover that can avoid obstacles along its path (Fig. 3). Input from the ultrasonic module is processed by Arduino and turned into output for the motor shield, which regulates the wheel direction and speed.

5 Conclusions

This project started in the 2015/16 academic year, with a first bulk of experiments built around the Arduino platform. Typical examples are in mechanics (a free-falling body or the harmonic oscillator), in thermodynamics (the heat-work equivalence), and in electronics (observing the charge and discharge of a *RC* circuit, the characteristic curve of an LED, …). The fruitful interaction between teachers with different backgrounds was central to devising new solutions and exploring novel applications. Indeed, a microcontroller like Arduino can be interfaced to a variety of sensors (ultrasonic ranger, temperature probe, …), thus fostering creativity and a sense of discovery. Hence, Arduino can be regarded as a kind of micro-laboratory, that is able not only to collect data, but also to control other devices and actuators. Arduino naturally lends itself, therefore, to introducing students to programming, automation, and robotics. As soon as students are able to write, compile, and load their first *sketch*, they can immediately see their code at work. It might only be a blinking LED or a fancy prototype of a rover, nevertheless, students will always feel a great sense of wonder and ownership.

Acknowledgements This work was funded by the Italian Ministry of Education, University, and Research as part of the project "PLS—Progetto Nazionale di Fisica" PN157YP17B. We are grateful to F. Capodaglio and P. M. Tricarico, who helped implement the project. We also thank the teachers and students from I.I.S. "Leonardo da Vinci" Civitanova Marche, Liceo Scientifico "Galileo Galilei" Macerata, Liceo "Giacomo Leopardi" Recanati, and I.I.S. "Francesco Filelfo" Tolentino for their enthusiastic participation.

References

1. OECD: PISA 2015 Results (Volume I): Excellence and Equity in Education, PISA; OECD Publishing, Paris (2016)
2. Mostafa, T.: Why Don't More Girls Choose to Pursue a Science Career? PISA in Focus **93**. OECD Publishing, Paris (2019)
3. Haugen, A.J., Moore, N.T.: A Model for Including Arduino Microcontroller Programming in the Introductory Physics Lab. Eprint arXiv: 1407.7613 (2014)
4. Arduino home page https://www.arduino.cc
5. Pereira, N.S.A.: Measuring the *RC* time constant with Arduino. Phys. Educ. **51**, 065007 (2016)
6. Organtini, G.: Arduino as a tool for physics experiments. J. Phys.: Conf. Ser. **1076**, 012026 (2018)

Weturtle.Org: A Web Community for Teacher Training and Sharing Resources in Educational Technologies

Michele Storti, Elisa Mazzieri, and Lorenzo Cesaretti

Abstract In recent years, there has been a stronger than ever need in Italy for teacher training in digital skills and pedagogical and teaching innovations. This is also due to an increase in national funds and European resources available for innovation in the field of education. This paper first describes the main innovations in learning that are made possible by web-based and other technologies, and how they currently meet teacher training needs. Next, the authors present Weturtle.org, a practical example of a "Community of Practice" and the TPCK model, which enables an integrative view at the subject, pedagogical and technological levels, to face the challenge of learning innovation. In the middle section, Weturtle.org is described with a focus on the opportunities for teacher training and validation, not only as an active user of the community, but also as a trainer him or herself. Finally, the authors present browsing data from October 2018 to September 2019, final considerations and future developments for the platform.

Keywords Teacher training · Educational technologies · TPCK framework · Community of practice

1 Introduction

The evolution of digital products and services currently makes it easier to introduce technology to educational contexts for the purpose of fostering learning processes. This perspective has already been theorized in the pedagogical model of constructionism [1], in which learning is described as a process of constructing mental models, made possible by the collaborative construction of tangible objects (also called "public entities") through technologies.

M. Storti (✉) · E. Mazzieri · L. Cesaretti
TALENT Srl, via Bachelet 23, 60027 Osimo, AN, Italy
e-mail: michele.storti@weturtle.org

L. Cesaretti
Department of Information Engineering (DII), Università Politecnica Delle Marche, Via Brecce Bianche, 60131 Ancona, Italy

D. Scaradozzi et al. (eds.), *Makers at School, Educational Robotics and Innovative Learning Environments*, Lecture Notes in Networks and Systems 240,
https://doi.org/10.1007/978-3-030-77040-2_42

Weturtle.org, the web community described in this article, represents an attempt to translate, in concrete terms, the integration of educational technologies and pedagogical models, inspired by the constructionist approach, and academic subjects.

The platform was designed by the startup TALENT srl, which has been active in the field of educational technologies for about four years. The company's mission is to promote educational innovation in schools and the development of soft skills, critical thinking and other human qualities in students, through the use of technological devices.

In this article, after an introduction on the current status of this technology in Sect. 2, the platform itself will be introduced in Sect. 3, followed by browsing data from October 2018 to September 2019 (Sect. 4); the final section presents final considerations and future developments for the platform.

2 Current State of Affairs

2.1 Training Needs in Digital Skills Among Italian Teachers

In recent years, the increase in resources and opportunities for teacher training in digital skills and didactic innovation, made possible by national and European investments, has paved the way for several changes in Italy.

An important boost came with the "La Buona Scuola" [2] (Good School) law, and was followed by a strategic plan for teacher training for the 2016–2019 period [3]. In the international context, 2017 saw the publication of DigComp 2.1 [4], the European framework for national policies on digital skills for citizens, and DigCompEdu [5], aimed specifically at teachers, educators and education stakeholders.

However, there are still critical issues in Italy, where there is only partial penetration of digital skills in schools. The 2018 Teaching and Learning International Survey (TALIS) [6] showed that only 47% of secondary school teachers regularly allow students to use technology in projects or classroom work, which is below the average in OECD and TALIS countries (53%). This can be partly explained by the average age of Italian teachers, which is currently the highest among OECD countries (49 years vs. an average of 44) and by the fact that only 52.2% have used teaching technologies in their formal education, against an OECD average of 56% and a TALIS average of 60.3%.

Regarding continuing education, although 68% of teachers in the 12 months prior to the survey participated in training that included teaching technologies, this is still the area of professional development where they report the greatest need. Only 35.6% of Italian teachers (OECD avg.: 42.8%, TALIS avg.: 49.1%) feel sufficiently prepared to use educational technologies.

Also, the context makes it difficult to use technology in schools: 31% of school managers report that the provision of quality education in their school is hampered by a lack or inadequacy of digital technologies for education (OECD avg.: 25%).

2.2 *Weturtle.Org: Between Theoretical Models and Changes in Online Learning*

Global innovations in online learning and training. Weturtle.org has its roots in an international trend marked by recent experiments based on web technologies that are transforming learning into a widespread, accessible, personalized and participatory process within online learning communities.

One example is the global Open Educational Resources (OER) movement, which, in many countries, has provided the impetus for individuals and institutions to share educational resources on thousands of educational sites and e-learning platforms, such as Moodle. Another relevant project is eTwinning, a platform launched as part of the eLearning Program of the European Commission. Its purposes is to enable teachers and staff in schools in many countries to collaborate and carry out projects within a European community of teachers.

MOOCs (massive open online course) are another innovative online learning model. The first examples of MOOC platforms (edX, Udacity and Coursera) began in the United States in 2011. Similar platforms appeared in Europe at around the same time, including EMMA, FutureLearn, Alison and Iversity, with free MOOCs from several European universities. In Italy two examples are EduOpen, which began from a project co-financed by the Ministry of Education, University and Research (MIUR), and brings together many Italian universities, and Federica.eu, developed by the University of Naples. A resource for online teacher education is the European Schoolnet Academy, a MOOC platform that offers free courses for teachers and educators in different subject areas.

The TPCK reference model Consistent with the aim to stimulate knowledge sharing processes through the creation of online communities, Weturtle.org specifically addresses teachers' training needs in the field of educational technologies and innovative educational methods. The reference model for the site was the technological pedagogical and content knowledge (TPCK) framework proposed by Mishra and Koehler [7, 8], which describes how the technological, pedagogical and academic dimensions can be successfully integrated into teaching practice. The community is an attempt to translate the model practically. In all content, there is strong integration between the features of the selected technologies, the academic topics and the didactic methodologies. Furthermore, as described below, the Tutorials section is directly connected to the technological dimension of the TPCK model and the Publications section to the pedagogical one, whereas Projects, Courses and Ebooks propose didactic paths in which the three dimensions are integrated.

Weturtle.org as a Community of practice. In recent years in Italy teachers' need for training in the field of digital and didactic innovation has led to informal initiatives by teachers and educational institutions, such as educational blogs, audio/video playlists, repositories and groups on social networks for sharing news and resources. These solutions, however, generally present obstacles including difficulty finding the resources (often not well organized and lacking search functions), and communications among users.

Weturtle.org was begun with the aim to create an online "Community of practice" for teachers—with the meaning proposed by Lave and Wenger [9, 10]—offering a user-friendly space with ease of access, research and resource sharing. Subscribers to the platform can publish projects subject to approval by the administrators, comment on uploaded content and rate it.

The community's mission is also to encourage production and sharing processes, to allow teachers to emerge as content authors in their professional community, thus qualifying their professional identity as trainers of other teachers. Teachers with specific skills are given the opportunity to offer free or paid training products (webinars, projects and ebooks) depending on their preference. This helps to expand and innovate the professional identity of the teachers who want to put their skills to use on the knowledge market.

3 Weturtle.Org

3.1 Browsing Weturtle.Org

Weturtle.org's home page has a top menu bar giving access to: the "Community" page showing all users who actively participate in the site by offering resources and training products; the "About us" page with information about the staff; the Store; and the Login area. In addition, it contains links to the platform's social media pages (YouTube, Twitter Facebook).

A bar below gives access to the following sections:

- PROJECTS: this section contains all lesson plans, which can be found by filtering by educational technology, subject and specific kit/software. There are tags for each project allowing users to find projects that have certain features (e.g., they use a kit). This area has two types of content: brief ideas for individual lessons (marked with a light bulb icon) and full projects with multiple lessons and a progression of activities (marked with a page icon);
- COURSES: this area contains all online courses and webinars, future and past. For each course, users can view details about the author, an introduction, a list of contents and objectives, the cost (if it is available for sale) and instructions on how to take part. One recent development being offered is free online courses to give teachers complete and high-quality training in certain areas of educational technologies (e.g., educational robotics, digital storytelling, videogame making), including video lessons, exercises and reading materials, followed by paid courses for more advanced learners. Teachers can get a recognized training certificate for all courses, which they can register on the Sofia platform as proof of fulfilment of their continuing professional development obligations;
- TUTORIALS: this section hosts training videos that explore the basic and advanced features of the technologies used in the projects; the area targets users who need basic training in the tools or who want to acquire more advanced skills;

- PUBLICATIONS: scientific and educational papers by international experts can be consulted on this page. A short description in Italian is provided for each article, along with a link to the online resource where it can be found;
- E-BOOKS: the area contains a collection of e-books dedicated to technology and didactics; some of these resources are provided free of charge, others are for purchase;
- BLOG: this space contains informative articles on topics of interest to teachers (e.g. accounts of workshops led by trainers, news on educational events in Italy and abroad, focus on noteworthy people in the field of educational technologies).

There is also a search function on the menu bar.

Further down the home page, eight quick buttons related to the technologies discussed in Weturtle.org give immediate access to the projects, which are already filtered according to the icon clicked on. Just below that, the home page is divided into areas that contain previews of the latest content added. The spaces between these areas contain calls to action for visitors, such as registering with the site, sharing a project and offering a training course of their own.

4 Analysis of User Browsing Data

The browsing data from Weturtle.org shown in Fig. 1 refer to the period from September 30, 2018 to September 30, 2019. The term "Sessions" refers to the number of single interactions (e.g., browsing pages) with the site. "Users": the number of unique visitors who visited the site; "New users": the number of users visiting the site for the first time; "Page views": the total number of pages viewed; "Pages/session": the average number of pages opened during a session; "Average session duration":

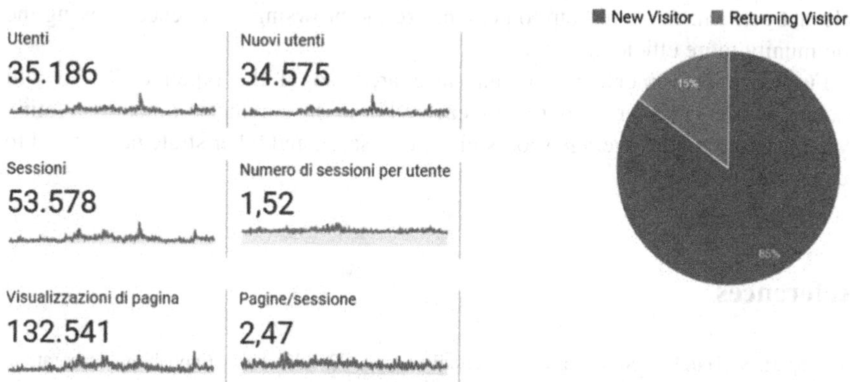

Fig. 1 User browsing data on Weturtle.org between 30 Sep. 2018 and 30 Sep. 2019

the average time the user is active on the site; "Bounce Rate": the percentage of visits where the user only sees one page.

Unlike the period analyzed previously from November 2016 to March 2017 [11], during which promotion of the website was mainly informal, with no investment in marketing, the data reported correspond to a period when the first investments were put into the communication (especially on social media and in newsletters) and marketing strategies. Compared to the previous period, there was a significant increase in the number of visitors, sessions and overall page views, whereas the average number of pages visited per session (2.47 versus 2.67), the bounce rate (70.62% versus 64.28%) and the average session time (00:02:14 versus 00:02:34) remained stable.

Most users visiting the platform (over 60%) are in the youngest age group of 18 to 34 years. The distribution of users by geographical area shows that the platform is used throughout Italy, but particularly in the large urban centers (Rome and Milan, followed by Naples and Turin).

5 Final Considerations and Future Developments

This article introduced Weturtle.org, a platform that aims to respond to the training needs of teachers in digital skills and educational innovation. Weturtle.org is also a "hybrid" model that seeks to combine the philosophy of OERs with the value of qualifying teachers as trainers on the online knowledge market.

Regarding future developments of the content, all the sections are in a growth phase, partly because the network of collaborating teachers is expanding, as is the total number of users. The most significant development concerns the forthcoming launch of online courses on the main themes of educational innovation.

A feed system will be implemented to update registered users regularly on new, relevant content: this will help to personalize the browsing experience, making the community more efficient.

Further improvements to the interface are also in the pipeline. These will encourage users to interact more with each other in the comments function, a rating system (based on the average scores given by users), and other strategies related to custom notifications.

References

1. Papert, S., Harel, I.: Situating Constructionism, pp. 193–206. Ablex Publishing Corporation, Constructionism (1991)
2. MIUR.: Piano Nazionale Scuola Digitale. (2015) http://www.istruzione.it/scuola_digitale/allegati/Materiali/pnsd-layout-30.10-WEB.pdf. verified on 08 Oct 2019
3. MIUR.: Piano per la formazione in servizio dei docenti 2016–2019. http://www.istruzione.it/piano_docenti/. Verified on 08 Oct 2019

4. Carretero, S., Vuorikari, R., Punie, Y.: DigComp 2.1. The digital competence framework for citizens with eight proficiency levels and examples of use. (2017) http://publications.jrc.ec.europa.eu/repository/bitstream/JRC106281/web-digcomp2.1pdf_(online).pdf. Verified on 08 Oct 2019
5. Redecker, C., Punie, Y.: European Framework for the Digital Competence of Educators: DigCompEdu. (2017) https://ec.europa.eu/jrc/en/publication/eur-scientific-and-technical-research-reports/european-framework-digital-competence-educators-digcompedu. Verified on 08 Oct 2019
6. TALIS 2018 Results (Volume I)—Teachers and School Leaders as Lifelong Learners. http://www.oecd.org/education/talis-2018-results-volume-i-1d0bc92a-en.htm, verified on 08 Oct 2019
7. Mishra, P., Koehler, M.: Technological pedagogical content knowledge: a framework for teacher knowledge. Teach. Coll. Rec. **108**(6), 1017–1054 (2006)
8. Koehler, M.J., Mishra, P.: What is technological pedagogical content knowledge? Contemp. Issues Technol. Teacher Educ. **9**(1), 60–70 (2009)
9. Lave, J., Wenger, E.: Situated Learning: Legitimate Peripheral Participation. Cambridge University Press., first published in 1990 as Institute for Research on Learning report 90-0013 (1991)
10. Wenger, E.: Communities of Practice: A Brief Introduction. (2011) https://wenger-trayner.com/introduction-to-communities-of-practice/. Verified on 15 April 2017
11. Cesaretti, L., Storti, M., Mazzieri, E., Galassi, A., Screpanti, L., Scaradozzi, D.: "Weturtle.org: una web-community per la formazione dei docenti e per la condivisione di risorse", Didamatica 2017, (2017)

Good Educational Robotics Practices in Upper Secondary Schools in European Projects

Marco Cantarini and Rita Polenta

Abstract In the fields of MINT (mathematics, ICT, natural sciences, technology), there is an increasing lack of young talent throughout Europe. It is clear that early exposure to scientific experiences is the key to motivating young people, especially girls, to develop an interest in these fields. The Erasmus + MINT "Kits for Kids" project is a current initiative for the design and production of integrated learning units as open educational resources for primary school students, of which there is a real demand. The main themes are learning media for primary education, such as simple mechanical machines, computer software, electrical appliances and related learning material. Our product, R4G—Robot for Geometry, is a device that can be used to teach mathematics, geometry and fractions. The aim is to improve performance and motivation from primary school level, using interactive and innovative teaching methods and tools. A second project, named EUWI—European Waste Investigation, is currently in progress, whose aim is to investigate water and pollution in the countries involved in the partnership.

Keywords Education · Teaching · Technology · Mathematics · Robotics · Environment

1 Introduction

It has become evident in all European countries, that interest in MINT subjects, and consequently the competences in these fields, and the number of MINT teachers, is falling across Europe. Thus, innovative methods are needed at the pre-school and primary levels, especially in the areas of science and technology [1].

M. Cantarini (✉) · R. Polenta
Istituto Istruzione Superiore "Volterra-Elia", Ancona, Italy
e-mail: m.cantarini@iisve.it

R. Polenta
e-mail: r.polenta@iisve.it

D. Scaradozzi et al. (eds.), *Makers at School, Educational Robotics and Innovative Learning Environments*, Lecture Notes in Networks and Systems 240,
https://doi.org/10.1007/978-3-030-77040-2_43

It is at this level that learning processes and individual competences must be initiated and nurtured, if future generations are to be successful in tomorrow's ever-more-technological world. It is a recognized fact in education that what we learn at an early age is more firmly anchored in our minds and can be built on successfully in subsequent learning phases. Any inhibitions that young children have, especially girls, can be reduced.

The learning units we developed in our Erasmus + MINT "Kits for Kids" project [2] are particularly innovative, because they provide playful access to science. It is our intention to assist in counteracting some of the deficiencies in the current educational situation at primary level. So far, initiatives in this field are still in their infancy, so our project will play a leadership role.

This project also helps to promote alternative learning and teaching concepts for primary schools.

2 R4G—Robot for Geometry

The idea suggested by a primary school teacher is to build a small moving/self-propelled device that can draw geometric shapes on large sheets of paper. The sheets of paper can be used to create posters to be hung on the wall, to get the children cut out geometric figures to help with memorization, to create complex geometric figures or explain other mathematical concepts like fractions.

2.1 Mechanics

During the first phase of the project, students in the electronics and mechanics departments of our school designed and built a prototype, using software and tools available in our school laboratories.

After the first prototype, partially created by cutting and modelling Plexiglas with machine tools, all the parts were designed in SolidWorks [3] software and 3D-printed [4]. During the project, thanks to the simple and rapid production of the pieces in the laboratory, the students were able to make changes to the drawings and optimize the products (Fig. 1).

One very important part of the device is the mechanism that moves the pen: it has to draw while the robot is moving, and stop and lift up when the robot stops.

For the second prototype, students used new motors with different characteristics, which required them to design new supports.

Later in the project, students developed a new, smaller and lighter prototype, and produced it entirely using a 3D printer.

The frame is a single piece and can contain all electronic control components (control board and batteries) (Fig. 2).

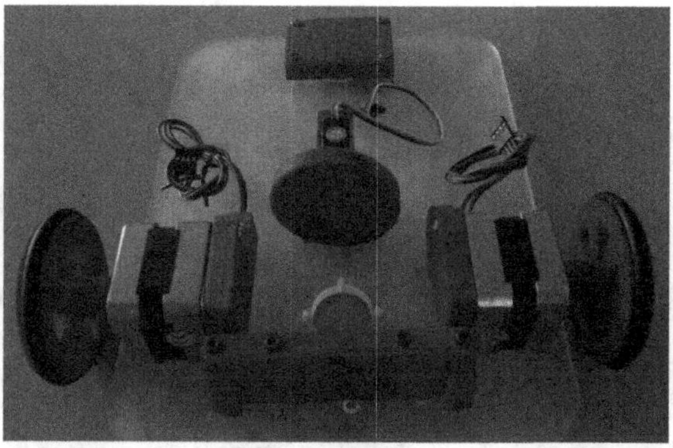

Fig. 1 First vehicle chassis

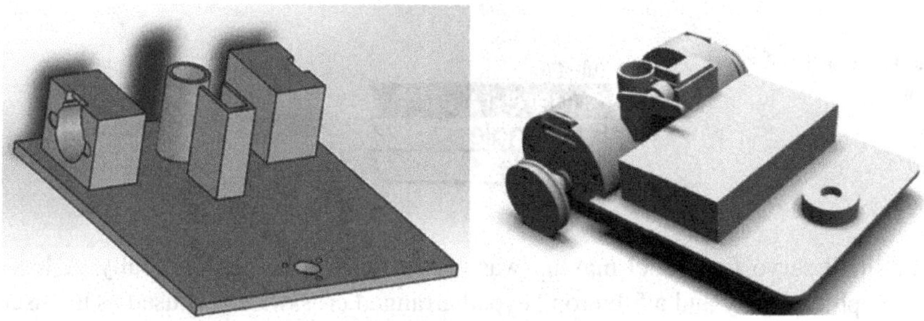

a) b)

Fig. 2 **a, b** 3D frame model

One very important part of the project was the cover design. R4G robot had to be fun-looking and attractive to young students, like a toy. With this in mind, the students designed and printed two different 3D covers, R4G Turtle and R4G Ladybug, both in bright colors (Fig. 3).

2.2 Electronics

The robot's control electronics went through various phases of development which were highly educational for the students. They got to experience all the steps in design of the electronics, from rapid prototyping to engineering for small series production. The first prototype was made using several electronic boards available on the market: a microcontroller board (Arduino), and a stepper motor driver like the one used in small 3D printers. The two main bipolar precision stepper motors connected to the wheels are also from the world of 3D printers.

Fig. 3 R4G Turtle and R4G Ladybug

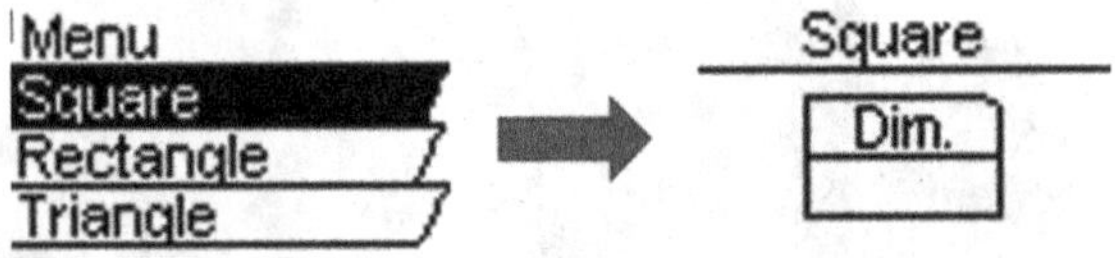

Fig. 4 Example of a text menu

A small servo for model-making was used to move the pen vertically. A low-cost graphic display and a 5-button keypad arranged crosswise was used as the user interface for selecting functions and modifying parameters.

The software was developed using a standard Arduino environment (IDE) [5, 6]. Some students were involved in developing various portions of the code, making ease of use a priority. Algorithms were developed for the design of squares, rectangles, triangles of different types, circles, and circular sectors. All geometric shapes could be selected via a text menu. All figures can be parameterized to generate drawings ranging in size from 10 cm to 1 m (Fig. 4).

This first prototype was successfully tested in several primary school classes (Fig. 5).

After the first prototype, given the interest and the soundness of the project idea, a second, a third and a fourth prototypes were developed. New functions were added to each new prototype, and the engineering was improved for possible small-scale production (Fig. 6).

In terms of hardware, the aim was to design a single board that would contain all the control and power components. An industrial-style PCB was created for the last prototype.

The same microcontroller on the Arduino Uno board was used for each prototype. This ensured the software was compatible from one prototype to another; therefore it was not necessary to change the development environment. Power was supplied by standard lithium batteries (18,650). We moved to low-cost unipolar motors fitted

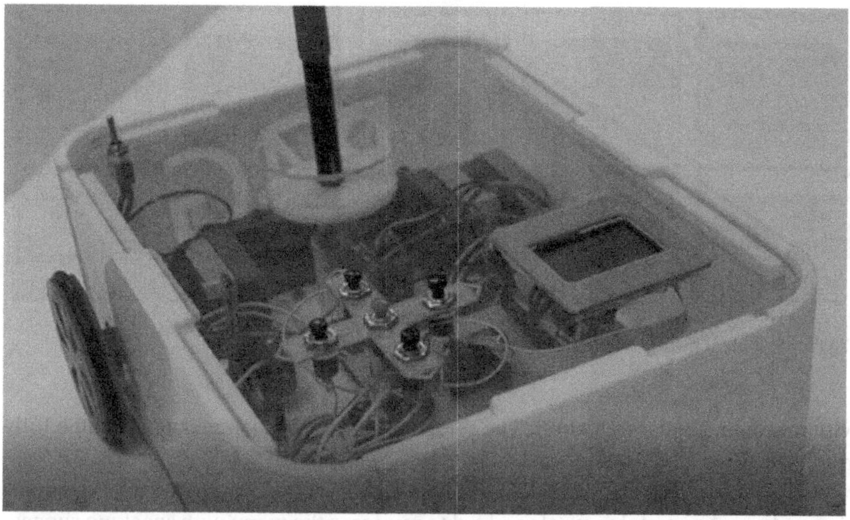

Fig. 5 First complete prototype

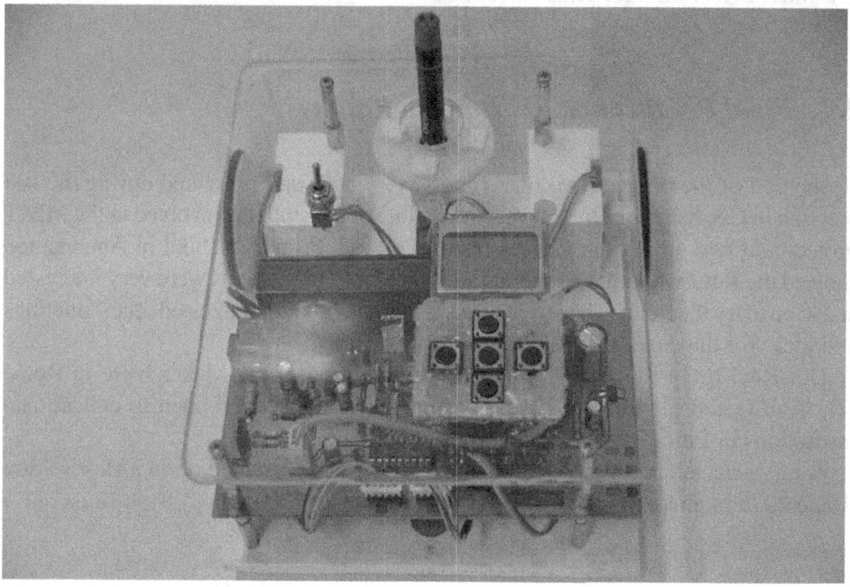

Fig. 6 2nd prototype with single PCB

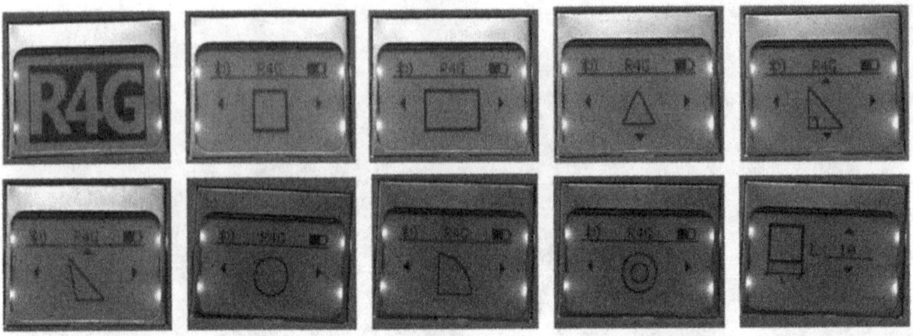

Fig. 7 Graphical menu

with mechanical gear reduction. The selection menu was changed and became fully graphical (Fig. 7).

A Bluetooth interface was added for interfacing with a computer, tablet or smartphone. An audio module was also added to reinforce the menu with speaking suggestions. This was tested successfully in a primary school in Germany during one of the Erasmus "MINT Kits for Kids" meetings.

2.3 Final Product and Sharing the Results

At the end of the project, two final robots were built and presented during the last meeting in Germany. Before this meeting, some of the students involved in the MINT project met two grade-four classes from "Collodi" Primary School in Ancona and showed the R4G robot to the students and teachers. The students were very interested and keen to try the robot. They tried guessing the shapes and suggested other functions and ideas for the cover (Fig. 8).

Our R4G robot took part in the European edition of the Maker Faire in Rome [7], from October 14 to 16, 2016. It competed in the section open to educational institutions in European Union countries (14–18 age group).

A jury selected 55 of the most innovative projects presented during Maker Faire Rome 2016 in an area dedicated entirely to schools (Fig. 9).

3 EUWI—EUropean Waste Investigation

This is a new Erasmus + project, which began in 2018 and is still in progress. Three other European schools are involved in the partnership. The aim of the project is to study local waters in each of the countries involved, by testing samples, and also by creating a scuba diving robot.

a)

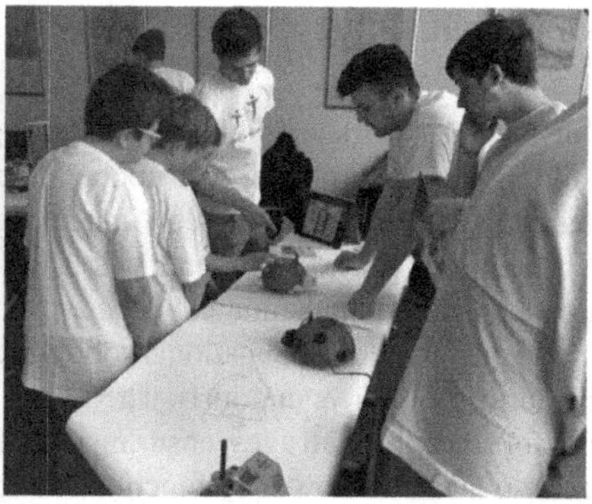

b)

Fig. 8 **a** Presentation at "Collodi" Primary School, Ancona (I); **b** Presentation at Berufskolleg Tecklenburger Land des Kreises Steinfurt, Ibbenbüren (D)

Fig. 9 Maker faire Rome—October 2016

A team of students specializing in chemistry, electronics and mechanics are working on these topics in collaboration with Dipartimento di Ingegneria dell'Informazione (DII) of Università Politecnica delle Marche.

4 Conclusions

Our students worked in groups, acquiring soft skills like problem-solving, team building, leadership, and peer cooperation. Students with different specializations shared their knowledge, thereby increasing their technical and subject skills.

The primary school students who tried the robot appreciated how easy it was to use, and learned mathematical concepts interactively while having fun.

In general, all the educational robotics experiences in our school originated from the curiosity of boys and girls. The use of robots proved to be an effective learning tool for students of all ages, ensuring they are not mere users of these instruments, but are also aware of how they function. Educational robotics enables them to identify constructive procedures to find solutions to concrete problems. It is learner-centered teaching, meaning that, instead of being passive recipients of concepts, learners learn effectively through experience, error, interaction with the environment and with others.

This type of learning is particularly productive, since knowledge comes from being active and doing, which begins with curiosity, from a question, and passes through trial and error, hypotheses, in pursuit of suitable solutions. Students will be motivated, involved, passionate because, after all, robotics is a game, a "serious game", which stimulates creativity, the imagination, and the ability to solve problems [8].

References

1. Buttolo, Robotica educativa. La didattica STEM (Science, Technology, Engineering and Mathematics), Sandit Libri
2. Erasmus+, https://ec.europa.eu/programmes/erasmus-plus/node_en
3. Solid Works, https://www.solidworks.com/it
4. Anna Kaziunas France, Stampa 3D, Ed. Tecniche Nuove
5. Cerri, Marcianò, Robol@b, Ed. Hoepli
6. Banzi, Shiloh, Arduino, la guida ufficiale, Ed. Tecniche Nuove
7. Maker Faire 2016, https://2016.makerfairerome.eu
8. Antonella Bonavoglia, Il Sole 24 Ore, November 2017

Assistive Robot for Mobility Enhancement of Impaired Students for Barrier-Free Education: A Proof of Concept

Alessandro Freddi, Catia Giaconi, Sabrina Iarlori, Sauro Longhi, Andrea Monteriù, and Daniele Proietti Pagnotta

Abstract Smart wheelchairs are in the category of assistive robots, which interact physically and/or non-physically with people with physical disabilities to extend their autonomy. Smart wheelchairs are assistive robots that enhance mobility, and can be especially useful for improving access to university premises. This paper proposes a smart wheelchair that can be integrated with an academic management system to enable students who have serious leg problems and cannot walk on their own to reach any academic building or room on a university campus autonomously. The proposed smart wheelchair receives information from the academic management system about the spaces on campus, the lesson schedule, the office hours of lecturers, and so on. Students can select the desired task from the user interface. The smart wheelchair can then guide the student autonomously to the desired point of interest, while planning the best barrier-free route inside the campus/building and, simultaneously, avoiding fixed and moving obstacles. The assistive robot has localization and navigation capabilities, which allow students to move about campus freely and autonomously, and benefit from a barrier-free education.

A. Freddi (✉) · S. Iarlori · S. Longhi · A. Monteriù · D. P. Pagnotta
Department of Information Engineering, Università Politecnica Delle Marche, Via Brecce Bianche, 60131 Ancona, Italy
e-mail: a.freddi@univpm.it

S. Iarlori
e-mail: s.iarlori@pm.univpm.it

S. Longhi
e-mail: s.longhi@univpm.it

A. Monteriù
e-mail: a.monteriu@univpm.it

D. P. Pagnotta
e-mail: d.proietti@pm.univpm.it

C. Giaconi
Department of Education, Cultural Heritage and Tourism, University of Macerata, Via Crescimbeni 30/32, 62100 Macerata, Italy
e-mail: catia.giaconi@unimc.it

D. Scaradozzi et al. (eds.), *Makers at School, Educational Robotics and Innovative Learning Environments*, Lecture Notes in Networks and Systems 240,
https://doi.org/10.1007/978-3-030-77040-2_44

Keywords Assistive robot · Smart wheelchair · Mobility enhancement · Education

1 Introduction

Assistive technology is a broad and active field of research [1], which has been experiencing a noticeable growth in recent years, thanks to massive economic investments from both the private and the public sectors. As a result of these investments, the gap between the current and potential market share of assistive technologies is gradually narrowing [2, 3]. Traditional wheelchairs are among the most common assistive devices, and represent one of the first solutions to the problem of mobility for impaired users [4]. Recently, electric wheelchairs have been turned into smart wheelchairs with integrated sensors and computational capabilities [5]. The market currently offers smart wheelchairs with several functionalities, such as, autonomous behaviors, interaction with domestic environments and parameter measurement, just to name a few [6, 7].

This paper proposes a smart wheelchair for use as a semi-autonomous mobility support for students who have serious leg problems and cannot walk on their own. The assistive robot has localization and navigation capabilities and, thanks to an integrated academic management system, it can guide the user between locations on a university campus, making it easier for them to attend lectures and reach common spaces, such as cafeterias and toilets.

2 Architecture

The proposed integrated system is mainly composed of the following elements:

1. smart wheelchair;
2. database containing information about the university campus.

2.1 *The Smart Wheelchair*

The smart wheelchair is based on a commercial electric powered wheelchair, namely the Quickie Salsa R2 produced by Sunrise Medical. This wheelchair has integrated sensors and computational power, which permit semi-autonomous navigation in partially-known environments. Its compact size and its low seat-to-floor height (starting from 42 cm) gives it flexibility and easy access under tables, making it suitable for indoor settings. The mechanical system is composed of two rear driving wheels and two forward caster wheels. The latter are not actuated wheels, but are able to rotate around a vertical axis. The wheelchair is equipped with an internal control module, the OMNI interface device, manufactured by PG Drivers Technology. This controller has the ability to receive input from different SIDs (standard input devices) and convert them to specific output commands compatible with the R-net control system. In addition, an Arduino MEGA 2560 microcontroller, a Microstrain 3DM-GX3-25 inertial measurement unit, two Sicod F3-1200-824-BZ-K-CV-01 encoders, a Hokuyo URG-04LX laser scanner and a Logitech C270 webcam make up the remaining equipment of the smart wheelchair. The encoders, inertial measurement unit and the OMNI are connected to the microcontroller, while the microcontroller itself and the other sensors are connected via USB to a computer running ROS (robot operating system). Signals from the Sicod and Microstrain devices are converted by the Arduino and sent to the ROS localization module. The information provided by the Hokuyo is used by the mapping module and by the path planning module for obstacle avoidance. Once a waypoint is chosen by the user, the path planning module creates the predefined path.

The smart wheelchair is navigated in three different steps [8–10]: localization, map building and path planning. The steps are described as follows:

- *Mapping*: representation of the environment where the robot is operating, which should contain enough information to enable the robot to accomplish its task; it mainly relies on the laser rangefinder, which can measure distances in the space.
- *Localization*: estimation of the robot's current position in the setting; the localization method is based on AMCL (adaptive Monte Carlo localization), which exploits recursive Bayesian estimation and a particle filter to determine the actual robot position.
- *Path Planning*: choosing the route to a desired goal location from the robot's position; this takes into account possible obstacles detected by the laser rangefinder.

All the steps described above are performed via ROS modules. The wheelchair software is essentially composed of ROS packages and nodes, which acquire data from the sensor sets, elaborate the information and command the wheels accordingly. A description of these functionalities can be found in [11], and Fig. 1 illustrates the working prototype under development in our laboratory.

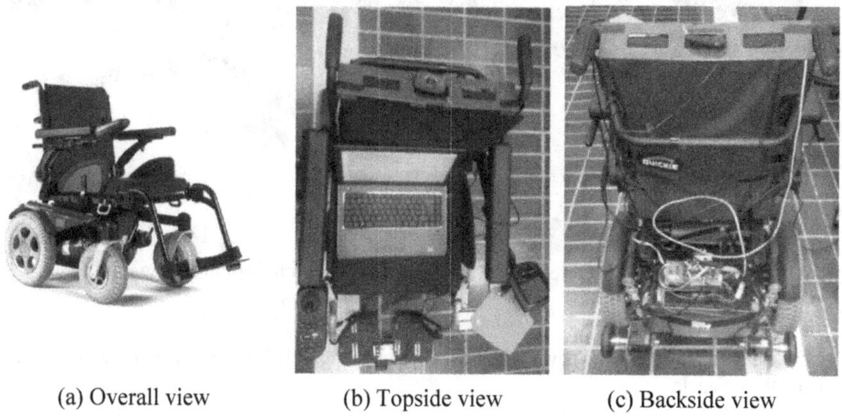

(a) Overall view (b) Topside view (c) Backside view

Fig. 1 Smart wheelchair test setup

2.2 *The Database*

The database is the second key part of the system. It contains all the relevant information about the university and is responsible for making the smart wheelchair adapt to the specific environment. In detail, the database contains:

Maps—Maps of the entire campus must be captured digitally, measured by laser and catalogued by floor.

Course Program—All available lessons must be included in the database, together with the timetable and the room where the lesson is to take place.

The user can simply choose the place or the lesson on a graphical user interface, and the wheelchair will take her/him straight to the desired location, choosing the shortest available route and guaranteeing obstacle avoidance.

3 Simulation Testing

Both the environment and the smart wheelchair, including its sensors, have been simulated in Gazebo. A "virtual user" was positioned in front of a computer running the simulator and chose an end point (i.e., a new classroom, or a space on the campus). After the path was generated, additional obstacles were positioned on the screen to simulate an obstacle avoidance situation. During path planning, once the laser scanner detected the presence of an obstacle, the wheelchair's speed was automatically decreased by the system, and eventually set to zero to give it time to select a new path (see Fig. 2). The simulation was repeated with different "virtual users" and "obstacles" in order to test the capabilities of the path planning algorithm.

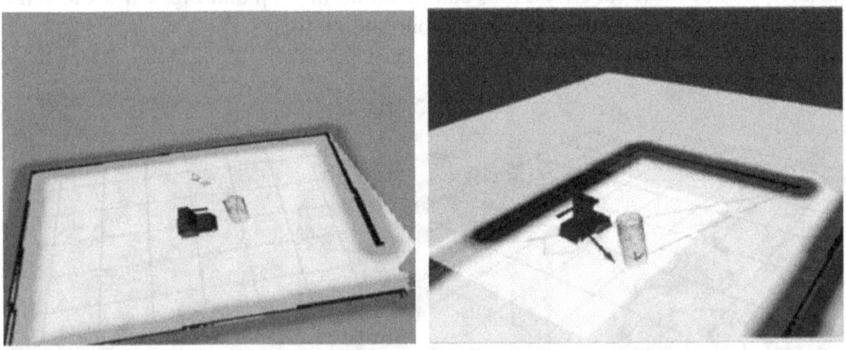

(a) Wheelchair with an obstacle in its way

(b) Wheelchair plansa new path to avoid the obstacle.

Fig. 2 The first image shows a new cylindrical obstacle. After it is identified, a new path is planned to avoid the obstacle

4 Conclusions

This paper has explored a solution that provides support to students with physical disabilities in the form of a smart wheelchair with an integrated academic management system containing relevant information about the university. This integrated solution allows the student to move through different areas of the campus. The algorithms related to the wheelchair for localization, mapping and path planning have been tested in a Gazebo simulator environment and performed well when it came to reaching the point selected on the map. A preliminary version of the database was organized by separating the information section about the maps and different sites on campus (e.g., classrooms, auditorium) from the course program and the schedule of lessons or particular events.

5 Future Works

Various aspects relating to navigation, interaction with the mobile robot and the user interface still need improvement to increase the smart wheelchair's functionalities. First, if there is an elevator on the planned route, the mobile robot could communicate with the elevator system and require the highest priority. This would avoid wait times for impaired students. Also, the wheelchair could be fitted with a system to recognize that people are moving around it, to avoid collisions and autonomously modulate the speed of the mobile robot in crowded spaces. The user interface is under development, and should provide an overall view of all possible actions. Finally, the possibility of integrating the smart wheelchair with the database and the management system is currently under investigation, as it is a key step in getting from simulation to real-life validation.

References

1. Benetazzo, F., Ferracuti, F., Freddi, A., Giantomassi, A., Iarlori, S., Longhi, S., Monteriù, A., Ortenzi, D.: AAL technologies for independent life of elderly people. In: ser. Biosystems & Biorobotics, Ambient Assisted Living: Italian Forum 2014, vol. 11, pp. 329–343. Springer International Publishing (2015)
2. Lenker, J.A., Harris, F., Taugher, M., Smith, R.O.: Consumer perspectives on assistive technology outcomes. Disabil. Rehabil. Assist. Technol. **8**(5), 373–380 (2013)
3. Ward, G., Fielden, S., Muir, H., Holliday, N., Urwin, G.: Developing the assistive technology consumer market for people aged 50–70. Ageing Soc. **37**(5), 1050–1067 (2017)
4. Benetazzo, F., Ferracuti, F., Freddi, A., Giantomassi, A., Iarlori, S., Longhi, S., Monteriù, A., Ortenzi, D.: AAL technologies for independent life of elderly people. In: Ambient Assisted Living, pp. 329–343. Springer (2015)
5. Fioretti, S., Leo, T., Longhi, S.: A navigation system for increasing the autonomy and the security of powered wheelchairs. IEEE Trans. Rehabil. Eng. **8**(4), 490–498 (2000)

6. Simpson, R.C., Poirot, D., Baxter, F.: The hephaestus smart wheelchair system. IEEE Trans. Neural Syst. Rehabil. Eng. **10**(2), 118–122 (2002)
7. Leishman, F., Monfort, V., Horn, O., Bourhis, G.: Driving assistance by deictic control for a smart wheelchair: the assessment issue. IEEE Trans. Hum. Mach. Syst. **44**(1), 66–77 (2013)
8. Bonci, A., Longhi, S., Monteriù, A., Vaccarini, M.: Navigation system for a smartwheelchair. J. Zhejiang Univ. Sci. **6**A(2), 110–117 (2005)
9. Cavanini, L., Benetazzo, F., Freddi, A., Longhi, S., Monteriù, A.: Slam-based autonomous wheelchair navigation system for AAL scenarios. In: MESA 2014—10th IEEE/ASME International Conference on Mechatronic and Embedded Systems and Applications, Conference Proceedings (2014)
10. Ippoliti, G., Longhi, S., Monteriù, A.: Model-based sensor fault detection system for a smart wheelchair. IFAC Proc. Volumes (IFAC-PapersOnline) **38**(1), 269–274 (2005)
11. Ciabattoni, L., Ferracuti, F., Freddi, A., Longhi, S., Monteriù, A.: Personal monitoring and health data acquisition in smart homes. In: Human Monitoring, Smart Health and Assisted Living: Techniques and technologies, pp. 1–22. IET (2017)

How Innovative Spaces and Learning Environment Condition the Transformation of Teaching: Good Practices and Pilot Projects

UP School: Motion, Perception, Learning

Lino Cabras and Fabrizio Pusceddu

Abstract The design strategy common to the educational spaces for the "Up School" based in the metropolitan area of Cagliari aims to frame a flexible learning space open to experimentation and the active exploration of places. Indeed, learning does not merely mean collecting and memorizing information; it also requires the ability to select, connect, understand and integrate, first by acquiring self-awareness and by developing perceptual abilities. Space—as experienced in its dynamic dimension—plays a crucial role in this process. The principles of the dynamic perception of space established by the most important investigations in neuroscience of recent years, were declared by the experimentations of the Bauhaus workshops, ahead of their time, as being strongly related to space, body and mind. Beginning with this premise, the "Up School" project—nursery, preschool and primary school—integrates an innovative educational program with the spatial layout of its environments. These spaces are conceived as a fluid sequence of "*affordances*" where, from an early age, children can shape their world within a perspective guided by good sustainability practices, enabling technologies and psychomotor equilibrium. Thus, the school system changes by being more conscious of its fulcrum: namely, the psychosomatic dimension of the individual.

Keywords Exploration · Active perception · Sharing · Dynamism · Community · Flexible spaces

The very etymology of the word knowledge contains an *active* meaning, which links the concept of testing reality with our need for choices, decisions and having awareness of the world we live in.

L. Cabras (✉) · F. Pusceddu
Department of Architecture, Design and Urban Planning, University of Sassari, Sassari, Italy
e-mail: lcabras@uniss.it

F. Pusceddu
e-mail: fapusceddu@uniss.it

D. Scaradozzi et al. (eds.), *Makers at School, Educational Robotics and Innovative Learning Environments*, Lecture Notes in Networks and Systems 240,
https://doi.org/10.1007/978-3-030-77040-2_45

Whether it is the Italian *conoscere* (from the Latin *cognoscere* cf. "recognize", "gnosis") or the English "knowledge" ("gnosis"), and whether it is voluntary or involuntary, knowledge is the result of an action.

Awareness is the factor that explains reality as a complex entity with which we interact by establishing relationships between physical objects and mental objects. Thus the experiential condition becomes the foundation for understanding the world and recognizing how reality is strongly affected by the way we describe, observe, and perceive it sometimes subjectively, sometimes collectively.

Knowledge is no longer a cognitive concept, but it is directly linked to the need for action. In order to act we need knowledge, but the only way we can have access to knowledge is by acting. According to Berthoz, the reason for this process is the intrinsic need of human beings to correlate the perceptive component with the motor component, that is, "perceiving something as a function of" and "making something as a function of". The brain is the core of these operations and it acts as a convergence tool between the two above-mentioned components, and also as an information processing center and action simulator [1]. Even learning activities in school environments are based on the same mechanism: the body is conceived as the essential component in relations with the immediate surroundings and with other individuals; motion is no longer considered a distracting element, but it has become the part of a process that we know to be intrinsic to our cognitive system. The most recent research in neuroscience appears to demonstrate a correlation between motor patterns and superior cognitive functions. Action is no longer understood to be a consequence of a perceptual phase—with a consequent interpretation—but rather constitutes the essential part of the process where all the components work simultaneously. This process is not structured in clearly distinguishable phases, but rather in the actual, or simulated, accomplishment of "motor actions" in which behavior is not seen as "mere movements" [2]. As a driver of processes and actions, space plays a revealing role in its surroundings and spontaneously suggests how they should evolve in the future. From this point of view, it is evident that spatial control and its design are foundational to the way human beings live, not only in terms of the functional need of shelter, protection, and comfort, but rather as a new structure of shapes and possible ways of inhabiting.

Today, we are keenly aware of the importance of the architecture of learning spaces, which should enhance sharing and cooperation processes, and the psychophysical well-being of individuals.

Pedagogical research confirms the need for architecture that is not conceived merely as closed off areas that are dedicated to specific settings—schools—but as spaces that can generate closer ties with the surrounding world, and that are modelled on complex interactions.

One hundred years after the founding of the Bauhaus, its educational principles have been largely confirmed by contemporary neuroscientific investigations, showing the similarity with the most contemporary learning theories. Thus, according to the program of the *Staatliche Bauhaus Weimar* described by founder Walter Gropius, laboratory practice determines a synesthetic experience within a social and didactic community. The aim of the school was to educate the natural abilities of individuals to understand the whole of existence, as a cosmic entity [3].

In the Bauhaus learning experience, it is even clearer that the study of form, color and space is the main tool for understanding reality. The pedagogical approach involves students starting a new learning process on a *tabula rasa* of their previous experiences. Johannes Itten, master of the *Vorkurs*—the school's preliminary course—was one of the teachers who applied a new radical learning method for art, aimed at reaching a new equilibrium. Itten's references took in the theories of Franz Cižek[1] on the stimulation of creativity and Dewey's principles of *Learning by doing,* the foundations of the entire Bauhaus teaching program. The goal of his course was to train man as a creative being, invoking the synergy of the energy of the body, mind and spirit.[2] Students' tactile skills were developed through specific exercises involving perception.[3] A new dynamic equilibrium in which individuals can see another way of inhabiting and perceiving reality is sought in both two-dimensional and three-dimensional terms. Paul Klee' tool of investigation was perspective in motion, or the *wandering viewpoint* [5], where "man is not a species, but a cosmic point" [6]. For Oskar Schlemmer, on the other hand, who was appointed director of the school's theatre workshop in 1924, the laws between the human body and space were investigated through abstract choreographies. From a pedagogical point of view Schlemmer reminds us that theatre, in its fundamental component of abstract choreography, can be a precious tool for achieving self-awareness of the body [7]. The modernity of these concepts, which focus on the dynamic nature of reality can easily be found in the contemporary definition of space, where motion and sensorial experimentation are the fundament of the act of perception.[4]

With this in mind, the Up School design views architectural space as a *continuum*, where learning takes place through a dynamic perceptual experience aimed at guiding children towards autonomy.

The Up School of Cagliari has a preschool and a primary school located in the historic "Villino Campagnolo" building. The main unit is in the villa itself and there are two small accessory buildings, a former store and kitchens. The outdoor space includes a large monumental terraced garden where children can play, discover nature and grow vegetables. Both the mobile and the fixed components of the didactic space

[1] Founder of the *Kunsterweberschule* of Wien, an art school exclusively for children, where he developed a new teaching technique aimed at promoting the free expression of students, drawing inspiration from Maria Montessori's theories.

[2] Itten used a series of exercises taken from the *Mazdaznan* cult he had belonged to, an exoteric doctrine widespread in Europe in the early twentieth century, founded by Otto Hanisch.

[3] To perceive means to be moved, and to be moved means to form. [...] Without movement—no perception, without perception—no form, without form stance. Substance—form. Form = movement in time and space; thus, substance = movement in time and space 4.

[4] Space occurs as the effect produced by the operations that orient it, situate it, temporalize it, and make it function in a polyvalent unity of conflictual programs or contractual proximities [...] In short, space is a practiced place. Thus the street geometrically defined by urban planning is transformed into a space by walkers. In the same way, an act of reading is the space produced by the practice of a particular place: a written text, that is, a place constituted by a system of signs [8].

provide multiple opportunities for children to interact during their daily activities. The furniture itself is designed and assembled to support children as they grow and to help them get the most out of making together and sharing knowledge. Thus, the conception of space is based on a radical re-assessment of action, enabling creative and active use. Inhabiting a space requires knowledge of how to interact with it, that is, knowledge is intrinsic to the space itself [9]. The pre-existing inner partitions of the villa were demolished to make way for flexible areas where groups of children of different ages can work and study in a shared space: classrooms were turned into learning environments and provided with functional and movable furniture designed for each different activity. Children learn in a "home" environment, made up of informal spaces, such as the reception and entertainment area for children, laboratories for experimental activities and water spaces for psycho-motor wellness. All the learning environments communicate with each other and overlook the large central hall—sharing space—which opens onto the terraced garden. The creativity room has a free, multifunctional configuration which depends on the activities taking place there. Paint stands and horizontal worktops are mobile and adaptable for plastic and manual arts. The fab lab space is the result of opening up four former rooms, whose original configuration can still be made out on the floor and in the way the walls are cut. The purpose is to facilitate flexible use of the space, as required by the multiplicity of the activities that take place in the fab lab: 3D printing, construction of small prototypes, coding classes and video/photo sets. The fab lab space is directly connected to the science laboratory and the augmented reality laboratory. The central space of the basement, which is directly linked to the terraced garden, features the "action area", dedicated to physical activity, relaxation and body awareness, and the water room, one of the central elements of Up School's educational program. Situated in the former utility spaces, the preschool is divided into two rooms by a movable, modular wooden wall, which can be opened when necessary. The wall also contains storage compartments and shutters connecting the two sections. A projector installed in the wooden roof generates an "interactive carpet" on the recycled vinyl floor, where children can go barefoot.

Similarly, the focus of the pedagogical program of the Up School preschool in Quartu Sant'Elena, namely, the development of basic motor skills in water, has spatial coordinates that are respectful of the pre-existing setting. Located in a traditional courtyard house, the preschool has a "water room" in the former storeroom, which has been redesigned as a glazed shell linked directly to the garden through a system of sliding doors, to achieve a continuous inside-outside space. The space contains three pools arranged in no particular layout and which the children use according to their age-related motor patterns. The different shapes are designed for specific needs: a rectangular pool, with a section increasing from 0 to 120 cm, with variable water flows; a shallow geodetic pool that has different pressure levels; a water floor that is 10 cm deep, with a soft surface coated with different materials—smooth, uneven, rough—for tactile stimulation. A wooden walkway takes the children from the water space to the organic garden, where a system for collecting rainwater is connected to the house's ancient water tank, in turn a learning tool on nature's cycles and sustainability (Fig. 1).

Fig. 1 The water room, Up School, Quartu Sant'Elena (photo© Stefano Ferrando)

In conclusion, Tagliagambe's [10] definition of the Up School is as a place of balance: balance between the body and the mind, between awareness and emotions, knowledge and expertise, tradition and innovation, imagination and sense of reality. Balance for the education of full, complex individuals, who can live and work in contemporary society.

References

1. Berthoz, F.: Il senso del movimento. Mc Graw-Hill, Milan (1998)
2. Rizzolatti, G., Sinigaglia, C.: So quel che fai. Cortina, Milan (2006)
3. Gropius, W.: Il progrmma dello Staatlische Bauhaus. In Wingler, H.: Il Bauhuas. Weimar, Dessau, Berlino 1919–1933, Feltrinelli, Milan (1987)
4. Itten, J.: Analisi dei maestri del passato. In Wingler, H.: *Op cit*
5. Klee, P.: Quaderno di schizzi pedagogici. Abscondita, Milan (2002)
6. Klee, P.: Diari, 1898–1918. Il Saggiatore, Milan (2004)
7. Schlemmer, O.: "Akademie und Bühnestudio". In Bistolfi, M., ed.: Oskar Schlemmer Scritti sul teatro, Feltrinelli, Milan (1982)
8. Certeau (De), M.: L'invenzione del quotidiano, Edizioni Lavoro, Rome (2001)
9. Emery, N.: L'architettura difficile -filosofia del costruire-. Marinotti, Milan (2007)
10. Tagliagambe, S.: Idea di Scuola, Antonio Tombolini editore, Ancona, (2016).

Landscapes of Knowledge and Innovative Learning Experiences

Massimo Faiferri and Samanta Bartocci

Abstract In these historic times, when there is a crucial shift in the way we consider the cultural and architectural aspects of learning spaces, it is important to investigate the role this spatial resource plays in the urban context. This, so we can understand the need to break with the outdated ideas about school that are deeply rooted in our society. There is common ground between architecture and pedagogy, a possible dialogue between space and knowledge, which can generate new explorations into the ordinary meaning of educational spaces and landscapes of knowledge, as a chance to expand the concept of inhabiting a space and how that impacts the world, and to devise a new urban condition. By first considering cities as a broad, extended learning space, we provide a chance and an incentive to reflect on the role of spatial design. The city is an important educational tool, since it represents a space of discovery, growth, socialization, tension, conflict and adventure. It is also where autonomy, adaptive intelligence and relational skills are developed. A new relationship between school and the city defines the future of learning and civilized co-existence.

Keywords Exploration · Active learning · Urban learning · Urban relations · Landscape of knowledge · Educational spaces

If we leave behind the comfort zone that brought us to the preordained conclusions we are used to, and we open up to reality without any prejudice, we will find that our perception of the world does not have an abstract point of view in the background, but is embodied in the observer. In this way, we take possession of the visible, meaning we can get rid of the dual concept of school–city, by devising a school that can be inhabited [1], or the concept of making school [2]; we can do this by moving away from the physical dimension of the building to the urban space, thus finding many similarities between how they are organized (streets/corridors, classroom/home, plazas/agora etc.…).

M. Faiferri (✉) · S. Bartocci
Department of Architecture, Design and Urban Planning, University of Sassari, Sassari, Italy
e-mail: faiferri@uniss.it

D. Scaradozzi et al. (eds.), *Makers at School, Educational Robotics and Innovative Learning Environments*, Lecture Notes in Networks and Systems 240,
https://doi.org/10.1007/978-3-030-77040-2_46

Much more important is the change in direction towards children and the city, that is, growing up in an urban environment [3], the city as a classroom [4] and the educating city [5].

Instead of school there is the city; it can enhance educating spaces and become a connecting place and a platform for open learning. Another important aspect to highlight on this premise is the following: if it is necessary to place the content of knowledge into specific settings that are expressly designed for that purpose, and structured according to a complex system and "intelligence" that grows with children's ability to "inhabit" and "use" them to learn how to learn, then experience, consisting of actions and choices, has a continuous relationship with the physical environment. For this reason, it is accepted that buildings express the values and the attitudes of their designer; but it is less evident, or less well known, the way in which buildings and space will affect users [6].

Psychological research demonstrates this need by recognizing space as an existential and experienced dimension [7], considering it a reference point of the identity of human experience [8]; even proxemics research has proven the strong influence of space on individuals' behavior [9], identifying empathy as the cognitive ability that makes people perceive other individuals' sensations, experiencing their resonance, and, finally, activating cognitive processes [10].

We look, touch, listen and measure the world with our whole bodies, and we structure the perceivable world around our bodies, which we consider to be the central element.

The body plays a fundamental role in the construction of the mental representation of ourselves, which will never be definitive, but will change continuously for the rest of our lives. We constantly understand and learn through the senses, defining the feeling of belonging to a space, a place, or a community.

It is a scientific certainty that the development of cognitive abilities occurs through the body, particularly through the experiences that children have with their body actions.

Action represents the fundamental element for attaining any kind of learning.

The relationship between single actions or more complex patterns of action creates the conditions for defining an organization of one self's acting, based on the experiences transferred from other situations or directed to the organization of new actions. Thus, all possible access routes to knowledge begin with the body, and body action is necessary for the development of cognitive processes.

Guiding space design towards a sense of possible action means being in the movement of choice, as an open process of subjective representation of reality. When we talk about "spaces inviting to action" we are thinking about a generative action as a new approach to design. A new connection between school and city defines the future of learning and civil co-existence and, as said above, evokes the metaphor of the "school as city", that is, a microcosm where corridors recall streets, classrooms recall buildings, spaces for socialization and leisure recall urban squares. This metaphor becomes more interesting and innovative when we reverse it: instead of a school there is a city, a city that turns into a school, animated by educational spaces, a place for learning that is open to a landscape of knowledge. This point of view leads us to consider the investigation of the *Ecourbanlab* research group[1] of the Department of Architecture, Design and Urban Planning, (DADU) of the University of Sassari, through the activities of the International Scientific School held since 2016 at the "Parco Scientifico e Tecnlogico di Porto Conte Ricerche" in Alghero, on the topic of Innovative Learning Spaces (ILS). In this research, the Scientific Schools[2] have become the field of investigation into how learning takes place (not only in schools) in contemporary society; every year the ILS scientific board has defined different objectives according to specific detailed topics, closely related to strategies of international research and regional planning.

The experiences over the last years have given rise to an executive board on which international professionals from all over the world have been involved in investigating the relationships between space design, pedagogy, energy, use of materials, computer science, physics, and new technologies. This provides an opportunity to experience the complexity of educational paths and, at the same time, find a way to renew the concept of school within an extended landscape of learning.

During the first edition of the scientific school, this approach was focused mainly on the design of school spaces, both new and existing ones.

In the second edition, the discussion was broadly on the synergy between the web and the possibility of integrating it with physical places and urban spaces, through augmented reality. The topic of the third edition introduced the principle that takes us from the school-based, urban and public learning space to "a city for everyone". This issue is not merely about disability, but rather it aims to explore, in technological and spatial terms, all the possible tools and facilitative solutions, in order to ensure the mechanisms of learning are spontaneous and inclusive for every citizen. The results of the school activities point out how the new ways of socializing and learning, supported by ICT and virtual reality, leads to inclusion being the main strategy that retains the ability to defend the local context [11]. Thus the urban context is the place where every inhabitant plays into an inclusive space based on sharing. The fourth edition of the International Scientific School, held in 2019, aimed to promote a high-value scientific and educational initiative to open up opportunities for dialogue among researchers, working every day in research spaces, and space

---

[1] Ecourbanlab is a research laboratory in the Department of Architecture, Design and Urban planning of the University of Sassari focused on space design as an investigative and testing tool of urban complexity. It analyzes and looks into several fundamental issues of the contemporary city, such as collective housing, urban re-use and renovation, and innovative learning spaces. Ecourbanlab is supervised by Professor Massimo Faiferri in collaboration with Samanta Bartocci, Fabrizio Pusceddu, Lino Cabras, Laura Pujia, and Rosa Manca.

[2] The Scientific schools *Innovative Learning Spaces* are organized by DADU (Department of Architecture, Design and Planning, University of Sassari within the SPIN-APP project); Innovative Spaces for Learning (Regional Law n.7, "Promotion of scientific research and technological innovation in Sardinia"), in collaboration with the INFN—Frascati (National Institute of Nuclear Physics, with the contribution of Sardegna Ricerche under the Scientific Schools project. The Scientific Schools are supervised by Prof. Massimo Faiferri.

Websites: http://ils2016.wixsite.com/uniss, http://ils2017.wixsite.com/uniss, https://ils2018school.wixsite.com/uniss, https://ils2019school.wixsite.com/uniss.

designers, while fostering a cross-pollination between the two fields of investigation. Research infrastructures are defined as facilities, resources or services used by researchers or companies for doing scientific research or enhancing knowledge and innovation. Research infrastructures can be great equipment, data archives, electronic devices or other kinds of structures. They can be precisely located, distributed in space or virtually defined: regardless of their nature, a research infrastructure is a form of excellence in terms of knowledge and innovation. The question we asked ourselves is: how can we find a connection between all these high-profile scientific activities and the local communities, in order to enhance local knowledge within a process of development for society and the economy? The dimension that relates to scale of knowledge aims to establish a reference range, by defining a set of problems, questions, needs and opportunities for which public policy and space organization can find an answer. Here the goal is to define the boundaries (both symbolically and geographically) of public action regarding space, where "boundaries" means "mobile borders" that can be changed as and when necessary. Thus, we can find a relationship between the scale of the abstract structure of knowledge and the organization of spaces for knowledge, from the formal institutions of learning to the territorial dimension. Therefore, we define a new urban and territorial morphology that assesses physical places according to their capacity for teaching, the interactions they promote, and the opportunities they offer for creating groups and enabling creativity. Ingold's [12] anthropological vision has provoked the transition from the concept of living in the world (*dwelling*) to one of living with the world (*inhabiting*). The phrase "interaction between perception and environment", often quoted as a way of describing potentially innovative learning spaces, presents a dichotomy between constructing and dwelling. This clarifies the vision of inhabiting as the sensation of feeling at home. This self-awareness opens up a pathway of re-imagination, which takes us from a conventional school space—or an unconventional one—to a wider vision and the spaces of research infrastructures. For the user, this awareness means choosing how to inhabit space, considering the idea that we do not move within a defined or static environment, but rather one that is "made" and is continuously refashioned.

The ILS Scientific school has worked locally to identify new learning territories, and has recognized public spaces and the public dimensions of the territory as learning spaces, for the areas identified in Sardinia Region's Smart Specialization Strategy (S3) for "Intelligent, sustainable and inclusive growth". A theater, a garden, a square, a church, a museum, a park, a former mine or an industrial building, for instance, retain certain features which, if properly reconsidered through space design, can transform ordinary, everyday places into devices for enhancing knowledge. Every place has its specific qualities, which are often not immediately or spontaneously readable. Hence, the research aims to use these traits to identify subjects for scientific publications and results, such as a network of "active spaces", directed at different levels of design (i.e., simply opening them up and making them available, making small changes to ensure they can actually be used, or setting up ad hoc installations). Augmented reality can link these places so they can share experiences and create a "connective urban learning platform".[3] These experiments—conducted in the form of workshops—have been an opportunity to test and adjust the ongoing scientific research. The outcome of these experiments is that they later formed the basis of important research projects, like those funded by the PRIN program,[4] and framework agreements between institutions. Furthermore, the research results have been published at exhibitions, conferences and in international scientific publications. In conclusion, the focus of research on architecture and learning is whether or not these spaces can be used as platforms of dialogue between public administrations and citizens, across different generations, cultures and institutions; whether they can share the same spaces and the same dynamics of innovation as museums, universities, research centers, and cultural and sports centers.

References

1. Ingold T.: The Perception of the Environment: Essays on Livelihood, Dwelling and Skill, Routledge, London (2000)
2. Weyland, B.: Fare scuola. Un corpo da reinventare, Guerini, Milano (2014)
3. Ward, C.: Il bambino e la città. Crescere in ambiente urbano, L'ancora del mediterraneo, Napoli (2000.)
4. Brusa, A., Borri, D., Porsia, F.: La città come aula, Edipuglia, Bari (1985).
5. Clemente, F.: I contenuti formativi della città ambientale, Pacini, Pisa (1974)
6. Plummer, H.: L'esperienza dell'architettura. Einaudi, Torino (2016)
7. Iori, V.: Lo spazio vissuto, Firenze, La nuova Italia, Firenze (1999)
8. Hillman J.: L'anima dei luoghi, Rizzoli, Milano (2004)
9. Costa, M.: Psicologia ambientale e Architettonica, Franco Angeli, Milano (2009)
10. Sinigaglia C., Rizzolati G.: So quel che fai. Il cervello che agisce e i neuroni specchio, Raffaello Cortina Editore, Milano (2006)
11. Tagliagambe, S.: Idea di scuola, Tombolini, Ancona (2016)
12. Ingold, T.: Ecologia della cultura, Meltemi, Milano (2001)

[3] A demo web platform has been developed from this concept, which looks at the city's broad potential as a learning space.

[4] See the PRIN project (Research Projects of National Interest) titled "PROtotypes of Schools to be lived—PROSE new architectural models for the construction, the renovation and resilient recovery of school buildings and to build the future in Italy." (Scientific Director Professor Massimo Faiferri, with the research laboratory *ecourbanlab*—partner of Frascati Scienza), and the "European Researchers' Night" project funded by the European Commission Horizon 2020|Marie Skłodowska–Curie (GA No. 818728).

Child Friendly Architectures. Design Spaces for Children and Adolescents

Marco d'Annuntiis and Sara Cipolletti

Abstract This paper presents the educational and laboratory experience of the course entitled "Child Friendly Architectures", taught during the 2019 academic year at the School of Architecture and Design (SAAD) of the University of Camerino, in collaboration with UNICEF Italia. The training course is the first in Italy to build a dialogue between the discipline of architecture and the protection and promotion of children and adolescents' rights. The course was offered to the university's students and was structured as two modules. In a series of training seminars, the first module, Teaching Activity, addressed the design of spaces for children and adolescents while looking closely at good practices and case studies. The second module, Application Activity, was a practical laboratory which guided students in a participatory process of planning. The students experimented with reading and planning a specific context in which they live, using specially structured tools and methods. The Child Friendly Architectures training course theorizes a way of thinking about the design of spaces for children and adolescents, taking into consideration their rights, and promoting the learning of tools, design techniques and new technologies. The competences involved in participatory planning—which can be learned—strengthen team work through important networking and listening opportunities. This helps young people to develop a critical awareness of children and adolescents' rights, and the quality of the spaces dedicated to them.

Keywords Design spaces for children and adolescents · Children and adolescents' rights · Laboratory · Innovative didactics · Participation

M. d'Annuntiis (✉) · S. Cipolletti
University of Camerino, "Eduardo Vittoria" School of Architecture and Design (SAAD), Viale della Rimembranza, 63100 Ascoli Piceno, Italy
e-mail: marco.dannuntiis@unicam.it

S. Cipolletti
e-mail: sara.cipolletti@unicam.it

D. Scaradozzi et al. (eds.), *Makers at School, Educational Robotics and Innovative Learning Environments*, Lecture Notes in Networks and Systems 240,
https://doi.org/10.1007/978-3-030-77040-2_47

1 Introduction

Every year, UNICEF Italia organizes multidisciplinary university courses on development education and education in rights. These courses are part of the university program, one of UNICEF's long-running activities.

UNICEF works closely with host universities, which provide the necessary professional skills and resources, to give young people the tools to read and analyze the issues that affect children and adolescents in Italy and the rest of the world. The Convention on the Rights of the Child is the main tool on which training courses are built and remains the key to understanding these matters but also to identifying possible solutions [1].

The diversity of the universities hosting the UNICEF university program brings a wealth of readings and possible applications of the Convention, directed at the individual subject matters and professions. Organized by the University of Camerino's "Eduardo Vittoria" School of Architecture and Design and UNICEF Italia during the 2019 academic year, *Child Friendly Architectures. Designing spaces for children and adolescents* is the first training course in Italy to build a dialogue between the discipline of architecture and protection and promotion of children and adolescents' rights.

Other UNICEF Italia programs conducted throughout the country are of significant importance for the project. Some of them include *"Child Friendly Schools"*, *"Child Friendly Cities"* [2] and other international projects like *"Malnate. The city of children and adolescents"*, which have set the framework for this project, guiding the actions and activities organized.

The aim of these programs is to recognize the subjectivity of children and young people, in order to build a world that is suitable for children and adolescents, using an integrated approach and a specific methodology. In keeping with the activities of UNICEF Italia and testing laboratory activities with a "maker" approach, the Child Friendly Architectures project delivers knowledge about children and adolescents' rights. It ensures the perspective of children and adolescent is made a priority in local government structures. Also, it promotes the active involvement of children in the issues that concern them, so their opinions are heard and they are at the center of considerations during decision-making (Fig. 1).

2 Construction of the Training Course

The *Child Friendly Architectures. Designing spaces for children and adolescents* training course has two important objectives. The first and most immediate is to increase graduates' skills on the theme of children and adolescents' rights, engagement with and the listening and participation of children and young people in matters relating to their study programs and their future professions. With regard to architecture, the training course aims to encourage a sense of awareness of the design

Fig. 1 Some of the UNICEF programs conducted in Italy, http://www.unicef.it

of spaces that are suitable for children and adolescents, by studying issues related to design devices, invested spaces, quality, project scales, technologies applied, and including a Teaching Module.

The second objective is to listen to the needs and proposals of children and young people, with the aim of improving the specific contexts in which children and adolescents live, by considering the relationships between city/neighborhood and the educational/school context. An Application Module was formulated with the direct involvement of the school children in a participation experience. The Child Friendly Architectures training course was implemented by a specially created network of people who offer their skills and professionalism in favor of recognition of the rights of children and adolescents. In addition to the two group leaders, namely "Eduardo Vittoria" School of Architecture and Design and UNICEF Italia, "Pericle Fazzini" Secondary School of Grottammare and the Municipality of Grottammare (province of Ascoli Piceno) were also involved in the project.

Collaboration between the university, the school and the public administration is fundamental for constituting a territorial context in which the rights of children and adolescents to be heard and to participate are guaranteed.

The speakers in the two modules are university lecturers, UNICEF experts, external consultants, and experts in participative methodologies (Fig. 2).

2.1 Teaching Module

Teaching Module 1 is for 20 students on degree courses from Camerino University's School of Architecture and Design. It takes place over six seminars exploring the themes of the Convention on the Rights of the Child. An in-depth study is made of the four general principles and the Rights of the Child according to the general United Nations' Commentary, with a particular focus on the rights to be heard and

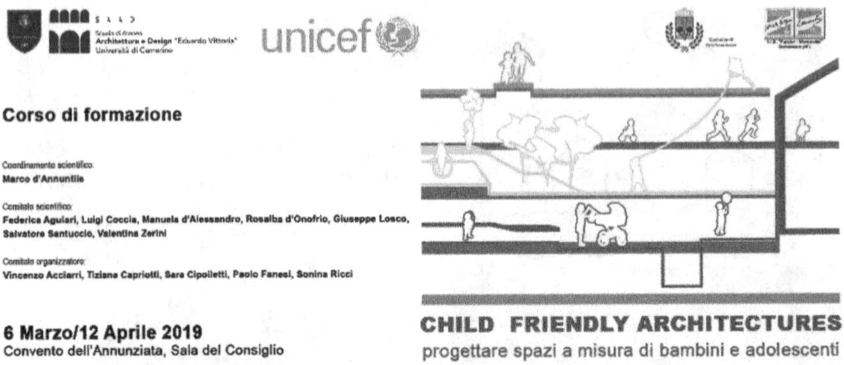

Fig. 2 The poster of the Child Friendly Architectures teaching course, edited by Sara Cipolletti

to participate and the right to education. How the spaces of the city [3], spaces of learning, spaces for leisure time and play can positively affect the quality of life and educational experiences is the subject of seminars on architectural topics. The relationship between architecture and spaces for children is analyzed through good practices and case study projects.

Different topics are investigated from historical examples to the contemporary spaces: for example, the summer *camp* in Misano (province of Rimini) designed by Giancarlo De Carlo; the playgrounds by Aldo Van Eyck in Amsterdam; the revolution of school spaces and the New Vertical Communities in Toronto.

One of the final stages of the course is dedicated to experimenting new technologies, such as 3D printing, laser cutting for the design and creation of games, teaching materials and child-friendly objects (Fig. 3).

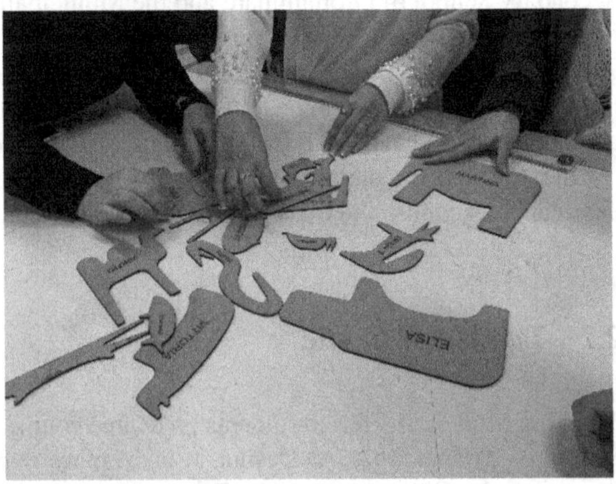

Fig. 3 Creating a laser-cut game in the laboratory Lab#Prototype, Unicam SAAD, ph C. Vagnozzi

2.2 Application Module

The activities in Application Module 2 involved "Pericle Fazzini" secondary school in Grottammare and the Municipality of Grottammare. A laboratory was created for secondary school adolescents with teachers and students from the University course acting as facilitators. The activity guides children through the use of specifically structured tools and methodologies to read an aspect of the context where they live: a route linking the historic city to the coastal settlement. Students conducted a survey to understand how adolescents use the space and their idea of the city. The students then organized round tables in order to discuss the findings of the survey and the developments in the area. The students developed special indicators—accessibility, pedestrian traffic, the reuse of disused buildings, green and water resources—to identify scopes where 'the teenagers' desires and vision intersect with and could improve interventions.

The adults involved in the activity offered their skills and guidance to the children/teenagers along the way (Figs. 4 and 5).

Fig. 4 Round table, ph. G. Sabatini

Fig. 5 Visit to the historical route, ph. A. Guaiani

3 Conclusions

The combination of a teaching module and an application laboratory with a "maker" approach offers students and schoolchildren a way to become familiar with and experience their rights. The laboratory, in particular, meets one of the Convention rights: the creation of a physical space and an ideal context in which children and adolescents can express themselves and can be heard.

In the final reports, the university students said that the course applies the rights of children and adolescents to the design of space, giving them a new perspective [4]. The teaching experience draws the attention, among academics and professionals, but also among young people themselves, to the fact that protecting children and adolescents' rights is pivotal to the choices made for the organization, planning and management of public and private spaces.

References

1. Saulle, M.R.: La Convenzione nel ventesimo anniversario, Comitato italiano per Unicef (2009)
2. AAVV: Costruire città amiche delle bambine e dei bambini, Comitato Italiano per Unicef (2005)
3. Krishnamurthy S.: Reclaiming spaces: child inclusive urban, design. Cities & Health, on line Journal homepage, https://www.tandfonline.com/loi/rcah20, (2019)
4. d'Annuntiis M., Cipolletti S: Child friendly architectures, Quodlibet, Macerata, (2020)

Multipurpose Learning Environments for Flexible Didactics

Anna Celeste Rubino

Abstract This paper investigates some of the issues and good practices in which 21st-century educational requirements are met by innovative design solutions for learning spaces, and their implications for education.

Keywords Learning environment · Pedagogy · Space design · Multifunction · Flexibility

1 Introduction

In recent decades, there has been much debate in the fields of architecture and pedagogy on the conception and design of new learning spaces, with an attempt to opportunely encompass all the elements that should contribute to an effective learning process.

The main considerations for such design include pedagogy, the technological tools used in teaching, the need for flexible and adaptable educational spaces in the different urban and social contexts of contemporary society.

What is the role of contemporary architecture in rethinking learning spaces? Can it influence learning processes or should it support pedagogical needs?

According to Gislason, most studies show that the changes introduced in the field of teaching and learning in the last two centuries have influenced the architecture of schools and the design of classrooms. He outlines two main transformations in the relationship between school design and school culture in western countries: the transition from single-grade classrooms to multi-grade school-rooms, and the spread of innovative educational practices, as an alternative to frontal lessons, such as cooperative education and informal learning activities [1].

A. C. Rubino (✉)
DICEA—Dipartimento di Ingegneria Edile Architettura, Università Politecnica delle Marche, Ancona, Italy
e-mail: a.c.rubino@pm.univpm.it

D. Scaradozzi et al. (eds.), *Makers at School, Educational Robotics and Innovative Learning Environments*, Lecture Notes in Networks and Systems 240,
https://doi.org/10.1007/978-3-030-77040-2_48

These variations have required a rethinking of learning spaces, involving space design experts and paving the way for interactions between the pedagogy and architecture sectors.

2 Pedagogy and Space Design in 21st-Century Learning

2.1 Critical Issues and New Requirements

It has recently become widespread practice in different European contexts for school designers to ask teachers and students what their needs are for new spaces, in an attempt to implement participatory planning.

Nevertheless, the link between architecture and the educational community is still characterized by distance and criticalities, and the word architecture is not widely used in the context of contemporary educational theories. The preference in educational environments is to refer to the building in its entirety, by calling it a "structure", or to specific elements of the learning environment, such as the tools used daily for teaching, referred to as resources [2].

One critical element is the difficulty teachers have designing and implementing non-traditional didactics, and adequately using the tools and spaces designed for active and collaborative teaching [3].

Another open issue is the inadequate level of communication and mutual understanding between the world of architecture and the world of education.

Teachers' requests regarding the arrangement of spaces for functional learning environments that are psychologically comfortable for students are not always properly interpreted and converted into architectural spaces.

The influence of social media and digital technologies is also a significant factor in the radical changes to the 21st-century learning environment.

Since the 1980s, Meyrowitz has illustrated that television and electronic media have altered social and communicative behavior, by dematerializing the boundaries between childhood and adulthood, and between private and public space [4].

An aptitude for using social media and the portability for students of smartphones have contributed to the breaking down of barriers between adults and children, determining the idea of *deterritorialization and disembodiment of space in the digital landscape* [5].

The traditional class space is seen as increasingly porous and precarious, owing to the ease and speed with which information can be shared between the worlds inside and outside school [6].

Demand in the twentieth century for child-centered pedagogy and the diversification and personalization of learning, as well as the use of visual technologies *are altering the experience and aspirations of learners* [7].

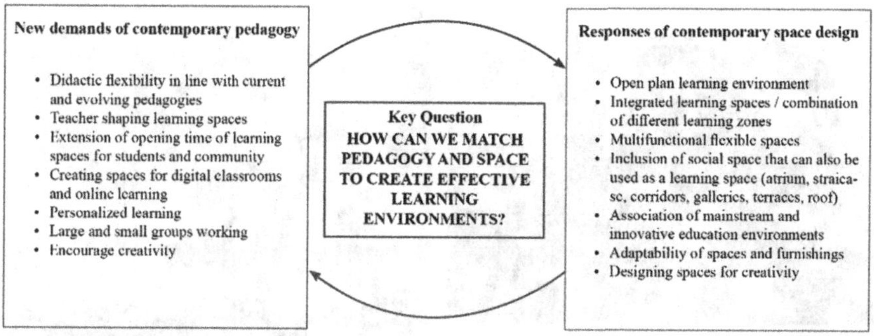

Fig. 1 The diagram summarizes recent demands made with respect to the architecture of learning environments within schools (author's diagram)

2.2 *Combining Pedagogies and Spaces*

The research focused on investigating how the configuration of contemporary educational spaces meets changing educational needs, from both a pedagogical and an architectural point of view (Fig. 1).

The classroom is the traditional teaching space and has been the subject of various experiments over time. However, there has been a recent trend in contemporary European schools to expand the educational space to the entire school building, including closed and open spaces, pre-existing buildings and newly built spaces.

The connective spaces also tend to be considered suitable for teaching, so much so that there is a growing trend for the hybrid use of learning spaces. These might be, for example, wide staircases used in multiple ways, as a space for taught lessons or collaborative lessons, as a lecture theatre for conferences and meetings, or simply as a socialization space for students.

In the current scenario of school architecture, there is a need to expand the physical learning space to include a variety of informal and social areas [3].

3 Multifunctional Learning Spaces in Contemporary Education Environments

Recent projects in different grades of school and university spaces include design solutions of this kind and the possibility for a flexible use of space both vertically and horizontally.

At Lorentz School in Leiden, Netherlands, designed by the architectural firm atelier PRO, the auditorium space was used for multiple purposes, in addition to its traditional function. These include: a theatre, a documentation center and library, and a space for computer lessons, as it contains workstations [8] (Fig. 2).

Fig. 2 **a** The documentation center also used for computer lessons at Lorentz School, Leiden, Netherlands (photo: Jannes Linders). **b** The auditorium used for performances at Lorentz School (photo: Jannes Linders). **c** Mediatheek on the stairs at Lorentz School (photo: Jannes Linders)

An area of experimentation of different learning spaces is the Bartlett Real Estate Institute (BREI), a new Centre for Learning Environments of University College London (UCL) located on the first floor of Here East university campus in East London, in a converted, multifunctional building erected for the 2012 London Olympics.

The variety of learning spaces is based on a holistic vision, to provide a 21st-century collaborative and inclusive environment for students, academics, designers and professionals involved in space investigation and design. In addition to open areas for panel debates and networking, fitted with a display screen, there are seminar rooms, a lecture theater and private study booths.

On the ground floor, wide spaces support flexible functions, such as the auditorium space, used for conferences and events, as an individual and group study space for university students, and as a laboratory space for secondary school students during the summer school.

An uncommon example of school architecture is the Bridge Academy, a public high school in the north-east London borough of Hackney.

The unusual location of the school on the banks of Regent's Canal led BDP architects to build vertically, distributing the classrooms and the open learning spaces

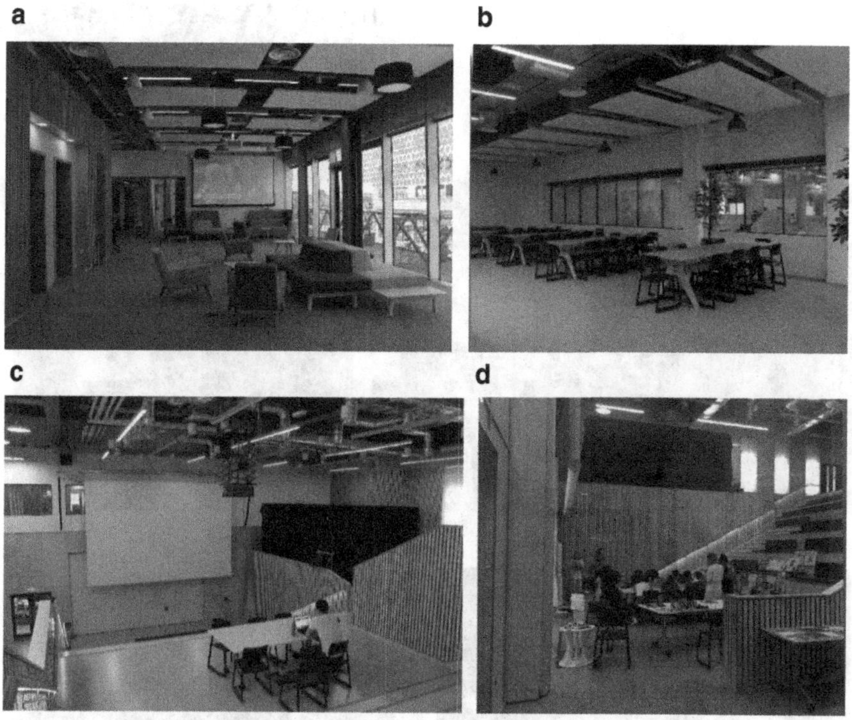

Fig. 3 **a** The break-out area in Bartlett Real Estate Institute (BREI) at UCL Here East, London (photo: Anna Celeste Rubino). **b** The lounge in BREI (photo: Anna Celeste Rubino). **c** The auditorium on the ground floor of UCL Here East (photo: Anna Celeste Rubino). **d** The auditorium used by secondary school students (photo: Anna Celeste Rubino)

over different floors, and connecting them physically via open balcony corridors (Fig. 3).

Owing to the building's geometry, most classrooms and learning zones in the galleries have irregular shapes, providing formal and informal learning environments [9] (Fig. 4).

The six floors face onto a ground floor "central square", a social heart space used for assemblies and recreation, above which the library is suspended. Both constitute centralized areas of focus within the school where learning and social activities can take place. The architect Papa of BDP, part of the Bridge Academy design team, said about the fundamental aspects of the project: *"We get inspiration and some specific concepts of architecture on the part of... designing the school... from one particular architect who influenced us both in this school and others that we've designed, which is Herman Hertzberger. The Montessori School in Amsterdam is very interesting particularly for two key concepts: first it is very focused around a space or a unifying idea that the whole school is one thing and you should really be able to stand in one part of the school see the rest of it in one glance. Second is*

Fig. 4 **a** One of the many open learning zones with workstation in the galleries of Bridge Academy, London (photo: Anna Celeste Rubino). **b** View of galleries on different floors at Bridge Academy (photo: Anna Celeste Rubino). **c** Ground floor hall, as a "central square" making an impressive spatial impact, Bridge Academy (photo: Anna Celeste Rubino). **d** The library floor overhanging the assembly hall (photo: Anna Celeste Rubino)

the staircase's location in different points of each level; it means you going... this staircase is... you're always heading towards another set of teaching and learning spaces and you never have that feeling of going round" [Interview transcript, Papa, Keith (interviewee), Rubino, Anna C. (interviewer), 2019].

4 Conclusions

The projects involving school and university spaces presented here contribute to defining a concept of contemporary learning spaces that responds to the needs of today's increasingly multicultural, dynamic and connected society. In light of the foregoing, the learning space is to be considered one that is socially constructed through student and community practices and multiple social relationships with variable layouts; the architectural dimension takes on multiple dimensions, with fluid internal and external boundaries, and incorporating different areas of proxemics. Designers and educational institutions should promote the creation of learning spaces of different sizes and with different equipment and furnishings. Such spaces might communicate with each other, favoring spatial fluidity, or be separated by walls or mobile panels. Furnishings should also be flexible and varied, to encourage autonomous use of the space by students, as well as an active willingness to adapt the conditions of comfort in the classroom and laboratories. There is a common denominator to many contemporary school and university projects: the presence of large halls and spaces for multiple uses both inside and outside buildings, for use by students, staff and local communities, can respond effectively to the increasing demand for social learning spaces.

The learning environments analyzed show a flexible and creative approach on the part of architecture, in an attempt to meet the needs of 21st-century education.

References

1. Gislason, N.: Building Innovation. History, Cases, and Perspectives on School Design, Backalong Books, Big Tancook Island, Canada (2011)
2. Dudek, M.: How Children Learn from Architecture and the Environment, Routledge, Taylor & Francis Group, UK (2019)
3. Mirchandani, N., Wright, S.: Future schools. Innovative Design for Existing and New Buildings, pp. 31–88. RIBA Publishing, part of RIBA Enterprises Ltd, Newcastle upon Tyne, UK (2016)
4. Meyrowitz, J.: No Sense of Place: The Impact of Electronic Media on Social Behaviour. Oxford University Press, Oxford, UK (1985)
5. Castells, M.: The Rise of the Network Society. Blackwell, Oxford, UK (2000)
6. Dussel, I.: Digital classrooms and the new economies of attention. In: Darian-Smith K., Willis J. (eds). Designing Schools. Space, place and pedagogy, Routledge, Taylor&Francis Group, London and New York (2017)
7. JISC Joint Information Systems Committee.: *Designing Spaces for Effective Learning. A guide of 21st century learning space design*, Copyright HEFCE Higher Education Funding Council for England (2006)
8. OECD Centre for Effective Learning Environments.: *Designing for Education. Compendium of Exemplary Educational Facilities 2011*, pp.163–165, OECD Publishing (2011)
9. Gough, P.: *BDP's Bridge Academy suggests an optimistic future for the design of school buildings*, Architecture Today n.194, monographic number (2009)

Adaptive Environments. New Spaces for Learning

Gianluigi Mondaini and Marco Rosciani

Abstract The architecture of schools can no longer be as rigid as it has been in the past. It needs to be spatially open, stimulating and, through its physical qualities, able to react and adapt to those who live and work there to use it as an active educational tool. However, architecture by itself is not enough to define new spatial models for education or for our society and its ever more complex problems. Thus, a plurality of forms of disciplinary participation is necessary. Pedagogy, with its close relationship with architecture, and technology, with the innovations it brings to teaching methods, are also involved. The dynamism in communication processes and educational practices that arises from new technologies demands a review of how school environments are organized, along with flexible, multifunctional and adaptable solutions for them. The most interesting models are those in which the conventional classroom "breaks down" physically, in favor of open-space learning environments, and flexible spaces. Such layouts help promote skills acquisition, thanks to the synergy between the technological elements and the physical qualities of the spaces that make learning more engaging.

Keywords Open space · Learning spaces · Technological environmental

1 Learning Environment

In recent years, we have witnessed a change in direction in the design of learning spaces within the scene of school architecture that is leading, also in Italy, to buildings that are designed with more attention on the relationship between the built space and learning. This reversal in trend stems largely from the will of institutions to keep pace with the changing needs of contemporary education and standards of quality of

G. Mondaini (✉) · M. Rosciani
Università Politecnica Delle Marche, Ancona, Italy
e-mail: g.mondaini@univpm.it

M. Rosciani
e-mail: m.rosciani@pm.univpm.it

D. Scaradozzi et al. (eds.), *Makers at School, Educational Robotics and Innovative Learning Environments*, Lecture Notes in Networks and Systems 240,
https://doi.org/10.1007/978-3-030-77040-2_49

school building, which still see us lagging behind especially in relation to Northern Europe.

The first steps in the initiative were the studies conducted by the Ministry of Education, University and Research (MIUR) with the support of the scientific research conducted by the INDIRE agency. In 2013, INDIRE produced guidelines for school buildings, with the goal—so far unattained, unfortunately—of updating the legal framework established in the Ministerial Decree of 18 December 1975. Hence, the strategic recommendations are for designing attractive and stimulating spaces, based on a concept of school as a multifunctional hub, through a shift in design towards spatial "generosity" and an innovative relationship with context [1]. A spatial openness that can interpret social change, leaving behind the typological standardization that has been repeated for many years and has led to the symbolic irrelevance of the building within the community.

The space will no longer be made of walls alone, but also by the people who animate, interpret and modify it with their presence and creativity in the didactic process. The architecture of schools can no longer be as rigid as it has been in the past. It needs to be spatially open, stimulating and, through its physical qualities, able to react and adapt to those who live and work there to use it as an active educational tool.

The main subject of discussion is certainly the classroom: a paradigm of the past tied to the system of mass literacy, that has hardly changed over the years and now it no longer suits the needs of a more advanced school system, where teaching methodologies are more focused on the personalization of programs and teaching. Yet, architecture by itself is not enough to define new spatial models for education, since the issues are becoming increasingly complex. What is needed is multidisciplinary participation: pedagogy, with its close relationship with architecture, and technology, with the innovations it brings to teaching methods. One interesting proposal is INDIRE's "1 + 4 learning spaces for a new generation of schools in Italy". The institute intends to use this space model to open up the classroom to a new configuration, in which four complementary spaces are placed on a par with the class's group space, rather than being subordinate to it: the individual area, which is the student's personal space, the informal space for relaxing and meeting others, the exploration space, where students can learn to make, and finally the agora, the community space [2]. Each of these spaces requires suitable technological equipment, from individual personal devices, such as tablets and notebooks, to interactive tools and immersive virtual environments for managing teaching; digital manufacturing laboratory equipment requires intensive, widespread technological infrastructure designed in an integrated way with architectural elements.

The models that are being pursued are those of the most avant-garde schools in northern Europe, where we are witnessing the physical disintegration of conventional classrooms in favor of open-space learning environments. One example of this is the work of Danish architect Rosan Bosch, who describes his own principles of learning landscape design, with the creation of open and flexible environments aimed at a better acquisition of skills, thanks to the synergy between the technological elements and the physical qualities of the spaces that make learning more engaging [3]. By

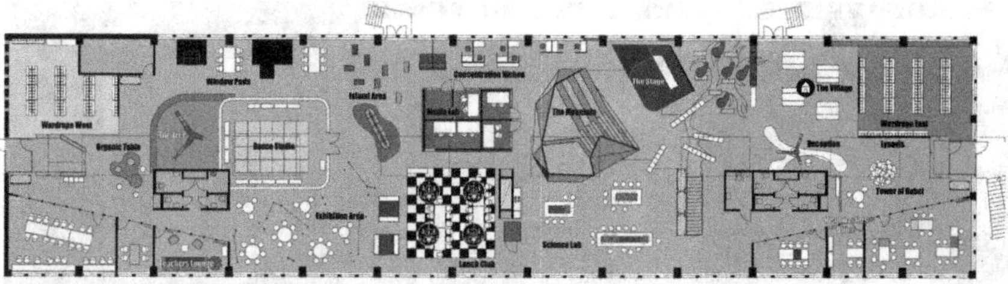

Fig. 1 Rosan Bosch Studio: Vittra Telefonplan, Stockholm (2011). From learning environments to learning landscape

analyzing planimetric developments in Swedish Vittra schools (Fig. 1), we see the physical freedom of the teaching spaces, in which the classroom disappears and the organization of space is governed by functional activities; spaces are equipped with digital technologies, which can adapt to the needs of students, following a logic of "reversed" didactics, the *flipped classroom* [4], or *cooperative learning* [5], aimed at improving educational interactions. Further design and technological developments can be found, with the built space evolving towards a more "nuanced" meaning. An architecture that is less physical and leaves room for technological interactions, and integration between real objects and virtual elements, with the potential to transform the learning landscape.

2 Research and Project Experiences

The workshops in the Scientific School Innovative Learning spaces, organized by the University of Sassari, produced several experiments, which led to a series of interesting reflections on the theme of innovative spaces for learning. These involved experts from various disciplines ranging from architecture and technology to pedagogy. Both of the experiences we present concern interventions on existing buildings, which came about from a desire to dismantle the static relationship between the classroom and the corridor, by using multifunctional architectural devices.

The first case concerned a secondary school in Sassari, which, although built only ten years earlier, was functionally, spatially and architecturally inadequate, far from the current needs of pedagogy. We came up with a functional review of the classroom that imagined a new relationship between the interior and the exterior of the building, but especially between the classroom and distributive space.

To do away with the fixity of the didactic spaces, partitions were replaced with sliding walls, equipped with interactive technology (Fig. 2). The classroom space thus leaves behind its original boundaries and becomes a dynamic setting, where learning experiences are also had in more informal areas, and much-needed dialogue is restored between different sections [6].

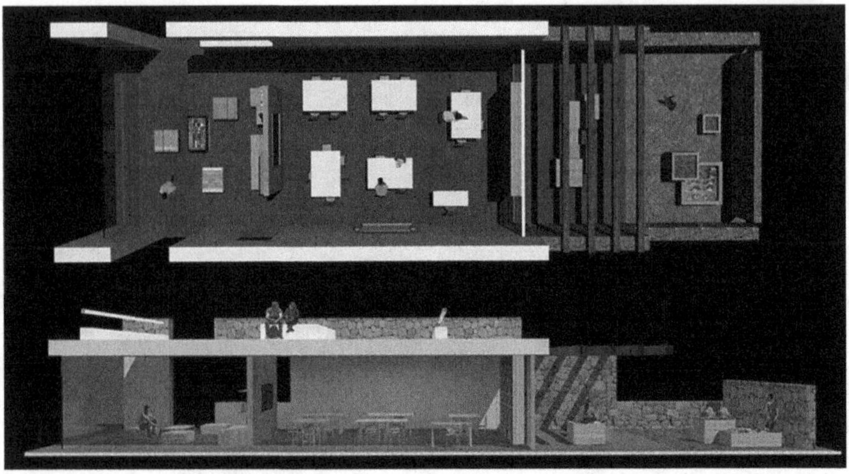

Fig. 2 Middle school extension project, Li Punti, Sassari. Team: M. Ferrari, C. Tinazzi, G.M. Dettori, M.M. Fiori, M. Lucesoli, A. Lucidi, D. Perrotta, M. Rosciani, (2016). Classroom concept, with extension of the teaching activities into the distributive space, thanks to a mobile wall system, and into the outdoor space

The second case concerned a historic, early twentieth-century building, with a rigid spatial layout, marked by the classroom–corridor sequence. The project seeks to recreate a less formal environment, aiming to hybridize learning and distribution spaces with the aid of a "parasite" element (Fig. 3). This is a spatial device that extends progressively from the horizontal plane onto the walls; a new "kitted out skin" that builds a continuous learning landscape that extends beyond the physical limits of the wall partitions, dislocating the various teaching functions over a wider area.

Another research experience that is similar to the one just described arose from the collaboration of the UNIVPM DICEA Department with Circolo Didattico "San Lazzaro" in Fano. This is a concrete opportunity for intervention within the National Plan for Digital Education, which envisaged the design of a creative workshop within a Montessori primary school.

The objective was to combine the old ceramic workshops and the adjacent science lab into a "maker space", one single creative environment, where experiential teaching can be applied to increase students' skills and competences in an engaging and participative way.

The first stage in the project involved demolishing a partition wall between two classrooms to create a single space. Next, we added modular furniture which, depending on how it is combined, can be used as workstations for computers and making activities, or can change configurations as needed to provide new spatiality to suit different didactic requirements, based on what work is being done and what devices are being used.

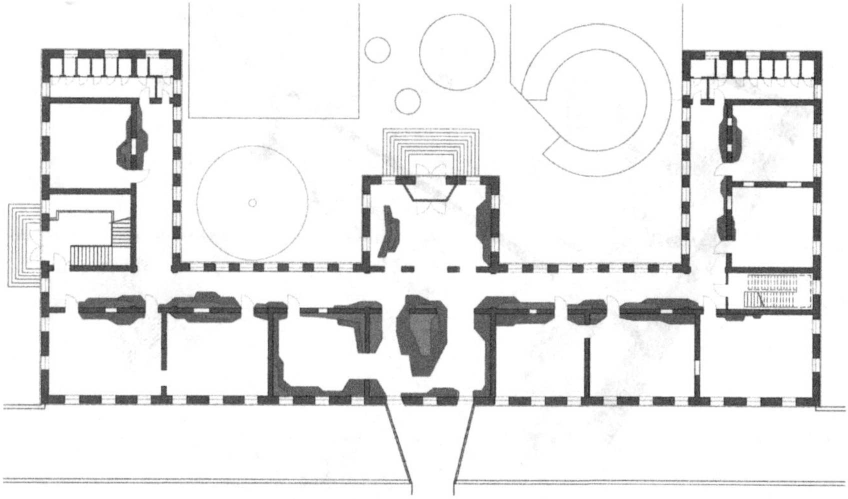

Fig. 3 Sacro Cuore School, Alghero. New learning landscape project. Team: G. Mondaini, L. Cabras, A. Dessì, E. Melis, M. Sotgiu, B. Zaru, (2016). The new architectural device for the hybridization of learning and distribution spaces

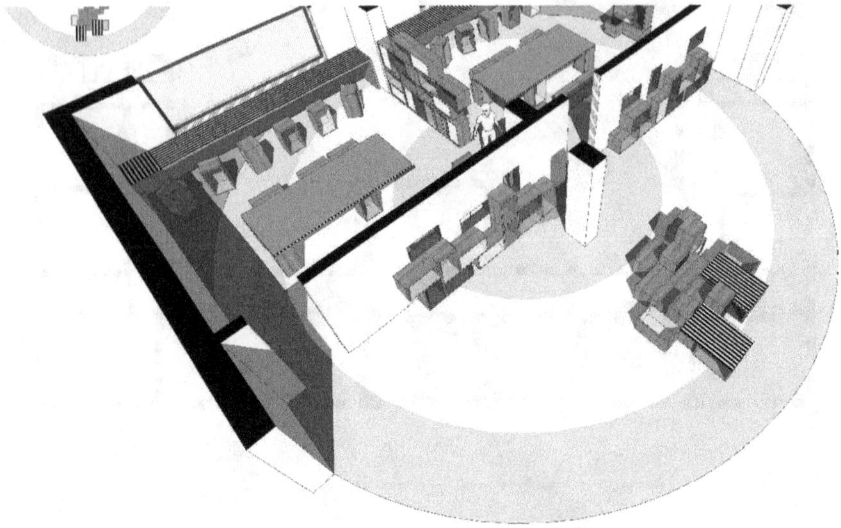

Fig. 4 *Atelier Creativo*, Montessori Primary School, Fano. Team: G. Mondaini, M. Rosciani, M. Tellechea, (2017). Design concept and images of the implementation

We used the same module for the wall that divides the classroom from the corridor; this time it was a case of subtraction, to create transparent openings and visual continuity between the interior and the exterior of the new atelier. Outside the laboratory, in the wide corridor, the furnishing elements can be used to display the works produced and can also act as "islands" for gathering and relaxing (Fig. 4). The workshop is also used by the community during non-school hours, particularly for preparations for the town's carnival.

References

1. Mondaini, G.: Spazio e formazione, 1st edn. Edilstampa, Rome (2018)
2. Borri, S.: 1+4 Learning spaces for a new generation of schools in Italy. In: The Classroom has Broken, 1st edn. Ediguida, Cava de' Tirreni (SA) (2018)
3. Bosch, R.: https://rosanbosch.com/en/approach/learning-spaces-need-enable-and-motivate-every-learner. Last accessed 25 Oct 2019
4. Mosa, E.: Lo spazio e il tempo nella scuola delle competenze. In: Biondi, G., Borri, S., Tosi, L.: Dall'aula all'ambiente di apprendimento. 1st edn. Altralinea Edizioni, Florence (2016)
5. Ferri, P., Moriggi, S.: Destrutturare l'aula, ma con metodo: spazi e orizzonti epistemologici per una didattica aumentata dalle tecnologie. Journal of Educational, Cultural and Psychological Studies **13**, 143–161 (2016)
6. Tinazzi, C.: Muri da ascoltare. In: ILS_Innovative Learning Spaces. LISt Lab, London (2016)

Conclusion

David Scaradozzi, Lorenzo Guasti and Paulo Blikstein

The chapters in this book, the FabLearn manifesto, and the associated research reveal a thriving community working on school innovation. We enjoyed reading fascinating contributions from many different countries aimed at students and young people of all ages. There are experiences involving young students from kindergarten and also research about vocational training. Each study, in its originality, refers to the cornerstones of the FabLearn Italy 2019 Conference: the innovation of laboratory teaching, the integration of robotics, making and coding in regular courses, and the rethinking of school spaces to make them more dynamic and suitable for innovation.

Education is vital to empowering our youth with critical competencies to understand and solve our future problems. Knowing how a student learns (and what he or she needs to learn) is essential for enhancing teachers' ability to guide students in new and complex learning environments, such as makerspaces, and new technologies can help this process by gathering information seamlessly and providing support.

Another critical theme in the FabLearn Italy community was recognizing the importance of the "craft" dimension of making, as it gives meaning and perspective for understanding the contributions of multiple cultures. It was fascinating to see how the maker movement has begun to include issues around ecology, environmental sustainability, and gender equality. Many of the projects presented concerned solutions to real-life problems, especially regarding historically underprivileged groups.

D. Scaradozzi (✉)
Dipartimento di Ingegneria dell'Informazione (DII), Università Politecnica delle Marche, Ancona, Italy
e-mail: d.scaradozzi@univpm.it

L. Guasti
Istituto Nazionale di Documentazione, Innovazione e Ricerca Educativa (Indire), Florence, Italy
e-mail: l.guasti@indire.it

P. Blikstein
Teachers College, Columbia University, New York, USA
e-mail: paulob@tc.columbia.edu

D. Scaradozzi et al. (eds.), *Makers at School, Educational Robotics and Innovative Learning Environments*, Lecture Notes in Networks and Systems 240,
https://doi.org/10.1007/978-3-030-77040-2

This allowed teachers and students to immerse themselves in real-life problems by engaging with the world around them.

From what we have learned from reading the contributions and in face-to-face interactions with the researchers in the FabLearn Italy community, we realize that schools will only change if our research for school innovation also changes. The entire educational ecosystem should be treated holistically, with all disciplines, components, and stakeholders working in synergy.

We need new models for the learning environments of the future. To make them happen, we need a vision, rigorous research, and diligent implementation. The FabLearn Italy community is making an exciting contribution to this crucial and complex endeavor.

Printed in the United States
by Baker & Taylor Publisher Services

Printed in the United States
by Baker & Taylor Publisher Services